Conversion of Système International to Imperial Units

VARIABLE	SI UNIT	IMPERIAL EQUIVALENT
Length	kilometer	0.621 mi
	meter	1.094 yd
	meter	39.37 in
	meter	3.281 ft
Area	hectare	2.471 A
	square kilometer	0.39 mi^2
	square meter	1.196 yd^2
	square centimeter	0.15 in^2
Volume	cubic meter	1.308 yd^3
	cubic meter	35.3 ft^3
	cubic centimeter	0.06 in^3
	cubic kilometer	0.24 mi^3
	liter	1.06 U.S. qt
	liter	0.88 imperial qt
	liter	0.264 U.S. gal
	liter	0.22 imperial gal
Velocity	kilometers/hour	0.621 mph
	meters/second	3.281 ft/sec[1]
	meters/second	2.24 mph
Pressure	standard atmosphere	14.7 psi
	bar	14.5 psi
Mass	kilogram	2.205 lb
	gram	0.035 oz
Temperature	Celsius	°C = 5/9(°F − 32)

Common abbreviations used in the Imperial System

mi	mile	mph	miles/hour	gal	gallon
yd	yard	oz	ounce	qt	quart
in	inch	lb	pound	A	acre
ft	feet	psi	pounds/in^2		

Climatology
An Atmospheric Science

Second Edition

John E. Oliver
Indiana State University

John J. Hidore
University of North Carolina at Greensboro

PRENTICE HALL
Upper Saddle River, New Jersey 07458

Library of Congress Cataloging-in-Publication Data

Oliver, John E.
    Climatology: an atmospheric science/John E. Oliver, John J. Hidore.--Rev. ed.
        p.   cm.
    Original ed. John J. Hidore was the author.
    Includes bibliographical references and index.
    ISBN 0-13-092205-6
    1. Climatology.   2. Atmospheric physics.   I. Hidore, John J. Climatology.   II. Title.

QC981.H5917 2002
551.5--dc21

2001033228

Executive Editor: Dan Kaveney
Assistant Editor: Amanda Griffith
Editorial Assistant: Margaret Ziegler
Production Editor/Composition: Prepare, Inc.
Executive Managing Editor: Kathleen Schiaparelli
Assistant Managing Editor: Beth Sturla
Marketing Manager: Christine Henry
Managing Editor, Audio/Video Assets: Grace Hazeldine
Art Director: Jayne Conte
Cover Designer: Kiwi Design
Manufacturing Manager: Trudy Pisciotti
Assistant Manufacturing Manager: Michael Bell
Vice President of Production and Manufacturing: David W. Riccardi

 © 2002 by Prentice Hall, Inc.
Upper Saddle River, New Jersey 07458

Printed in the United States of America

10 9 8 7 6 5 4 3 2 1

ISBN 0-13-092205-6

Pearson Education Ltd., *London*
Pearson Education Australia Pty., Ltd., *Sidney*
Pearson Education *Singapore*, Pte. Ltd.
Pearson Education North Asia Ltd., *Hong Kong*
Pearson Education Canada, Inc., *Toronto*
Pearson Educación de Mexico, S.A. de C.V.
Pearson Education — Japan, *Tokyo*
Pearson Education Malaysia, Pte. Ltd.

Brief Contents

Contents

11 Tropical Climates   206

12 The Mid-Latitude Climates   223

13 *Polar and Highland Climates*   *241*

PART III   Past and Future Climates   259

14 *Reconstructing the Past*   *260*

15 *Natural Causes of Climatic Change*   *280*

Preface to the Second Edition

The years since the publication of first edition of *Climatology* have been of major significance in the development of the discipline: It has now passed from a rather academic pursuit to a widely known area of study. *Climatic change* and *global warming* have become common terms, and there are few global processes that have received as much media coverage as El Niño and La Niña. Associated with the wide coverage of climatic phenomena is the remarkable growth of the World Wide Web (WWW) and its attendant multitude of home pages dealing with weather and climate.

Although these developments have brought climate to the forefront of popular science, such an awareness is fraught with potential problems, not the least of which is sorting out fact from speculation in the printed word and website. Unfortunately, the ready availability of climatic data often results in the use of data without the necessary rigor of appropriate statistical methods and misconceptions unwittingly arise.

It is within such a framework that we have prepared this second edition. Although part of the original content has been retained, the following changes are noted.

- A chapter on statistical methods has been included. This is located at the end of the content (chap. 20), yet it can be introduced and used at any juncture. The study of climatology does require statistical analysis, and various basic methods are described.

- We have added a section on applied climatology. In the limited confines of an introductory text, it is not possible to cover all components of a given area of study, and we have made no attempt to be comprehensive. However, the addition of this section will, we hope, provide the basis for further studies in the ways in which climate influences everyday life.

- An expanded section on oceans and atmosphere/ocean interactions is included in a new chapter. Both historic and recent events are dealt with.

- The treatment of recent climatic change and global warming is expanded to include the basic chemistry involved and the potential impacts of change.

- End of chapter material has been added. Key terms are listed and defined in an expanded glossary. Straightforward review questions are also given.

- We have made a conscious effort to retain both the literary and scientific components of the study of climate. To complete the latter without interrupting the text flow, some quantitative expressions are included as boxes in selected chapters. These provide a ready reference should readers require a more formal explanation of aspects described in the text.

- Although the authors encourage the use of the Internet, with but a few exceptions (such as the National Climate Center), we do not promote any particular sites. However, it is of consequence to use climate data and knowledge, and we recommend use of the Laboratory Manual designed to accompany this text. Selected web data sources are located in that source.

The authors wish to acknowledge all those who have helped determine the content and provided useful input into this revised work such as Richard W. Dixon, Southwest Texas State University; Mark L. Hildebrandt, Southern Illinois University; Karl K. Leiker, Westfield State College; David McGinnis, University of Iowa; Donald L. Morgan, Bringham Young University; Jan Westerik, Concord College. We are especially indebted to our wives, Loretta and Suzanne, for both their patience and support.

John E. Oliver
John J. Hidore

Climatology
An Atmospheric Science

Second Edition

PART I

Physical and Dynamic Climatology

Blue Snow, The Battery by George Wesley Bellows,
Columbus Museum of Art, Ohio: Museum Purchase, Howald Fund

CHAPTER
1

The Basis
of Modern Climatology

CHAPTER OUTLINE

The earth's atmosphere is a fundamental and critical part of the environment of planet Earth. Together with the lithosphere (the solid earth), the hydrosphere (water in all its forms), and the biosphere (living organisms), it makes the earth the habitable, hospitable place that it is. The atmosphere serves several major functions on the planet. Beyond providing the reservoir of gases required for life, it also plays a critical role in the distribution or redistribution of energy over the planet. The atmosphere provides an insulating layer around the earth that raises the mean temperature of the surface from −23°C (−9°F) to near 15°C (59°F). It shields the surface from the large doses of ultraviolet radiation that would destroy most life forms. It also serves as a major transporter of heat horizontally across the earth's surface and vertically away from the surface. Tropical regions receive far more energy than polar regions. Atmospheric circulation helps equalize this imbalance by moving heat from the warmer to the colder areas. In addition to transferring energy, the atmosphere transfers water from low to high latitudes and from ocean to land.

THE ATMOSPHERIC SCIENCES

Because of the complexity of the gaseous envelope that surrounds the planet, atmospheric scientists often divide its study into specific areas of interest. One such division identifies the three fields of aerology, meteorology, and climatology. **Aerology** (or aeronomy) is essentially the study of the free atmosphere through its vertical extent. Initially, the major goal of aerology was the identification of atmospheric structure and the amount and distribution of its parts. Today aerology deals mostly with the chemistry and physical reactions that occur within the various atmospheric layers. The word *aerology* is less widely used now than it once was, and its content is frequently considered as part of meteorology. We deal with this aspect of the atmosphere only briefly in later chapters.

     Meteorology is the science that deals with motion and phenomena of the atmosphere, with the view to both forecasting weather and explaining the processes involved. It deals largely with the status of the atmosphere over a short period and uses principles of physics to reach its goals. **Climatology** is the study of atmospheric conditions over periods of time measured in years or longer. It includes the study of the kinds of weather that occur at a place. It is concerned not only with the most frequently occurring types, the average weather, but the infrequent and unusual types as well. Dynamic change in the atmosphere brings about variation and occasional extremes that have long- as well as short-term impacts. As a result, climatology is the study of all the weather at a given place over a given period. Within this definition, climatology is partly meteorology. Yet because it also analyzes climatic conditions at locations on the earth's surface, it is also geographical. Thus, as the British climatologist E. T. Stringer noted, climatology does not belong entirely within the fields of meteorology or geography. It is a science—really an applied science—whose methods are strictly meteorological but whose aims and results are geographical.

CLIMATOLOGY: A BRIEF HISTORY

The study of climatology began at least as early as the ancient Greek culture. In fact, the word *climate* comes from a Greek word meaning slope. In this context, it refers to the slope, or inclination, of the earth's axis. It specifies an earth region at a particular place on that slope—that is, the location of a place in relation to parallels of latitude. This mathematical derivation represents one of many contributions to mathematical geography by philosophers such as Eratosthenes and Aristarchis. The Greek search for knowledge about the world resulted in written works on the atmosphere. The first climatography was *Airs*, *Waters*, and *Places* written by Hippocrates in 400 B.C. In 340 B.C., Aristotle wrote *Meteorologica*, the first treatise on meteorology.

The academic world did not resume the Greek interest in the atmosphere for many hundreds of years. Although Arab scholars of the 9th and 10th centuries expanded on the Greek writings, the renewed interest came in the middle of the 15th century with the Age of Discovery. Extended sea voyages and development of new trading areas led to descriptive reports of climates outside of Europe. Many of these descriptions were quite fanciful; they provided the basis for misconceptions about parts of the world that prevailed for centuries.

Scientific analysis of the atmosphere began in the 17th century with the design of instruments to measure atmospheric conditions. These provided data from which laws applying to the atmosphere were derived. Galileo invented the thermometer in 1593, and Torricelli invented the barometer in 1643. In 1662, Boyle discovered the basic relationship between pressure and volume in a gas.

Instruments were improved and standardized during the 18th century, and extensive data collection and description of regional climates began. Explanation of phenomena through the study of the physical processes began in the 19th century. Climatic data are the measurements of the earth's climate system. The most widely recorded data are surface temperature and precipitation. The longest complete climatic records exist for temperature and precipitation. These range in length from some 325 years for measurements in Central England to about 200 years for stations in Europe and the United States. Most of the stations that make up today's observational network have records of less than 100 years. In 1817, von Humboldt constructed the first map that showed temperatures using isotherms. Soon after, in 1827, Dove explained local climates using polar and equatorial air currents.

In the 20th century, contributions became more frequent and ideas essential to understanding the atmosphere evolved. The significant contributions of individuals since the beginning of the 19th century are too many to present here, although Table 1.1 lists some key developments.

An important event in the development of climatology was the National Climate Program Act. Signed into law in September 1978, this act has the stated purpose "to establish a national climate program that will assist the nation and the world to understand and respond to natural and man-induced climate processes and their implications." The program, as shown in Table 1.2, is a statement of the importance of climate and our understanding of it.

THE CONTENT OF CLIMATOLOGY

Figure 1.1 shows the diversity of approaches available in climatic studies. Climatography consists of the basic presentation of data in written or map form. As the names imply, physical and dynamic climatology relate to the physics and dynamics of the atmosphere. Physical climatology deals largely with energy exchanges and physical processes. Dynamic climatology is more concerned with atmospheric motion and exchanges that lead to and result from that motion. Some scientists consider the study of past and future climates to be a subgroup of climatic studies. However, methods used in studying past and future climates are the same as those in other aspects of climatology.

The approaches suggested in Figure 1.1 are largely self-explanatory with the possible exception of synoptic climatology. The object of the synoptic approach is to relate local and regional climates to atmospheric circulation. The scale of studies shows that climatic investigation covers areas ranging from very large (macroclimates) to very small (microclimates).

Table 1.1   Significant Events in the Development of Climatology

Date	Event
ca. 400 B.C.	The influence of climate on health is discussed by Hippocrates in *Airs, Waters, and Places*.
ca. 350 B.C.	Weather science is discussed in Aristotle's *Meteorologica*.
ca. 300 B.C.	The text *De Ventis* by Theophrastus describes winds and offers a critique of Aristotle's ideas.
ca. 1593	Galileo describes the thermoscope. (The first thermometer is most likely attributed to Santorre, 1612.)
1662	Francis Bacon writes a significant treatise on the wind.
1643	Evangelista Torricelli invents the barometer.
1661	Boyle's law on gases is propounded.
1664	Weather observations begin in Paris; although often described as the longest continuous sequence of weather data available, the records are not homogeneous or complete.
1668	Edmund Halley constructs a map of the trade winds.
1714	The Fahrenheit scale is introduced.
1735	George Hadley writes his treatise on trade winds and effects of Earth rotation.
1736	The Centigrade scale is introduced. (It was first formally proposed by du Crest in 1641.)
1779	Weather observation begins at New Haven, CT, the longest continuous sequence of records in the United States.
1783	The hair hygrometer is invented.
1802	Lemark and Howard propose the first cloud classification system.
1817	Alexander von Humboldt constructs the first map showing mean annual temperature over the globe.
1825	August devises the psychrometer.
1827	Beginning of the period during which H. W. Dove developed the laws of storms.
1831	William Redfield produces the first weather map of the United States.
1837	Pyrheliometer for measuring insolation is constructed.
1841	Movement and development of storms are described by Espy.
1844	Gaspard de Coriolis formulates the "Coriolis force."
1845	Berhaus constructs first world map of precipitation.
1848	Dove publishes the first maps of mean monthly temperatures.
1862	Renou drafts first map (showing western Europe) of mean pressure.
1879	Supan publishes a map showing world temperature regions.
1892	Beginning of the systematic use of balloons to monitor free air.
1900	The term *classification of climate* is first used by Köppen.
1902	Existence of the stratosphere is discovered.
1913	The ozone layer is discovered.
1918	Beginning of the development of the polar front theory by V. Bjerknes.
1925	Beginning of systematic data collection using aircraft.
1928	Radiosondes are first used.
1940	Nature of jet streams is first investigated.
1960	United States launches the first meteorological satellite, Tiros I.
1978	U.S. Congress passes the National Climate Act.
1978	The United States bans the use of CFCs as aerosol propellants.
1987	The Montreal Protocol limiting the production of CFCs is signed by more than 30 nations.
1989	The state of Vermont bans the use of CFCs in auto air conditioners.
1990	Doppler radar network introduced in the United States.
1992	Rio de Janeiro Earth Summit leads to "Framework" Convention on Climate Change.
1998	Kyoto Protocol on climatic change.
2000	Climate change concerns lead to disruption of UN Climate Conference in The Hague.

Table 1.2   Elements of the National Climate Program Act of 1978

The programs shall include, but not be limited to, the following elements:

1. assessment of the effect of climate on the natural environment, agricultural production, energy supply and demand, land and water resources, transportation, human health, and national security

2. basic and applied research to improve the understanding of climatic processes, natural and human-induced, and the social, economic, and political implications of climatic change

3. methods for improving climate forecasts

4. global data collection on a continuing basis

5. systems for the dissemination of climatological data and information

6. measures for increasing international cooperation in climatology

7. mechanisms for climate-related studies

8. experimental climate forecast centers

9. submission of five-year plans

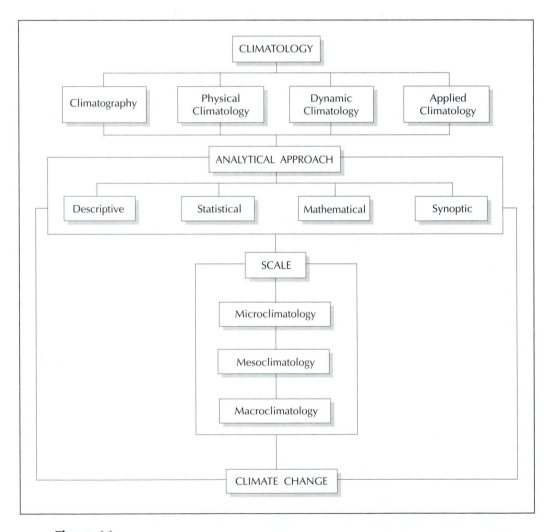

Figure 1.1
Subgroups, analytic methods, and scales of climate study (from Oliver J.E., *Climatology: Selected Applications*, © 1981, John Wiley and Sons, New York).

Within each of the areas identified in Figure 1.1, specialized analytic methods have been derived. For example, the exchanges of energy are examined using an energy budget approach. The various pathways and exchanges of energy in various forms are accounted for. Similarly, a water budget analysis examines the phases of water and the concomitant energy exchanges.

ATMOSPHERIC VARIABLES AND DATA ACQUISITION

Table 1.3 lists the variables measured most frequently today. Almost 10,000 stations throughout the world record at least one of these variables. They are part of a primary land-based system of observational stations. Unfortunately, the distribution is not even over the earth. Large parts of the world have but a sparse network of recording stations.

Since the oceans occupy almost three quarters of the earth's surface, climatic data over the oceans is of prime importance. Archives of surface data from the earth's oceans began in 1854. Through international agreement, the major maritime nations began a regular program of recording atmospheric and oceanic data from merchant and military ships. Nonetheless, long-term data from the oceans exist only for the popular sea lanes, and for large oceanic areas only limited data are available.

In the early 1800s, the only data representing conditions at high altitudes came from mountain observatories. In 1885, balloons became more widely used to monitor air currents and temperatures. During World War I (1914–1918), the use of balloons, kites, and aircraft multiplied rapidly. A similar impetus occurred in World War II (1939–1945). The need for upper air observations prompted a group of countries to establish a worldwide network of upper air observation stations. Balloon carried instruments, called **radiosondes**, are released into the atmosphere at specified times to simultaneously sample atmospheric conditions. Today, a network of about 1,000 stations using radiosondes routinely measures air temperature, dewpoint temperature, and wind direction and velocity.

Besides the parameters listed in Table 1.3, observations of other variables take place at selected weather stations. These include solar radiation and sunshine, soil temperature and evaporation, air pollution data, and water quantity and quality. At present there are about 100 stations measuring solar radiation and 150 monitoring sunshine duration. Many agriculture research stations measure evaporation.

Of importance in recent years is the development of satellite imagery to derive climatic data. **Remote sensing** provides quantitative values for temperature, humidity, and winds. It is also important in the study of the distribution and variation in cloud and snow covers. A later chapter includes methods of using satellite images.

Table 1.3   Commonly Observed Atmospheric Variables

Air temperature
Barometric pressure
Cloud type, height, and amount
Current or prevailing weather
Dew point temperature
Precipitation
Sunshine
Wind velocity and direction

Although weather data are gathered at many places, the data are of little value unless they are readily available to potential users. Fortunately, there are agencies and organizations that control the collection and processing of original data. The international agency responsible for worldwide climatic data is the World Meteorological Organization (**WMO**) located in Geneva, Switzerland. This United Nations (UN) agency publishes *Meteorological Services of the World*—a directory of the 150 member nations and the agency in each country responsible for climatic data. There are several scientific groups, both national and international, that maintain collections of data. These are readily obtained via the National Climate Data Center web page.

The official library for records generated by government weather services in the United States is the National Climatic Data Center in Asheville, North Carolina. This center publishes summaries of national and world climatic data that are available in printed form, at their website, and on compact disks. Beyond the national and international agencies, general climatic data are available in many textbooks and almanacs while several companies have assembled daily climatic data for most U.S. stations.

A new dimension in the study of the atmosphere was added when satellites were placed in orbit with the express purpose of sensing the atmosphere. The first was TIROS in 1960. This was followed in 1966 by the launching of the geostationary ATS-1 (Applications Technology Satellite), the first in a series of weather satellites. A geostationary (or earth-synchronous) satellite orbits the earth at a height of about 35,900 km (22,300 mi). At this altitude, the velocity necessary to maintain orbit is equal to the rotational velocity of the earth. Although the earth and satellite are both in motion, the effect is that the satellite appears to remain in place over a given point on Earth. The working satellites are the Geostationary Operational Environmental Satellite (GOES)/Synchronous Meteorological Satellite (SMS) series. The significance of satellite imagery in climatology is the subject of a later chapter.

Other developments in climatology are international programs for weather and climate studies. An appropriate example is the Global Atmospheric Research Program (GARP; it is also called the Global Weather Experiment). It is a concerted research effort by more than 140 countries, the World Meteorological Organization (WMO), and the International Council of Scientific Unions. The goal of this program is to test the practical limits of weather forecasting and determine the statistical properties of the general circulation of the atmosphere. This will lead to a better understanding of the physical basis of climate.

Data Representation

As part of the geographic aspect of climatology, maps play a significant role in the depiction of climates and climatic data. Many of the maps use isolines. These are lines joining locations of equal value to show distribution of the elements. The name given to the isoline depends on the climatic element shown on the map (Table 1.4).

The presentation of numeric values is often most clearly seen in graphs and diagrams. Throughout this text, as in most climatology works, graphs are used to show trends and patterns because they are easier to comprehend than a long list of numbers. But a graph is a statistical tool based on a mathematical grid that has one or more numerical data sets, and care need be used in its preparation. Most graphs are made up of two axes that need be scaled with calibrated graduations and clearly numbered. By convention, a graph consists of a horizontal axis (x-axis or abscissa) and a vertical axis (y-axis or ordinate). The x-axis is associated with the independent variable, that which influences the variable on the y-axis. The variable on the y-axis is dependent on that on the x-axis. The x-axis on many climate graphs shows time.

Some graphs may actually be classed as diagrams, with histograms and climagrams providing apt examples. A histogram is used to illustrate the relative frequency of particular values. A climagram, sometimes referred to as a hythergram, relates mean values of two climatic variables as a loop. The examples shown in Figure 1.2 use the average monthly temperature and precipitation values at Terre

Table 1.4 Isolines Used in Climatology

Isoline	Lines of
Isallobar	Equal pressure tendency showing similar changes over a given time
Isamplitude	Equal amplitude of variation
Isanomaly	Equal anomalies or departures from normal
Isobar	Equal barometric pressure
Isocryme	Equal lowest mean temperature for specified period (e.g., coldest month)
Isohel	Equal sunshine
Isohyets	Equal amounts of rainfall
Isokeraun	Equal thunderstorm incidence
Isomer	Equal average monthly rainfall expressed as percentages of the annual average
Isoneph	Equal degree of cloudiness
Isonif	Equal snowfall
Isophene	Equal seasonal phenomena (e.g., flowering of plants)
Isoryme	Equal frost incidence
Isoterp	Equal physiological comfort
Isotherm	Equal temperature

Haute, Indiana. Figure 1.2a is a continuous graph with the months of the year on the x-axis and temperature on the y-axis. The bar graph in Figure 1.2b is average monthly precipitation. The intersection of monthly average temperature and monthly precipitation is plotted in Figure 1.2c. This climagraph is typical of a midlatitude continental station; other climate regimes show different shapes and patterns.

Units

The historical development of science has provided a legacy of units in which variables may be measured. In some cases this has led to a confusing array, which is well illustrated by the units used for temperature. The Kelvin, Celsius, and Fahrenheit scales used to indicate temperature provide an appropriate example. (Note: Conversion factors for this and other units are given in the Appendix while each unit is defined in the Glossary.)

The influence of historical development on the use of units is well illustrated using length. Among the earliest relationships established in the English system was that three barleycorns laid end to end or eight laid side by side equal one inch. Twelve of these measurements were then equal to one foot. The latter, however, was open to question because it is said that during the reign of James I, the length of his foot became the official length. The use of barleycorns to measure an inch is still reflected in present day measure. A number 7 shoe is 1/3rd inch (1 barleycorn) longer than a number 6.

In such a way, the English system developed over many years and was most prevalent until about 1790, when the metric system was introduced in France. In much of Europe, this became the most widely used system. Based on decimal relationships, basic units (meter, gram, liter) were given appropriate prefixes based on increments of 10.

Clearly, for any system of units to be valuable, there must be a consistent set of standards. On a worldwide basis, the International Bureau of Weights and Measures, located in Paris and established by the Treaty of the Meter in 1875, sets

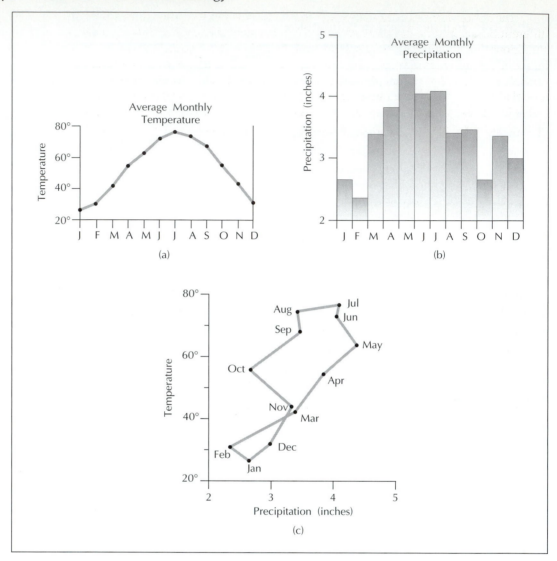

Figure 1.2
Some ways in which climatic data are represented (a) line graph (b) histogram or bar graph (c) climagram or hythergram.

the standard. This treaty called for International Conferences to be meet periodically. In 1960, the Conference adopted the International System of Units (or SI for Systeme International). SI units comprise the metric system based on fundamental physical quantities. Initially, six quantities were designated: mass (kilogram), length (meter), time (second), temperature (Kelvin), current (ampere), and luminous intensity (candela), with a seventh, the mole (the number of grams to which the molecular weight is numerically equal), adopted in 1971. Units frequently used in atmospheric studies—such as temperature, pressure, and energy—are derived from the basic quantities. These are dealt with as each is encountered in later chapters.

THE STANDARD ATMOSPHERE

Earlier in this chapter, it was stated that climatology is partially meteorologic. Weather is the basic ingredient in climate. Therefore, it is essential to begin the study of climate with an introduction to the atmosphere and its workings.

The atmosphere is continually changing and is never exactly the same at any two different points in time or space. Although it is constantly changing, there is a set of conditions that describes the atmosphere most of the time. It is helpful to know the most frequent conditions in a vertical column through the atmosphere to use these conditions as a basis for examining the extent of change. Therefore, scientists constructed a model atmosphere. In the United States, this model is the U.S. **Standard Atmosphere** (Table 1.5). This standard atmosphere represents an idealized, steady-state representation of Earth's atmosphere from the surface to 1,000 km (600 mi) because it is assumed to exist

Table 1.5   The Standard Atmosphere

Altitude (km)	Temperature (°C)	Pressure (mb)	P/P_0*	Density (kg/m^3)	D/D_0*
30.00	−46.60	11.97	0.01	0.02	0.02
25.00	−51.60	25.49	0.03	0.04	0.03
20.00	−56.50	55.29	0.05	0.09	0.07
19.00	−56.50	64.67	0.06	0.10	0.08
18.00	−56.50	75.65	0.07	0.12	0.09
17.00	−56.50	88.49	0.09	0.14	0.12
16.00	−56.50	103.52	0.10	0.17	0.14
15.00	−56.50	121.11	0.12	0.20	0.16
14.00	−56.50	141.70	0.14	0.23	0.19
13.00	−56.50	165.79	0.16	0.27	0.22
12.00	−56.50	193.99	0.19	0.31	0.25
11.00	−56.40	226.99	0.22	0.37	0.30
10.00	−49.90	264.99	0.26	0.41	0.34
9.50	−46.70	285.84	0.28	0.44	0.36
9.00	−43.40	308.00	0.30	0.47	0.38
8.50	−40.20	331.54	0.33	0.50	0.40
8.00	−36.90	356.51	0.35	0.53	0.43
7.50	−33.70	382.99	0.38	0.56	0.45
7.00	−30.50	411.05	0.41	0.59	0.48
6.50	−27.20	440.75	0.43	0.62	0.50
6.00	−23.90	472.17	0.47	0.66	0.54
5.50	−20.70	505.39	0.50	0.70	0.57
5.00	−17.50	540.48	0.53	0.74	0.60
4.50	−14.20	577.52	0.57	0.78	0.63
4.00	−11.00	616.60	0.61	0.82	0.67
3.50	−7.70	657.80	0.65	0.86	0.70
3.00	−4.50	701.21	0.69	0.91	0.74
2.50	−1.20	746.91	0.74	0.96	0.78
2.00	2.00	795.01	0.78	1.01	0.82
1.50	5.30	845.59	0.83	1.06	0.86
1.00	8.50	898.76	0.89	1.11	0.91
0.50	11.80	954.61	0.94	1.17	0.95
0.00	15.00	1013.25	1.00	1.23	1.00

Source: J. M. Morgan and M. D. Morgan, *Meteorology*, 3d ed. (New York: Macmillan, 1991), 541.

*P/P_0 = ratio of air pressure to sea-level value; D/D_0 = ratio of air density to sea-level value.

in a period of moderate solar activity. The standard atmosphere does not represent an average condition, but a steady-state condition. This is a condition where the atmosphere is in balance with the processes that govern it. The standard atmosphere defines profiles of chemistry, temperature, pressure, density, and several other variables.

Atmospheric Chemistry

The atmosphere consists of a mixture of gases that in its normal state is colorless, odorless, and tasteless. The gases contained are governed by gas laws considered later. Some of the particles in the atmosphere are single atoms, such as argon and helium. Others are molecules consisting of atoms of two or more elements, like water vapor and carbon dioxide.

Knowledge that the atmosphere includes many gases began with the work of chemists in the 18th century. The first gas studied in detail was carbon dioxide (CO_2). Its discovery in 1752 is somewhat surprising since, compared with nitrogen and oxygen, there is only a small amount of carbon dioxide in the atmosphere. Gaseous nitrogen, discovered by Rutherford in 1772, at first was called *Mephitic air*. Shortly after this, Joseph Priestly isolated oxygen, which he called *dephlogistated air*.

Over time other gases were discovered, the last of which was argon, isolated in 1894. Eventually the gases were given their modern names, and their relative proportion by volume in the atmosphere was determined. As shown in Table 1.6, nitrogen and oxygen make up the bulk of the atmosphere. There are other gases present in very small, but important, amounts. Of singular importance in understanding the ways in which atmopsheric gases act is an understanding of the Gas Laws. These laws, which show the relationships among temperature, pressure, and volume, are outlined in Box 1.1.

The gases making up the atmosphere may be divided into two groups. They are the constant gases (those relatively constant by volume) and the variable gases.

Constant Gases

The constant gases remain in the same proportion in the atmosphere upward to an altitude of about 80 kilometers (49 mi). The three most important constant gases

Table 1.6 Chemical Composition of the Dry Atmosphere Below 80 km

Gas	Parts per Million	Percentage of Total
Nitrogen	780,840.0	78.1
Oxygen	209,460.0	20.9
Argon	9,340.0	0.9
Carbon dioxide	350.0	
Neon	18.0	
Helium	5.2	
Methane	1.4	
Krypton	1.0	
Nitrous oxide	0.5	
Hydrogen	0.5	
Xenon	0.09	
Ozone	0.07	

Note: Data from U.S. Department of Commerce, NOAA, *United States Standard Atmosphere* (Washington, DC: Government Printing Office, 1976).

QUANTITATIVE EXPRESSION: BOX 1.1

The Gas Laws

The Gas Laws provide a series of relationships among basic gas properties.

1. *Boyle's laws* describe relationships among pressure, volume, and density of gases. In each case, the equations assume temperature to be held constant.

The first law states

$$P_0 V_0 = P_1 V_1 = K,$$

where P_0 and P_1 and V_0 and V_1 are the pressure and volume at two times, 0 and 1, respectively. K is a constant. The relationship indicates that an increase in pressure results in a decrease in volume and vice versa.

The second law relates pressure (P) to density (D)

$$P/D = K \text{ (at a constant temperature)}.$$

As pressure increases, so does density.

2. If pressure is considered constant, the volume of a gas can be related to temperature by *Gay–Lussac's law:*

$$V_t = V_0(1 + t/273),$$

where V_t is the volume at a temperature t (in °C) and V_0 is the volume at 0°C.

3. *Charles' law* provides the relationship between pressure and temperature when volume is held constant.

$$P_t = P_0(1 + t/273).$$

Here P_t is pressure at a given temperature, t (in °C); and P_0 is pressure at 0°C.

4. The *combined gas law* uses a gas constant, r, to combine the relationships of gas, pressure, volume, and temperature.

$$PV = rT.$$

This *equation of state* can be written in a form that draws on the molecular weight of a mass to derive a universal gas constant, R, and the density of the gas

$$P = \rho RT,$$

in which P is pressure, ρ is density, R is the universal gas constant (2.87×10^6 erg/g°K), and T is temperature. This equation allows derivation of any one variable if the other two are known.

are nitrogen (78% by volume), oxygen (21%), and argon (0.93%). Nitrogen is by far the most abundant of the gases, but it is relatively inactive in the atmosphere. Argon is also inactive, but it is present in rather small amounts. Oxygen is present in large quantities and is very active in the chemical processes of the physical and biological environments. There are many other constant gases in lesser amounts.

Variable Gases

Variable gases, as the term implies, vary in proportion of total atmospheric gases from time to time and place to place. The most important variable gases are water vapor and carbon dioxide. Water vapor content varies from near zero to a maximum of about 4% by volume, but this small relative volume is extremely important. Carbon dioxide exists in amounts that average near 0.03%. Of course, water vapor is the source of all precipitation that falls on the earth. Water vapor, carbon dioxide, and dust all absorb solar radiation. Ozone, another variable gas, is in the lower atmosphere in small amounts, the average being about one part per million. Another variable element of the atmosphere, which in many ways acts like a gas, is the particulate matter suspended in the air (**aerosol**). This includes soil particles, smoke residue, ocean salt, bacteria, seeds, spores, volcanic ash, and meteoric particles (see Table 1.7). The primary source of the solid particles is the earth's surface. Particulate matter decreases rapidly with altitude. High-altitude particles occur from meteoric dust, volcanic eruptions, and nuclear explosions in the atmosphere. The

Table 1.7 Major Ingredients in a Steady-State Tropospheric Aerosol

Source	Total (tons)	Percentage of Total
Vegetation	1.7×10^7	25.8
Dust rise by wind	1.6×10^7	24.1
Sea spray	7.6×10^6	11.9
Forest fires	6.2×10^6	9.9
Sulfur cycle	5.5×10^6	8.6
$NO_x\ NO_3$	5.5×10^6	7.7
Nitrogen cycle (ammonia)	3.9×10^6	6.0
Combustion and industrial	1.7×10^6	2.6
Anthropogenic sulfates	1.7×10^6	2.6

overall amount of particulate matter in the atmosphere varies from as little as 100 parts per cubic centimeter to several million parts per cubic centimeter.

Both particulate matter and water vapor are quite important in the atmosphere. They are largely responsible for the day-to-day variation in solar energy reaching the surface of the earth. Solid particles, moreover, serve as nuclei for condensation of water vapor and thus are necessary for precipitation.

VERTICAL STRUCTURE OF THE EARTH'S ATMOSPHERE

There are changes in the atmosphere that occur with height. The atmosphere can be divided vertically into several different zones based on a variety of changes that occur with altitude (Fig. 1.3).

Troposphere

The lowest zone is the **troposphere**. The troposphere is a turbulent zone that has a rather uniform decrease in temperature with height. Worldwide, the average rate of change in temperature with altitude (the lapse rate) is 6.5 degrees Celsius per 1,000 meters (3.5°F/1,000 ft). Similarly, as shown in Figure 1.4, the vertical distribution of other variables shows a varied decrease with height.

The troposphere has two subzones—the lower and upper troposphere. The lower troposphere extends upward to about 3 kilometers above the surface; it is the zone in which there is maximum friction between the earth and the atmosphere. It is the zone most affected by the daily changes in surface conditions. Another characteristic of this zone is the frequent existence of temperature inversions. A temperature inversion exists when the temperature increases rather than decreases with height.

The upper troposphere extends to a mean height of 11 kilometers (6.8 mi). It is less affected by the daily changes that take place near the surface and from the effects of friction at the surface. The primary changes that take place in this zone result from the secondary circulation (atmospheric storms) and the seasonal change in energy.

The water vapor content of the atmosphere at any point in time and space depends on the temperature of the air and to a lessor extent on atmospheric pressure, the proximity to a moisture source, and the history of the air mass. Water vapor content is normally highest close to the surface for two reasons. One is that most of the water in the atmosphere gets there as a result of evaporation from the ocean and from evapotranspiration. The second reason is that normally air tem-

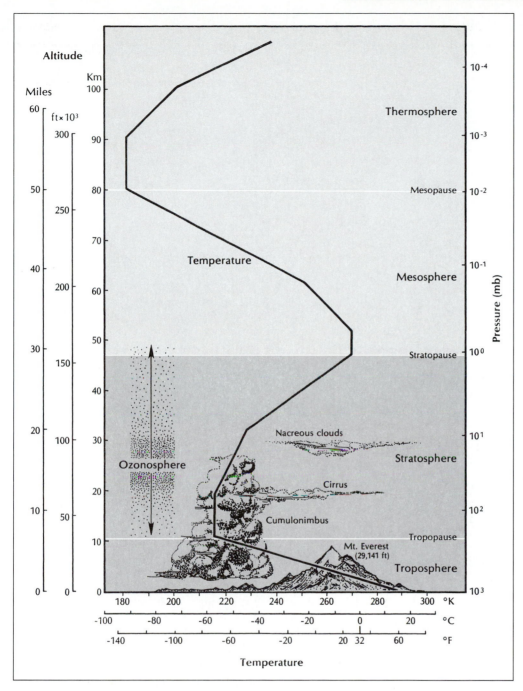

Figure 1.3
Vertical structure of the atmosphere.

perature is highest near the surface. The most water vapor recorded in the atmosphere is on the shores of the Red Sea where temperatures average 34°C (93°F). Water temperature reaches 35°C (95°F) here in summer. At air temperatures from 35°C down to −5°C (95°F–23°F), the amount of water vapor the air can hold decreases by a half for every 10°C (18°F) drop in temperature. The recorded range of water vapor in the atmosphere near the surface varies from a low of 0.1 ppm in the Antarctic and Siberia to a high of 35,000 ppm along the Persian Gulf. Due to the combination of atmospheric temperatures and altitude above the moisture source, the mean water vapor content decreases rapidly with altitude to a height of 12 to 15 kilometers (7.4 to 9.3 mi). Here it reaches a level of 2 to 3 ppm.

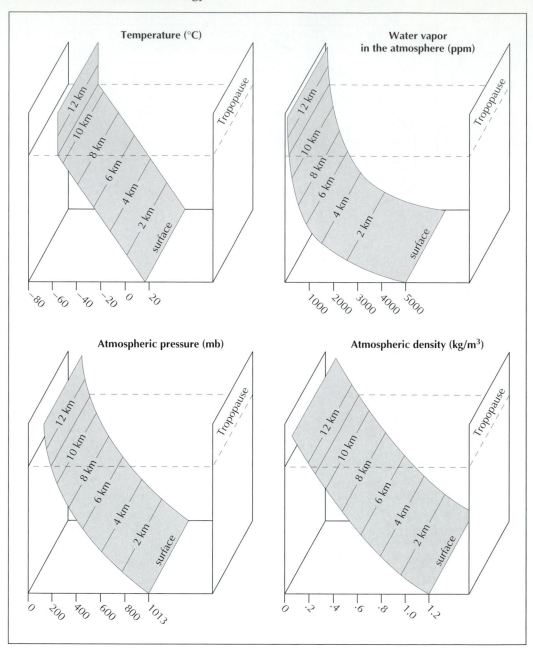

Figure 1.4
Vertical distribution of temperature, water vapor, pressure, and density in the lowest 14 km of the atmosphere.

The atmosphere is an unconfined gaseous fluid resting on the surface of the earth. As a result of the air having mass and being compressible, the mass and pressure of the air decrease with height. Atmospheric pressure at any point on the surface or in the atmosphere is the force per unit area exerted by the mass of the atmosphere above that point. Air close to the surface is subject to the mass of gases above. Hence, the greatest portion of the mass of the atmosphere lies near the surface. Atmospheric pressure decreases with height in a geometric fashion. That is, it decreases at a decreasing rate. The pressure data in Figure 1.4 show that surface pressure is on the order of 1,000 millibars. In the lower stratosphere, only 15 kilometers (9 mi) above sea level, the pressure is only 100 millibars. Of the total mass, about 50% lies below 5,500 meters (18,000 ft), 84% below 13 kilometers (8 mi), and 99% below 30 kilometers (18 mi). The density of the atmosphere decreases with height at a geometric rate, but the gases stay in roughly the same proportion up to heights of at least 80 kilometers (48 mi). The gases decrease in density away from the earth and extend for a distance of at least 1,000 kilometers (600 mi).

A distinct boundary layer known as the *tropopause* marks the upper reaches of the troposphere. There are actually a series of overlapping layers at different heights that make up the tropopause. This boundary zone is significant in several ways. Basically it marks the upper limit of most turbulent mixing started from the surface. It represents a cold point in the vertical temperature structure of the atmosphere. It also marks the upper limit of most of the water in the atmosphere. Very little moisture penetrates the tropopause except through severe thunderstorms, which will go as high as 20 kilometers (12 mi). The amount of water vapor in the atmosphere above the tropopause is very small. At the equator, the tropopause is at a height of 16 to 17 kilometers (9.9–10.5 mi). The temperature averages −70 to −85°C (−94 to −121°F), and the pressure averages only 100 millibars, only a tenth of sea-level pressure. Above the north and south poles, the tropopause is at a height of 9 to 12 kilometers (5.5–7.4 mi), with a temperature averaging −50 to −60°C (−58 to −76°F) and at a pressure of some 250 millibars.

Stratosphere

Above the tropopause is the stratosphere, which is named for the layered nature of the air at these levels. The **stratosphere** is relatively stable, relatively dry, and has relatively little vertical motion. Some high-velocity winds occur just above the tropopause; otherwise winds are noticeably absent.

The temperature at the tropopause is about −58°C (−72°F). The temperature remains nearly the same up to about 20 kilometers (12 mi). The upper region extending to some 50 kilometers (31 mi) has temperature increases with height that range up to as much as 4°C per kilometer (2.2°F/1,000 ft). However, the temperature at the top of the stratosphere is near that at the bottom.

Above the Stratosphere

The upper boundary layer of the stratosphere is the **stratopause**. As is the case with the tropopause, there may not be a single boundary layer, but instead a series of overlapping layers making up a transition zone.

Above the stratopause is a layer identified by a temperature decrease with altitude (**mesosphere**). Beginning at an elevation of about 48 kilometers (29 mi), the decline in temperature continues outward to the **mesopause** near 80 kilometers (49 mi). Up to the mesopause, the mixture of gases is about the same as at the surface. For this reason, the lower 80 kilometers of the atmosphere is termed the **homosphere**. Beyond the 80-kilometer altitude, the composition of the air begins to change. Above the homosphere atmosphere, gases tend to stratify on the basis of molecular weight. This layer is termed the **heterosphere**. To a height of 220 kilometers (130 mi), molecules of nitrogen largely make up the atmosphere. Layers of helium and oxygen exit outward from there (Fig. 1.5).

The mesosphere merges with a zone called the *ionosphere*. As the name implies, the ionosphere contains ionized gases and free electrons resulting from absorption of solar radiation. Short-wave radiation from the sun is absorbed, and the energy causes the electrons to split from the atoms of nitrogen and oxygen. The proportion of ionized particles is large because the density of particles is low and solar energy high. The higher in the ionosphere, the fewer particles, the more solar energy available, and the more ionization that takes place.

Incoming solar radiation begins to interact with the atmosphere at altitudes beyond 500 kilometers (310 mi). This outer region is also where gaseous particles

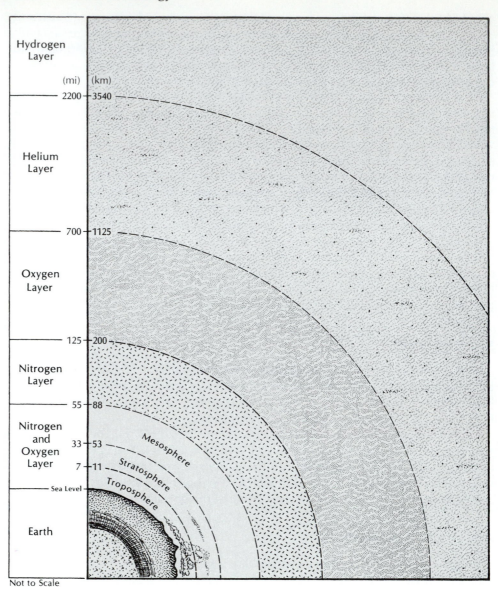

Figure 1.5
Chemical zones in the atmosphere.

escape the earth's gravity. Particles are far enough apart that some of them have a high enough velocity in a direction away from the earth to escape to space. The atmosphere is very thin at this altitude, but it is here that incoming space vehicles and meteorites begin to heat due to friction.

Above this, to perhaps 1,125 kilometers (675 mi), atomic oxygen is prevalent. Beyond this layer of atomic oxygen, helium is most common out to 3,540 kilometers (2,124 mi). Still farther out, hydrogen atoms predominate. The boundaries among the zones are not clearly defined. The heights represent the altitudes above which a different chemistry predominates.

SUMMARY

Climate is the total of weather at a place for a particular period. Study of climate is often divided into climatography, synoptic climatology, dynamic climatology, applied climatology, and climate forecasting.

The study of climatology began in ancient times, but received a considerable impetus from information gained in the great voyages of discovery. It was during this period that instruments were developed and the time when they made actual observations and recorded the data. The development of weather satellites now aids climatic interpretation and understanding. Similarly, the creation of the National Climate Program has promoted study.

Climatologists gather, store, and analyze climatic data. Most climatic data come from weather monitoring stations and are stored by both government and private agencies. Computers and computer techniques aid in data analysis, including summarizing and time series analysis. The modeling of past and future climates is growing rapidly in importance.

The atmosphere changes with time and from place to place. Important changes occur with altitude; to provide a baseline for atmospheric studies, the standard atmosphere was developed as a model. This model established attributes, such as pressure and temperature, as it is most frequently found.

KEY TERMS

Aerology	Mesosphere	Stratosphere
Aerosols	Meteorology	Tropopause
Climatology	Radiosondes	Troposphere
Heterosphere	Remote sensing	WMO
Homosphere	Standard atmosphere	
Mesopause	Stratopause	

REVIEW QUESTIONS

1. Explain the difference between weather and climate.
2. In what way is climatology a study of geography?
3. Why was the Age of Discovery important for climatology?
4. What variables are measured in atmospheric studies?
5. What types of data are available today?
6. What types of international cooperation are found in the atmospheric sciences?
7. What are the main areas of study of climatology?
8. What is the Standard Atmosphere? Why is it of value?
9. Differentiate between constant and variable atmospheric gases.
10. What are the major characteristics of the troposphere? The homosphere? The ionosphere?

CHAPTER
2

The Energy Balance

CHAPTER OUTLINE

The earth–atmosphere system is sustained by the supply of energy from the sun. Solar radiation is the ultimate source of energy that results in the varied atmospheric conditions experienced on Earth. It is fitting that a study of climatology is initiated by examining energy transferred and exchanged in the earth–atmosphere system.

Energy is the capacity for doing work. It can exist in a variety of forms and change from one to another (Table 2.1). The transfer of energy from place to place is of major importance in climatology. There are basically three ways in which this transfer can take place: conduction, convection, and radiation.

Conduction consists of energy transfer directly from molecule to molecule where the molecules are densely packed and contact one another. Energy always moves from an area of more energy to an area of less energy. In a sense, it is much as water flows from high places to lower places.

Convection involves the transfer of energy by the movement of an energy-laden substance from one location to another. Both convection and conduction depend on the existence of a physical substance in which to operate. This substance may be a solid, liquid, or gas.

Radiation is the only means of energy transfer through space without the aid of a material medium. The major source of energy on our planet is the sun. Between the sun and the earth, where a minimum of matter exists, radiation is the only important means of energy transfer. So important is this solar input that most climatological phenomenon are related to it.

To comprehend fully the role of solar energy in the functioning of the atmosphere requires three stages. First is understanding the nature of the energy emitted by the sun. Second is understanding the effect of solar radiation on the earth–atmosphere system. Third is a knowledge of changes that it undergoes in the system.

Table 2.1   Energy Forms and Transformations

	Forms
Radiation	The emission and propagation of energy in the form of waves
Kinetic energy	The energy due to motion: one half the product of the mass of a body and the square of its velocity
Potential energy	Energy that a body possesses by virtue of its position and that is potentially converted to another, usually kinetic energy
Chemical energy	Energy used or released in chemical reactions
Atomic energy	Energy released from an atomic nucleus at the expense of its mass
Electrical energy	Energy resulting from the force between two objects having the physical property of charge
Heat energy	A form of energy representing aggregate internal energy of motions of atoms and molecules in a body

Examples of Transformations

Atomic energy $\longrightarrow$ Radiation $\longrightarrow$ Heat $\longrightarrow$ Radiation
(Sun)   (Sunlight)   (Earth surface)   (Terrestrial)

Radiation $\longrightarrow$ Chemical energy $\longrightarrow$ Food chain
(Sunlight)   (Photosynthesis)

Potential energy $\longrightarrow$ Kinetic energy $\longrightarrow$ Heat
(Water vapor)   (Raindrop)   (Friction)

THE NATURE OF RADIATION

Every object above the temperature of absolute zero −273°C (−459°F) radiates energy to its environment. It radiates energy in the form of electromagnetic waves that travel at the speed of light. Energy transferred in the form of waves has characteristics depending on **wavelength**, amplitude, and frequency (Fig. 2.1). Using wavelengths as a criterion, radiant energy exists along a spectrum from very short to very long (Fig. 2.2). The characteristics of the radiation emitted by an object mainly depend on its temperature. Information relating to radiation laws is given in Box 2.1, while the following addresses the application of the laws to solar and earth radiation.

The amount of radiation varies as the fourth power of the absolute temperature (°K). The Kelvin scale is based on the concept of absolute zero. Absolute zero is the theoretical temperature at which all molecular motion would cease. It is equivalent to −273°C or −459°F. The hotter an object is, the greater the flow of energy is from it. Stephan–Boltzman's Law expresses this relationship as follows:

$$F = \sigma T^4,$$

where

> F = flux of radiation emitted per square meter
> σ = constant (5.67×10^{-8} W/m^{2}K^4 in SI units)
> T = Object's surface temperature in degrees Kelvin.

For example, the average temperature at the surface of the sun is 6000 K. The average temperature of Earth is 288 K (59°F). The temperature at the surface of the sun is more than 20 times as high as that of Earth. Twenty raised to the fourth power is 160,000. Therefore, the sun emits 160,000 times as much radiation per unit area as Earth. The sun emits radiation in a continuous range of electromagnetic waves. They range from long radio waves with wavelengths of 10^5 meters (62.5 mi) down to very short waves such as gamma rays, which are less than 10^{-4} micrometers (4^{-10} in) in length.

Another law of radiant energy (Wien's Law) states the wavelength of maximum radiation is inversely proportional to the absolute temperature. Thus, the higher the temperature, the shorter the wavelength at which maximum radiation occurs:

$$\lambda_{max} = 2897/T,$$

where

> T = Temperature in degrees Kelvin.

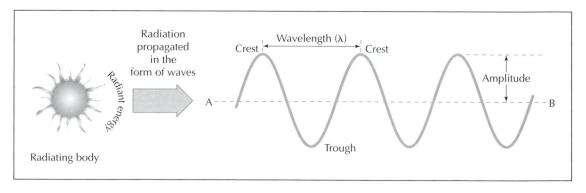

Figure 2.1
Radiant energy is transferred in the form of waves. The waves are defined in terms of wavelength, amplitude, frequency, and speed. Wavelength is the distance between wave crests. Amplitude is half the height difference between crest and trough. Frequency is the number of waves past a point in space per unit time. Speed is the distance a wave travels per unit time.

For the sun:
$$\lambda_{max} = 2897/6000 = 0.48\ \mu m.$$

For Earth:
$$\lambda_{max} = 2897/288 = 10\ \mu m.$$

At the temperature of the sun's surface, the maximum radiation is in the range from 0.4 to 0.7 micrometers. This is precisely the range of radiation that the human eye

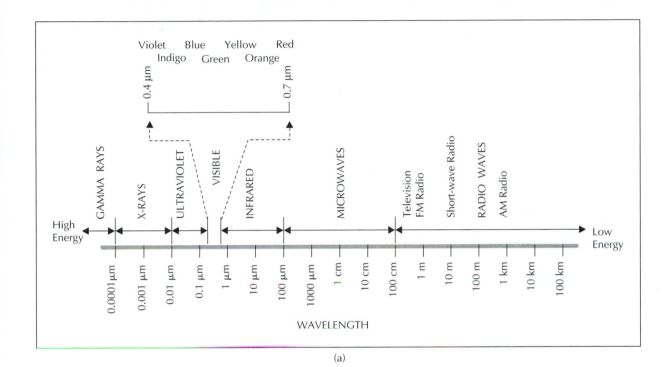

(a)

(b)

Figure 2.2
(a) The electromagnetic spectrum of solar radiation passes from shortwave high energy to longwave low-energy waves. In a range between 0.1 and 1 μm (10^{-4} and 10^{-5} cm) is a highly significant band. In this band, the energy is visible to the human eye. (b) The portion of the electromagnetic spectrum containing most of the energy emitted by the sun.

QUANTITATIVE EXPRESSION: BOX 2.1

The Radiation Laws

1. Radiant energy is characterized by wavelength (λ). The product of wavelength and frequency of the wave is a constant, the speed of light (c).

$$c = \lambda f,$$

where λ is wavelength, f is frequency, and c the speed of light (3×10^{10} cm/s). The direct relationship indicates that as λ increases, f will decrease. *Short waves are characterized by high frequencies, long waves by low frequencies.*

2. Basic energy laws apply to a black body; black bodies absorb all incoming energy and radiate at a rate determined by their temperature. The rate of radiation (F) is given by the *Stefan–Boltzmann law*:

$$F = \sigma T^4,$$

where F is the flux of radiation (W/m^2), T is temperature in degrees Kelvin, and σ is a constant (5.67×10^{-8} W/m^2/K^4). This equation shows that the flux of energy from a black body is related to its temperature. *The higher the temperature, the greater the flux.*

3. Radiators other than black bodies will not be perfect radiators. These gray bodies emit less radiation; the emissivity (ε) is thus less than 1 and is given by

$$F = \varepsilon \sigma T^4,$$

where F and σT^4 are components of the Stefan–Boltzmann equation and ε is the emissivity of the gray body at a particular wavelength.

4. *Kirchoff's law* relates emissivity to absorption (or absorptivity) of radiation at a particular wavelength.

$$\alpha \lambda = \varepsilon \lambda.$$

$\alpha \lambda$ is the fractional amount of energy that is absorbed (compared with that of a black body) at a given wavelength while $\varepsilon \lambda$ is emissivity at that wavelength. *This law shows that strong absorbers of radiation are also strong emitters.*

5. If the temperature of a black body is known, its wavelength of maximum emission (λ_{max}) is given by *Wien's law*:

$$\lambda_{max} = K/T,$$

in which K is a constant (2897) and T is temperature in degrees Kelvin. The inverse relationship means that *the higher the temperature of a black body, the shorter the wavelength of maximum emission.*

---

perceives, and we call it the *visible range*. This is a very good example of evolutionary processes. Human eyes have evolved to take maximum advantage of that part of solar radiation that is in greatest abundance. Individuals do not all see the same range of radiation as some see shorter and longer wavelengths than others.

THE SOLAR SOURCE

The sun is a gaseous mass with a diameter 109 times that of Earth. The source of the sun's energy is nuclear fusion produced in the core of the sun, where a nuclear reaction takes place that changes hydrogen into helium. Because the sun is a gaseous mix, no sharp boundaries exist within it, and different sections have different characteristics; it may be described using arbitrary divisions. Major parts are the core, **photosphere** (or visible surface), **chromosphere**, and **corona** (Fig. 2.3).

Photosphere

The photosphere is the bright outer layer of the sun that emits most of the radiation, particularly visible light. It consists of a zone of burning gases 300 kilometers (200 mi) thick. The photosphere is an extremely uneven surface. There are many small bright areas called *granules*. Their brightness is the result of bursts of extremely

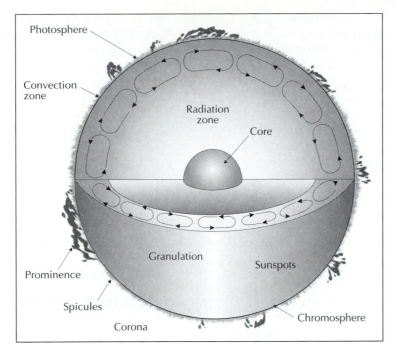

Figure 2.3
Internal structure and surface features of the sun (from Tarbuck E. J. & Lutgens F. K., *Earth Science*, 6e., © 1991, Prentice Hall, Upper Saddle River, N.J.).

hot gases formed in the photosphere that well up to the surface. There are darker areas around each granule that are cooler gases. The granules average 1000 kilometers (620 mi) in diameter. The hot gases spread out, cool on reaching the surface, and settle back below the surface. The net effect of all of these granules is convection of hot gases outward and cooler gases inward. It is much like the boiling of a large kettle. The elements in the photosphere are hydrogen (90 percent) and helium (10 percent). At the radiative surface, the effective temperature is on the order of 6000°K (11,000°F). The surface temperature determines both the amount of energy emitted and the wavelengths at which it occurs.

Chromosphere

Just above the photosphere is the chromosphere. It is a relatively thin layer of burning gases. This region of gases is under little pressure and extends away from the sun as much as a million kilometers (600,000 mi). It is here that the particles reach velocities high enough to leave the sun. The electrons and protons escape the sun in a stream called the *solar wind*. When they reach the earth environment, they meet the magnetic field that surrounds the earth. The magnetic field diverts most of these particles toward the two poles. One feature of solar activity that occasionally affects the earth are solar flares. They are sudden and explosive bursts of energy near sunspot clusters. They have a life span of only a matter of hours. They emit huge quantities of radiant energy and atomic particles. The atomic particles add to the solar wind. When these particles reach Earth, they disturb the ionosphere. This in turn interrupts radio and satellite communications and triggers the aurora.

Sunspots

One of the most prominent visible features of the sun's surface are dark areas known as **sunspots**. Sunspots appear as dark areas because they are about 1500°C (2700°F) cooler than the surrounding chromosphere. Science credits Galileo with reasoning that they were a feature of the sun and not clouds between the earth and the sun. They start as small areas about 1600 kilometers (1000 mi) in diameter. The individual sunspot has a lifetime ranging from a few days to a few months. Each spot has a black center, or umbra, and a lighter region, or penumbra, surrounding it.

Because records of sunspots have existed for many years and sunspots are relatively easy to see, they are one indicator of solar activity. The number of sunspots appears to follow an 11-year cycle. First, the sunspots increase to a maximum, with 100 or so visible at a given time. Over a period of years, the number diminishes until only a few or none occurs. Change from the maximum to the minimum number takes about 5.5 years, and thus the complete cycle is 11 years. We must stress that this is not an absolute periodicity as the interval between maxima cannot be forecast with certainty. The 11 years represent a mean interval.

Just how the sunspot cycle influences weather and climate is a matter of controversy. Some researchers claim to have found that weather patterns in a particular part of the earth follow the sunspot cycle closely. In other cases, no relationships exist. Even the physical process by which sunspots affect weather are not clear. The wavelengths of these emissions correspond to the wavelength of X-rays and they have different effects on different parts of the atmosphere. A further complication is that sunspot activity coincides with solar wind intensity, and it is difficult to separate the effects of each.

THE ATMOSPHERE AND SOLAR RADIATION

The energy emitted by the sun passes through space until it strikes some object. The intensity of radiation reaching a planet is in proportion to a basic physical law known as the *inverse square law*. This law states that the area illuminated, and hence the intensity, varies with the squared distance from the light or energy source. Thus, if the intensity of radiation at a given distance X is one unit, at a distance of 2X the intensity of the light will be one fourth that X. This law controls the intensity of solar energy intercepted by planets in the solar system. Earth only receives about 1/2,000,000,000 of the sun's energy output.

We know the amount of energy radiated by the sun and the mean Earth–sun distance of 149.5 million kilometers (93 million mi). Knowing this we also know the amount of radiation intercepted by a surface at right angles to the solar beam at the outer limits of the atmosphere. This quantity of radiation is approximately 1367 W/m^2 (1.97 calories per square centimeter per min) and is known as the **solar constant**. This value is often cited in another measure of solar energy, the Langley, which is 1 calorie per square centimeter per minute.

The solar constant is the basic amount of energy available at the outer limits of Earth's atmosphere. Several processes deplete the solar radiation as it passes into the atmosphere. These processes include reflection, scattering, absorption, and transmission.

Reflection

The earth and its atmosphere reflect part of the solar radiation back to space. There is considerable variation in reflection of natural surfaces. Reflectivity, or **albedo**, is expressed as a percentage of the incident radiation reaching the surface. Clouds are by far the most important reflectors in the earth environment. Reflectivity ranges from 40 to 90 percent depending on the type and thickness.

Water covers the largest area of Earth's surface. The reflectivity of water depends on the angle of the solar beam and the roughness of the water surface. Reflectivity of water decreases as the sun gets higher and higher in the sky. When the surface is smooth and the sun is near the horizon, reflectivity is high. People out in boats before 10 a.m. and after 2 p.m. can sunburn even when wearing broadbrimmed hats. Hats do not protect from radiation reflected from the water. Water reflects as little as two percent of the radiation when the solar angle is 90° and the water is choppy.

Table 2.2   Typical Albedos of Various Surfaces to Solar Radiation

Type of Surface	Albedo (%)
Fresh snow	75–95
Clouds	
Cumuliform	70–90
Stratus	60–84
Cirrostratus	44–50
Planet Venus	78
Old snow and sea ice	30–40
Dry sand	35–45
Planet Earth	30
Desert	25–30
Concrete	17–27
Savanna	
Dry	25–30
Wet	15–20
Grass-covered meadow	10–20
Tundra	15–20
Dry, plowed field	5–25
Asphalt road or parking lot	5–17
Green field crops	3–15
Deciduous forest	10–20
Coniferous forest	5–15
Moon	7
Water	
Angle of inclination of the sun:	
0°	99+
10°	35
30°	6
50°	2.5
90°	2

The land surface of the earth reflects only 40 to 50 percent of solar radiation. Reflectivity of land surfaces varies with the type of surface cover. Fresh snow reflects more than 75 percent and dry sand over 35 percent of the incident radiation. Table 2.2 provides examples of the reflectivity of various natural surfaces. Earth has an albedo of 30 percent, which represents the mean reflectivity from the ocean, land, and atmosphere.

Scattering

Scattering is the process by which small particles and molecules of gases diffuse part of the radiation in different directions. The process changes the direction of the radiation in a relatively random fashion. The English scientist Lord Rayleigh (1842–1919) developed the explanation for scattering, and the effect is called *Rayleigh scattering*.

The amount and direction of scatter depends on the ratio of the radius of the scattering particle to the wavelength of the energy. Furthermore, the amount of scatter is inversely proportional to the fourth power of the wavelength. This means that in a given set of conditions, the shorter wavelengths scatter more readily than

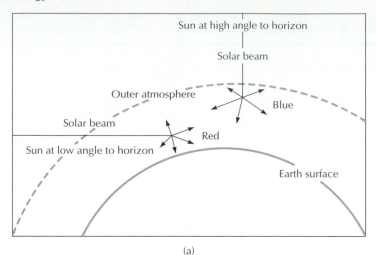

(a)

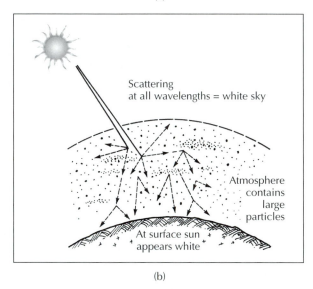

(b)

Figure 2.4
(a) Sky color depends on the lengths of the path solar radiation takes through the atmosphere. The shortest wavelengths are scattered first. Thus, when the radiation is perpendicular to the earth, the sky appears blue. When the sun is low in the sky, the blue is all filtered out and yellow, orange, and red predominate.
(b) Scattering of the full visible spectrum on a hazy day. The radiation is all scattered such that white light seems to come from all directions.

does long wave radiation. For example, radiation of wavelengths twice as long as that of another wavelength scatters only one sixteenth as much.

The most obvious effect of scattering in the atmosphere is sky color. The only reason our sky appears blue is because of scattering of radiation in the shorter wavelengths of visible light. As radiation in the visible range enters the outer regions of Earth's atmosphere, small gas molecules scatter the shortest wavelengths, the violet, first. Normally, the atmosphere scatters and absorbs the violet so the sky does not have a violet color. Blue scatters next. This randomly diffused radiation in the blue range scattered through the lower atmosphere gives it its color. Most of the radiation in the visible range has wavelengths greater than the diameter of particles in the dry atmosphere. This energy passes through without being changed in any way.

Scattering also explains the orange and red colors seen at dawn and sunset (Fig. 2.4). At these times, the radiation passes through the atmosphere at a very low angle. Thus, it passes a long distance through air close to the ground. The lower atmosphere contains not only the dry gases, but water vapor, solid particles, organic material, and salt. These larger particles scatter the longer wavelength radiation. In fact, the atmosphere absorbs most of the radiation in the shorter wavelengths (violet to green). Only the longest wavelengths—orange and red—pass through.

In early morning and late evening, the bottoms of clouds often have a red tint. This results from red light reflected from the clouds. The colors associated with shorter wavelengths scatter before the light reflects from the clouds. Spectacular sunsets occur when there is a lot of dust in the atmosphere. Dust scatters the radiation as do gas particles. Dust from storms on the surface or from volcanic eruptions causes unusually colorful sunsets. There were colorful sunsets following the eruption of El Chichon, Mexico, on April 4, 1985. Following the eruption of Krakatoa in 1883, there were brilliant sunsets for several years until the dust settled.

The sky appears black to astronauts above the earth's atmosphere and also as seen from the surface of the moon. Since there is no atmosphere on the moon to scatter the radiation, there is no color.

Rayleigh scattering is not the only scattering mechanism. Gustave Mie developed another theory for scattering in 1908. This theory states that molecules that have a larger ratio of diameter to wavelength than those that give rise to Rayleigh scattering results in the scattering of light at all wavelengths. More light scatters in a forward or continuing direction than in a backward direction. The sky is often the darkest blue shortly after a rain because the rain washes out the larger particles of debris and removes much of the moisture. There is then less scattering and a darker sky. On a clear day, the sky is a darker blue directly overhead than near the horizon. This is because there is less overall scattering in the direct path through the atmosphere overhead. Large particles scatter the light coming from near the horizon. The bright white color of the sky on a hazy day is due to the scattering of all the visible light as it passes through the haze. When the sky is cloud covered, only scattered radiation reaches the ground.

In polar regions, solar radiation scattered downward toward the earth (sky radiation) is a significant part of the total radiation. During the season when the sun does not come above the horizon, sky radiation is the main source of radiant energy.

Absorption

Absorption retains incident radiation and converts it to some other form of energy. Most often it changes to sensible heat, which raises the temperature of the absorbing object. For example, sunlight striking the side of a house is absorbed and heats the wall.

Gas molecules, cloud particles, haze, smoke, and dust absorb part of the incoming solar radiation. Such absorption is selective for gases absorb only in certain wavelengths. Each gas has a characteristic absorption spectrum and so different gases absorb different portions of the electromagnetic spectrum. The two most common gases in the atmosphere, nitrogen and oxygen, absorb ultraviolet radiation. Triatomic oxygen (ozone) absorbs shorter wavelengths than nitrogen and oxygen.

Nitrogen does not absorb much incoming solar radiation. The maximum solar radiation is at 0.5 μm, and nitrogen does not absorb well in this frequency. Oxygen (O_2) and ozone (O_3) absorb well at wavelengths below 0.3 μm, with most of the absorption occurring in the ionosphere. Water vapor absorbs fairly well in the infrared range but not in the range of maximum solar radiation. These three gases make up over 99 percent of atmospheric gases. Since none are good absorbers in the visible range, where most solar radiation occurs, there is a major window that lets in solar radiation.

Transmission

Reflection, scattering, and absorption deplete the solar beam as it passes through the atmosphere. **Transmissivity** is the proportion of the solar radiation ultimately passing through the atmosphere. Transmissivity depends on both the state of the atmosphere and the distance the solar beam must travel through it. The relative distance the solar beam travels through the atmosphere is the path length or optical air mass. The path length has a value of 1 when the sun is directly overhead or 90° above the horizon. The path length increases as the angle of the sun above the horizon decreases, and it can be calculated for any sun angle. For example, if the sun is 30° above the horizon, the path length has a value of two since the solar beam has twice the distance to travel through the atmosphere.

THE PLANETARY ENERGY BUDGET

The amount of solar radiation reaching a unit area of the surface—the insolation—is made up of energy transmitted directly through the atmosphere and scattered energy. These two radiation sources make up the global solar radiation. On days of thick cloud cover, no direct radiation reaches the surface. Solar energy reaches the surface only as scattered or diffuse radiation.

Figure 2.5 is a simple model of the planetary energy balance. This model is similar to that of the standard atmosphere. It represents a composite of the planet over a period of years. The model shows the percentages of the total annual energy inflow reflected, scattered, absorbed, and transmitted. Reflected energy plays no part in planetary heating. For the planet, a composite (or average reflectivity) of the atmosphere, ocean, and land masses is about 30 percent. This is termed the *planetary albedo*.

Clouds are the most important element in reflecting solar radiation back to space. The brilliant white of cloud tops you see when flying above them is reflected radiation of all wavelengths. Both Earth and Venus are very bright reflectors due to cloud cover. It is the cloud cover of Venus that makes it the brightest of the planets as seen from Earth. It is the second brightest object in the sky after the moon. While the moon appears bright on a clear night, it has an albedo of only seven percent. The moon lacks a cloud cover, and the rocks of the surface are relatively dark color. The earth appears much brighter from space than the moon does as it reflects more than four times as much sunlight.

Twenty-two percent of the incoming solar radiation is scattered, with most of it reaching the ground. However, of the 22 percent scattered, over a fourth goes back to space. Like reflected energy, this scattered radiation that goes back to space plays no part in heating the planet. In the model, the atmosphere absorbs 17 percent of the

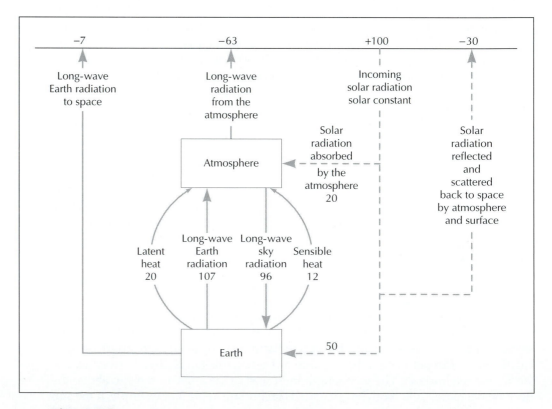

Figure 2.5
A model of the disposition of solar energy, Earth energy, and atmospheric energy.

incoming radiation. The stratosphere absorbs much of the shorter wavelengths, such as ultraviolet. Water vapor and carbon dioxide in the troposphere absorb the longer wavelengths. The remainder of the energy budget flow considers the long wave energy emitted by the earth and the interchanges of that energy in the atmosphere.

The Greenhouse Effect

Life is possible on Earth solely because of a process known as the **greenhouse effect**. It is more properly called the atmosphere effect for the analogy to a greenhouse is imperfect, but the name is firmly entrenched and its use is continued here. The process assumes that the solar energy passing through the glass (analogous to the atmosphere) is absorbed by materials (analogous to Earth's surface) inside. The radiated heat from inside the greenhouse cannot pass freely through the glass (analogous to greenhouse gases) and the air of the greenhouse (analogous to the lower atmosphere) is warmed.

If not for a complex interchange between the earth's surface and the atmosphere, as described by the greenhouse effect, the mean temperature of the atmosphere near the surface would be −20°C (−4°F). There would be no water in liquid or gaseous form. A selective absorption process raises the mean temperature of the atmosphere near the surface to 15°C (59°F). Solar radiation is of fairly short wavelengths, and the atmosphere is relatively transparent to this radiation. The solar energy received at the surface of the earth is processed and eventually radiated or otherwise released to the atmosphere.

Nitrogen and oxygen are poor absorbers of both short-wave solar radiation and long-wave Earth radiation. Several of the variable gases, including water vapor and carbon dioxide, are relatively transparent to solar radiation in the visible range but good absorbers of earth radiation. Clouds are even better absorbers of Earth radiation. Figure 2.6 shows the radiation curve for the earth and the bands in which the atmosphere absorbs this radiation. The earth radiates energy at wavelengths in the infrared band of 3 to about 30, 5 to 8, and beyond 13 micrometers in length. Carbon dioxide also absorbs radiation at 4 micrometers and from 13 to 17 micrometers. The result is that most Earth radiation that escapes to space does so in a narrow band from 8 to 13 micrometers. This is the atmospheric window for Earth

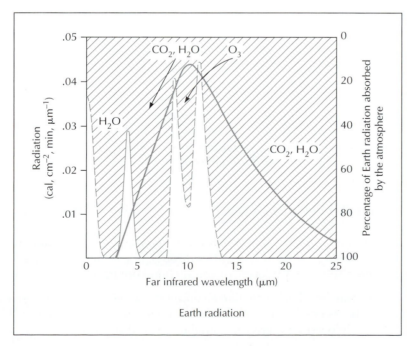

Figure 2.6
Earth radiation and atmospheric absorption.

QUANTITATIVE EXPRESSION: BOX 2.2

Energy Flow Representation

The exchanges and flows associated with energy inputs into the earth–atmosphere system are represented by a series of symbolic equations. Use of the equations permits easy calculation once values are input.

Shortwave solar radiation ($K\downarrow$) reaching the surface is made up of the vertical radiation (S) and diffuse radiation (D).

$$K\downarrow = S + D.$$

Some of the energy is reflected back to space ($K\uparrow$) so that net shortwave radiation (K^*) is the difference between the two

$$K^* = K\downarrow - K\uparrow.$$

Net longwave, terrestrial radiation (L^*) is comprised of downward atmospheric radiation ($L\downarrow$) less upward terrestrial radiation ($L\uparrow$).

$$L^* = L\downarrow - L\uparrow.$$

The amount of energy available at any surface is thus the sum of K^* and L^*. This is net all wave radiation (Q^*)

$$Q^* = K^* + L^*,$$

which may also be given as

$$Q^* = (K\downarrow - K\uparrow) + (L\downarrow - L\uparrow).$$

Q^* may be positive or negative.

High positive values will occur during high sun periods when $K\downarrow$ is at its maximum and atmospheric radiation, $L\downarrow$, exceeds outgoing radiation, $L\uparrow$.

Negative values require outgoing values to be greater than incoming. This happens, for example, on clear nights when $L\uparrow$ is larger than other values.

On a long-term basis, Q^* will vary with latitude and surface type.

---

radiation. Figure 2.6 also shows there is some absorption by upper atmospheric ozone at about 10 micrometers. Thick clouds are effective absorbers of Earth radiation over the entire range, even in the bands from 8 to 13 micrometers. So when layers of cloud over 100 meters (330 ft) thick cover the sky, the atmosphere is able to absorb most of the earth's radiation.

The gases and liquid water that absorb earth radiation are also good radiators of energy. The atmosphere radiates part of the energy absorbed to space and part back to the earth's surface. Nearly two thirds of atmospheric radiant energy is directed back to the surface. This provides an additional energy source to direct solar radiation. It is the largest source of radiant energy absorbed at the earth's surface. It is nearly double the amount of energy received directly from the sun. It is this additional atmospheric radiation that raises the mean temperature at the surface to 15°C (59°F).

The flows of energy described as part of the planetary energy budget may be represented by symbolic equations. Box 2.2, Energy Flow Representation, provides a summary of the symbols often used.

THE EARTH'S SURFACE AND SOLAR ENERGY

To maintain a steady-state temperature over a period of years, the earth disposes of the energy it receives. It does this through one of three processes: radiation, evaporation (with transpiration), or conduction-convection.

Radiation. The earth, like all objects with a temperature above absolute zero, radiates energy. The planet has a mean temperature of 15°C (59°F). This is a much lower temperature than the sun, and so the wavelengths of radiation are much longer. Earth radiation is at its maximum at 10 micrometers. Radiation from the surface of the earth is in wavelengths some 20 times as long as the incoming solar radiation. Energy gained by the earth that is not returned to the atmosphere by radiation is transferred by latent heat and sensible heat transfer.

Latent Heat Transfer. Changing liquid water to water vapor (evaporation) requires a considerable amount of energy. This absorbed energy is stored as latent heat in the water vapor until it changes back to liquid water. Condensation in the atmosphere then adds this energy to the air. (For an explanation of the evaporation-condensation process, see chap. 4.)

Sensible Heat Transfer. Conduction moves some energy from the earth to the air above. Because air is a very poor conductor of heat, the process only heats a few centimeters of air in this fashion. However, this warmed layer of air moves upward by convection and the conduction process continues at the surface. The conduction-convection process is the main heating process in the lower atmosphere and provides the environmental temperatures that people experience—the sensible temperature. This process is discussed in more detail in a later chapter.

These processes may be represented by

$$Q^* = H + LE$$

or net radiation (Q^*) is partitioned between sensible heat transfer (H) and latent heat transfer (LE). The ratio of sensible to latent heat is given by the **Bowen Ratio**, which is discussed further in chapter 3.

THE STEADY-STATE SYSTEM

The energy flow is a continuous process. Energy from the sun enters the earth–atmosphere system and ultimately returns to space. To remain at a steady-state temperature, incoming energy and outgoing energy must be in balance. If more energy comes in than goes out, the earth would get progressively hotter. If more energy goes out than comes in, it would get cooler.

The earth's surface receives energy from two main sources. One is direct and scattered solar radiation. Fifty percent of the solar radiation reaching the planet passes through the atmosphere to the surface. The atmosphere transmits 33 percent of incoming radiation, and an additional 17 percent reaches the surface as scattered energy. The second source of energy is atmospheric radiation. This is part of the greenhouse effect. The total amount of energy the earth receives from these two sources is 146 units. Each unit is equivalent to 1/100 of the total solar radiation reaching the outer atmosphere.

Radiation, evaporation, and conduction balance the incoming 146 units of energy. Radiation to the atmosphere removes most of the energy received at the surface. A small amount (7 units) is radiated directly back to space. Evapotranspiration removes another 20 units of energy. Conduction-convection removes the remaining 12 units. If we consider the net loss of heat beyond the exchange of the greenhouse effect, the earth loses nearly equal amounts of heat by radiation and evaporation of water. Conduction removes a smaller amount. Convection moves heat rapidly away from the surface. Convection and evaporation-condensation move heat upward from the surface. In greenhouses, air warmed by conduction cannot escape and so it continues to warm.

The atmosphere is also in balance. The atmosphere receives energy by absorbing direct solar radiation, absorbing earth radiation, conduction of heat at the surface, and heat of condensation. Seventy-seven percent of the energy received

Table 2.3   Energy Balances of the Earth Surface, Atmosphere, and Planet. Units Are in Hundredths of the Incoming Solar Radiation over a Year

Energy Balance of Earth's Surface			
Inflow		*Outflow*	
Solar radiation	50	Earth radiation	114
Sky radiation	96	Latent heat	20
Total	146	Conduction	12
		Total	146

Energy Balance of the Atmosphere			
Inflow		*Outflow*	
Solar radiation	20	Radiation to space	63
Condensation	20	Radiation to surface	96
Earth radiation	107	Total	159
Conduction	12		
Total	159		

Energy Balance of Earth			
Inflow		*Outflow*	
Solar radiation	100	Reflected and scattered	30
Total	100	Sky radiation to space	63
		Earth radiation to space	7
		Total	100

by the atmosphere comes from the earth. The largest single source of energy is Earth radiation. Only 13 percent of the total energy comes from the direct absorption of solar radiation. Therefore, we must understand that, as a result of the greenhouse effect, the prime heat source for the atmosphere is the earth's surface. Radiation to space and radiation back to the surface balance the energy received by the atmosphere.

Energy received by the earth is counterbalanced by an equal amount of radiation to space. The largest amount of energy released is by radiation from the atmosphere. This is primarily from cloud tops. Reflection and scattering of solar radiation and direct radiation from the surface remove the remaining energy. Table 2.3 shows the inflow and outflow of energy for the surface, atmosphere, and planet.

The earth is in a steady-state balance of incoming and outgoing energy. The temperature of Earth undergoes small changes, but the mean temperature stays nearly the same. It varies only slightly around 15°C (59°F) at the surface, although there are daily, seasonal, and year-to-year changes. There are mechanisms at work that prevent major change and keep the system in a steady state. For instance, if the earth receives more energy than usual, the temperature must go up. If the temperature goes up, the earth radiates away energy at a higher rate (Wien's law). By the same token, if the earth starts to cool, it will lose less heat by evaporation, conduction, and radiation. This offsets the reduction in incoming energy. Since the energy flow is in a steady state, the planetary climate is steady over a period of years. One of two events must occur for the steady-state temperature of the lower atmosphere to change, and hence for the climate to change. There must be a change in either the flow of energy to and from the planet or a change in the internal greenhouse effect. We see later in this book that such changes do occur.

SUMMARY

The flow of energy into, through, and out of the environment is dynamic and complex. The source for the energy that drives the atmosphere is solar radiation. The supply of solar radiation is steady but not constant. Changes occur in the amount of radiant energy ejected from the sun. It is steady enough to be called the solar constant. Energy reaching the atmosphere is disposed of in several ways. Some is reflected and scattered back to space, some is absorbed, and some is scattered and transmitted through to the surface. It is this latter energy that becomes the major heat source for the atmosphere. There is a built-in storage mechanism known as the greenhouse effect that raises the temperature such that life can exist. The energy balance of the planet is a steady-state system that maintains a mean temperature at the surface of about 15°C (59°F).

KEY TERMS

Albedo
Bowen ratio
Chromosphere
Conduction
Convection
Corona

Greenhouse effect
Latent heat
Photosphere
Radiation
Scattering
Sensible heat

Solar constant
Sunspots
Transmissivity
Wavelength

REVIEW QUESTIONS

1. How is energy transferred? Give example of processes and exchanges.
2. Why does the sun emit energy mostly in the form of light, whereas earth radiates at longer wavelengths?
3. How much energy does the sun emit? How do we know?
4. What are the main features of the sun's structure?
5. What process can result in colorful sunsets?
6. Differentiate between reflection and scattering in the atmosphere.
7. Outline the absorption characteristics of common atmospheric gases.
8. Describe the main features of the greenhouse effect on Earth.
9. By what processes is energy transferred from the surface to the atmosphere?
10. Why is the earth–atmosphere system considered a steady-state system?

CHAPTER
3

Atmospheric Temperatures

CHAPTER OUTLINE

The Seasons

Daily Temperature Changes
Daytime Heating
Nighttime Cooling
Daily Temperature Range

Seasonal Lag and Extreme Temperatures

Factors Influencing the Vertical Distribution of Temperature

Factors Influencing the Horizontal Distribution of Temperature
Quantitative Expression: Box 3.1 *The Heat Budget*
Latitude
Surface Properties
Aspect and Topography
Dynamic Factors

Temperatures over the Earth's Surface

Summary

Key Terms

Review Questions

T emperature is the most widely used atmospheric measurement. It is a measure of the quantity of energy, or heat, present in a substance. More basically, it is a measure of the kinetic energy of the motion of the molecules in the body. The temperature of a substance measures the amount of heat per unit volume. The total heat in a substance depends not only on the temperature, but on the mass. Raising the temperature of 25 grams (0.9 oz) of water from 20°C to 25°C (68°F–77°F) requires five times more energy than raising 5 grams (.15 oz) of water the same amount. The temperature in the container with 5 grams of water is the same as that in the container with 25 grams (0.9 oz), but the total heat is substantially different.

The temperature of substances, or within substances, determines the direction of the flow of energy. The flow of heat is always from the area of most heat to that of the least heat. Or the flow of heat is from the substance with the highest temperature to that of the lowest. The study of atmospheric temperatures is the study of the ebb and flow of energy at a place and the variation in energy from place to place.

THE SEASONS

Temperature varies with time over the earth's surface due to changes in radiation received at the surface. There are two periodic patterns of radiation influx and temperature that are due to the motions of the earth. One produces the **seasons**, and the other produces the daily changes in radiation and temperature. There are few places, if any, on the earth's surface that are truly without seasons. Seasons are most often thought of as summer, fall, winter, and spring. This is because most people live in the mid-latitudes where temperature change from summer to winter is very large. However, over large areas of the earth, the change from hot to cold is not as important as the change from a rainy season to a dry season. The most important differences in energy received at various locations on Earth result from basic motions of the earth in space. These are rotation on its axis and revolution around the sun. Figure 3.1 is a schematic diagram of these Earth–sun relationships over a period of a year.

In its revolution around the sun, the earth follows an elliptical orbit so that the distance from Earth to the sun varies. Earth is usually closest to the sun (**perihelion**) on January 4 and most distant (**aphelion**) on July 4. The date varies as a result of leap year. The amount of solar radiation intercepted by Earth at perihelion is about seven percent higher than at aphelion. This difference, however, is not the major process in producing the seasons.

The most important element in producing the seasons is the amount of radiation received at a place through the year. The amount of radiation varies as the angle of the sun above the horizon (intensity) and the number of hours of daylight (duration) change through the year. The intensity of solar radiation is largely a function of angle of incidence—the angle at which the solar energy strikes the earth. The angle of incidence directly affects both the energy received per unit area of surface and the amount of energy absorbed. Since the earth is nearly spherical, a curved surface is exposed to the radiation. Intensity of radiation is maximum in latitudes where solar radiation is perpendicular to the surface; this is the solar equator (Fig. 3.2). The intensity of radiation decreases north and south of the solar equator as the angle of incidence decreases.

The seasonal changes in energy result from the inclination of the earth on its axis. The axis of the earth is tilted 23°30′ from being perpendicular to the **plane of the ecliptic** (Fig. 3.3). As the earth revolves about the sun, the solar equator moves north and south through a range of 47°. The geographic equator (0° latitude) is the mean location of the solar equator. The solar equator moves north and south through the

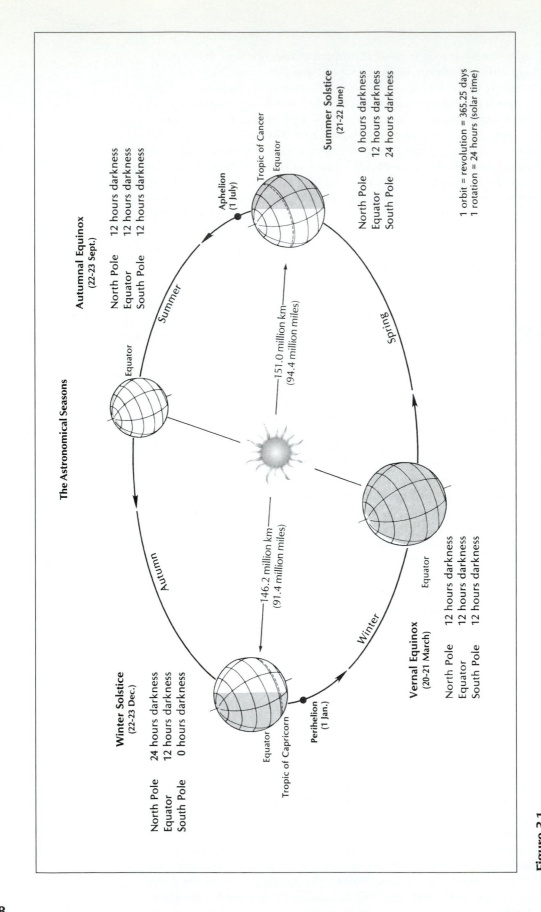

Figure 3.1
Orbit of the earth around the sun, the seasons, and change in length of day.

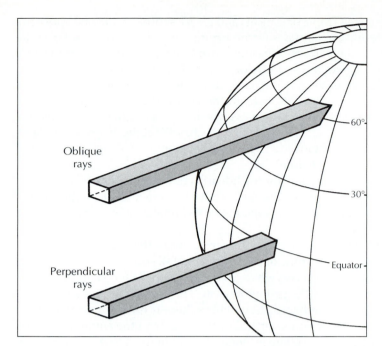

Figure 3.2
Angle of the rays of the sun and intensity of radiation. The same amount of energy is contained in both beams of radiation. The lower the angle of the beam, the larger the area over which the beam is spread.

Figure 3.3
Earth's tilted axis results in variations of the angle of overhead sun and length of daylight. Diagram shows conditions at the equinoxes and solstices.

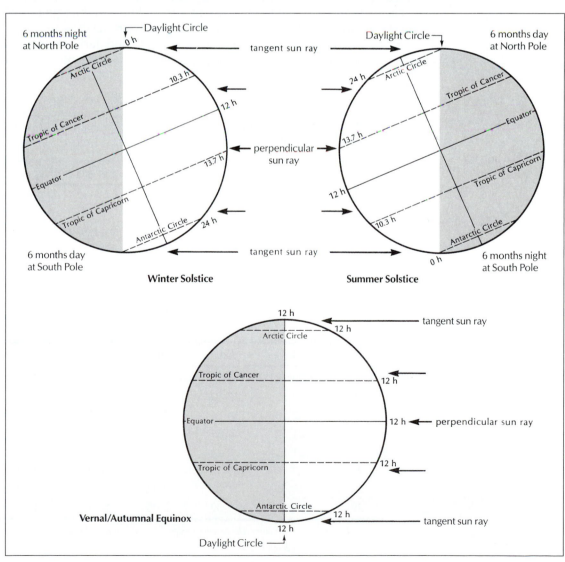

year between 23°30′ north (Tropic of Cancer) and 23°30′ south (Tropic of Capricorn). The two Tropics represent the latitudes farthest from the equator where solar radiation is perpendicular to the surface sometime during the year. As the earth travels around the sun during a year, the intensity of solar radiation varies at all latitudes.

The geographical equator has the least variation in the angle of incidence. Here the sun is never more than 23°30′ from the zenith (the point directly overhead). All latitudes between the two Tropics experience a variation in the angle of incidence between 23°30′ and 47°. The variation reaches 47° at the Tropics. All places between 23°30′ and 66°30′ of latitude experience a change of 47° in the angle of solar radiation through a year's time.

The primary factor responsible for the hot and cold seasons as well as the wet and dry seasons is the revolution of the earth about the sun and the inclination of the earth's axis to its orbital plane. This results in an imbalance of energy over the earth's surface through the year.

The change in the length of the daylight period strengthens the seasonal variation in temperature. Only at the time of the spring and fall equinoxes are the length of daylight and darkness equal everywhere over the earth. There is an imbalance between daylight and darkness the rest of the year. This imbalance is such that the daylight period is longer in the hemisphere where there is maximum intensity of solar radiation. Thus, when the sun's vertical rays are north of the equator, the daylight period in the northern hemisphere is longer than 12 hours. On the summer solstice, the length of daylight increases from 12 hours at the equator to 24 hours at the Arctic Circle (Tables 3.1, 3.2, and 3.3 provide data relating to latitude and various Earth–sun relations). The imbalance in daylight and darkness results in a greater accentuation of the seasons than the position of the solar equator alone would produce.

Table 3.1   Ephemeris of the Sun
(Declination of The Sun on Selected Days of the Year)

Day	Jan.	Feb.	Mar.	April	May	June
1	−23°04′	−17°19′	−7°53′	+4°14′	+14°50′	+21°57′
5	22 42	16 10	6 21	5 46	16 02	22 38
9	22 13	14 55	4 48	7 17	17 09	22 52
13	21 37	13 37	3 14	8 46	18 11	23 10
17	20 54	12 15	1 39	10 12	19 09	23 22
21	20 05	10 50	−0 05	11 35	20 02	23 27
25	19 09	9 23	+1 30	12 56	20 49	23 25
29	18 08	–	3 04	14 13	21 30	23 17

Day	Jul.	Aug.	Sept.	Oct.	Nov.	Dec.
1	+23°10′	+18°14′	+8°35′	−2°53′	−14°11′	−21°40′
5	22 52	17 12	7 07	4 26	15 27	22 16
9	22 28	16 06	5 37	5 58	16 38	22 45
13	21 57	14 55	4 06	7 29	17 45	23 06
17	21 21	13 41	2 34	8 58	18 48	23 20
21	20 38	12 23	+1 01	10 25	19 45	23 26
25	19 50	11 02	−0 32	11 50	20 36	23 25
29	18 57	9 39	2 06	13 12	21 21	23 17

Source: U.S. Naval Observatory, *The American Ephemeris and Nautical Almanac for the Year 1950.*

Table by R. J. List (Washington, DC, 1950).

Table 3.2   Length of Daylight for Intervals of 1° of Latitude*

	0°	1°	2°	3°	4°	5°	6°	7°	8°	9°
0°	12:07	12:11	12:15	12:18	12:22	12:25	12:29	12:32	12:36	12:39
	12:07	12:04	12:00	11:57	11:53	11:50	11:46	11:43	11:39	11:36
10°	12:43	12:47	12:50	12:54	12:58	13:02	13:05	13:09	13:13	13:17
	11:33	11:29	11:25	11:21	11:18	11:14	11:10	11:07	11:03	10:59
20°	13:21	13:25	13:29	13:33	13:37	13:42	13:47	13:51	13:56	14:00
	10:55	10:51	10:47	10:43	10:39	10:35	10:30	10:26	10:22	10:17
30°	14:05	14:10	14:15	14:20	14:26	14:31	14:37	14:43	14:49	14:55
	10:12	10:08	10:03	9:58	9:53	9:48	9:42	9:37	9:31	9:25
40°	15:02	15:08	15:15	15:22	15:30	15:38	15:46	15:54	16:03	16:13
	9:19	9:13	9:07	9:00	8:53	8:46	8:38	8:30	8:22	8:14
50°	16:23	16:33	16:45	16:57	17:09	17:23	17:38	17:54	18:11	18:31
	8:04	7:54	7:44	7:33	7:22	7:10	6:57	6:42	6:27	6:10
60°	18:53	19:17	19:45	20:19	21:02	22:03	–	–	–	–
	5:52	5:32	5:09	4:42	4:18	3:34	2:46	1:30	–	–

*The upper figure in each pair of figures represents the longest day, and the lower figure represents the shortest day. To find the length of daylight for 37 degrees, read down to 30° in the left column and across the row to 7°. The longest day at 37° is 14 hr, 43 min. Note that the two periods do not add up to 24 hr because daylight is measured when the rim of the sun, rather than the center of the sun, is visible.

There are two major aspects of the energy balance that distinguish tropical environments from those of mid-latitudes and polar areas. The first aspect is that the influx of solar energy in tropical areas is high throughout the year even though it does vary some. From this point of view, these are indeed environments without a winter. The intensity of solar radiation is high all year because the rays of the sun are always at a high angle. Also there is very little change in the length of daylight, ranging from 11 to 13 hours throughout the year.

Polar areas differ from those of the Tropics and mid-latitudes in the annual distribution of solar energy. During the winter months in polar regions, direct solar radiation may be absent. But some radiation, including some in the visible range, continues to find its way to the surface. In summer, solar intensity is low, but the duration is long in summer and so the total energy is quite large. Much of the energy goes to evaporate water or melt snow and ice.

The intensity of radiation in the polar regions is never very high for any length of time when compared with mid-latitude or tropical systems. At the Arctic and Antarctic Circles, the sun never climbs more than 47° above the horizon, and at the poles it is never more than 23 1/2° above the horizon. Both polar areas receive more hours of radiation during the year than other parts of the earth due to refraction of the sunlight. When the sun is below the horizon, reflection and refraction of the sunlight rays produce twilight. These processes bring some sunlight to the surface until the sun is 18° below the horizon (astronomical twilight). Thus, at high latitudes, there may be several months with no direct sunshine but almost continuous twilight. The stars, moon, and auroras are also sources of light for these areas so darkness is not so intense as might be and total darkness seldom exists.

The northern hemisphere has the most land area; it is 40 percent land and 60 percent water. The southern hemisphere is only 20 percent land and 80 percent

Table 3.3   Total Daily Solar Radiation Reaching the Ground (cals/cm^2, Transmission Coefficient = 0.6)

	Approximate Date							
Latitude	Mar. 21	May 6	June 22	Aug. 8	Sept. 23	Nov. 8	Dec. 22	Feb. 4
90° N		127	299	125				
80°	6	158	309	156	5			
70°	47	234	349	232	46			
60°	120	312	406	308	118	10		10
50°	202	376	450	372	199	58	19	58
40°	282	426	477	421	278	130	75	131
30°	350	453	481	449	345	213	152	215
20°	404	459	465	454	398	293	237	296
10°	436	444	428	439	430	366	323	370
0°	447	407	372	404	440	422	397	427
10° S	436	353	303	349	430	461	457	465
20°	404	282	222	279	398	475	497	480
30°	350	206	143	204	345	470	514	475
40°	282	125	70	124	278	441	509	445
50°	202	56	18	55	199	391	481	395
60°	120	10		10	118	323	434	327
70°	47				46	242	373	245
80°	6				5	164	330	166
90°						131	319	133

Source: Smithsonian Institution, Smithsonian Meteorological Tables, 1966. Values apply to a horizontal surface.

water. Temperatures outside the Antarctic continent vary much less on a seasonal basis in the southern hemisphere than in the northern hemisphere.

DAILY TEMPERATURE CHANGES

The rotation of Earth on its axis produces alternating periods of day and night and daily variations in temperature result.

Daytime Heating

After the sun comes above the horizon in the morning, the surface begins to heat. If the air is relatively calm, conduction rapidly moves heat from the surface to the boundary layer of air. This heat transfer takes place in a very limited laminar layer of air, often only a few millimeters deep. Gradually, the heat is distributed upward by diffusion of heated molecules of air. On a hot, clear, still day in summer, there may be a thermal gradient of 20°C (36°F) in the lower two meters of air. In tropical deserts, it may become impossible to see through a surveying instrument by 10 a.m. The upward movement of heat so disturbs the air as to make sighting impossible. If wind is blowing, the movement of heat upward is much more rapid and the gradient in the lower two meters is much less. Even with turbulence carrying heat away from the surface, daily changes in temperature normally do not extend above one kilometer.

As expected, temperatures are highest during the day when large amounts of energy are flowing to earth and lowest at night. However, there is not a one-to-one relationship in the time of highest solar input and highest temperature because air temperature largely results from absorbing earth radiation.

Highest daytime temperatures usually occur several hours after the time of maximum solar input, which is local noon. Despite the common notion that the hour of highest temperature occurs at the time when solar energy input equals the flow of outgoing energy, such is not true. Measurements show that equilibrium between incoming and outgoing radiation often occurs about an hour and a half before sunset and not at the time of maximum temperature. During the early afternoon, the earth receives a steady flood of incoming long-wave radiation from the lower atmosphere (the greenhouse effect). It is when this flow of long-wave radiation from the atmosphere reaches a maximum that highest temperatures occur.

The extent of the daily lag in maximum temperature varies. Under normal conditions, the lag will be greatest when the air is still and dry and the sky is free of clouds. In these instances, the maximum temperature of the day may not occur until an hour or so before sunset. When humidity is high or the atmospheric aerosol is unusually thick, the lag will not be as great. Minimum lags occur when there is a cloud cover such that the incoming and outgoing radiation are reduced.

Actual daily high and low temperatures can occur at any time of day if there is a shift in wind direction and cold or warm air is fed into the area. This is typical of temperature changes brought about by circulation around mid-latitude lows and also by the development of land and sea breezes.

Moist soil or a cover of vegetation also reduces the maximum temperatures. The process is that of evaporation or transpiration. Evaporation is a major cooling mechanism. Five hundred and ninety calories of heat are taken from the environment for each gram of water that evaporates. This is why it never gets as warm in summer over water as it does over land. Cities east of the Mississippi River generally have mean daily high temperatures and extreme highs, which are 10°C (18°F) lower than cities in the western desert. The higher atmospheric humidity and a surface covered by vegetation reduce the temperature of the lower atmosphere.

Nighttime Cooling

Once the sun passes the zenith, radiation intensity starts to decrease. Sometime near sunset on a clear day, the ground surface and the boundary layer of air begin to receive less energy than they emit and begin to cool. The ground surface cools most rapidly since it is a better radiator than air. Soon the coolest air is close to the ground and temperature increases with height. This is counter to the standard atmospheric state.

Earth radiation decreases through the hours of darkness and the surface cools. How much cooling occurs depends on how long the night is, how high the humidity is, and on wind velocities. Cooling is greatest on a night when humidity is low, the sky is clear, and the wind is calm. A wet surface retards cooling just as it retards warming. Dew, frost, and fog are major feedback mechanisms that prevent the atmosphere from getting still colder. When dew forms, 590 calories of heat are added to the atmosphere for each gram of water. This heat added by condensation offsets the heat lost by radiation. Frost is an even more effective means to control cooling. Six hundred eighty calories of heat are added to the boundary layer of air for each gram of water that sublimates as frost. Fog likewise is a major block to cooling. Not only does the condensation add heat to the air, but the fog acts as a blanket reducing radiative heat loss. In desert regions, dew and frost are very important in reducing nighttime cooling. Minimum daily temperatures occur in the early morning when incoming solar energy and outgoing earth radiation balance. Shortly after dawn, incoming solar rays provide enough energy to balance outgoing terrestrial rays.

Daily Temperature Range

The daily range in temperature is a function of both daytime heating and nighttime cooling. When conditions are good for rapid inflow of radiant energy in the daytime and outgoing radiation at night, the range will be large. The conditions that favor high daily ranges in temperature are clear skies, low relative humidity, and calm winds.

The magnitude of the daily change in temperature varies a great deal. Near the equator, the daily range exceeds the annual range. Near the two poles, the daily range is reduced to almost zero since there is generally only one daylight and one nighttime period each year.

Over the ocean, the daily range is also small. There are several reasons for the low range. First is the high specific heat of water. It heats slowly and cools slowly. Second, there is mixing of the surface water with the water below that modifies heating and cooling. Third, solar radiation penetrates deeper into the ocean than into the land. This distributes the heat more evenly with depth.

The daily range is a result of solar heating during the day and earth radiation at night. The daily variation decreases rapidly with height. The greatest changes are in the boundary layer and it decreases upward to about one kilometer (0.6 mi). Above 1 or 2 kilometers (0.6–1.2 mi), the daily variation is minimal.

In North America, daily ranges are a function of cloud cover and humidity. In the dryer regions, daily ranges may exceed 22°C (40°F), whereas in more humid regions, the normal range is nearer 17°C (30°F). The greatest known recorded daily range in temperature occurred in North Africa, where the temperature dropped from a high of 56°C (132°F) in the afternoon to 0°C (32°F) the following morning. The difference is an amazing 56°C (100°F).

SEASONAL LAG AND EXTREME TEMPERATURES

There is a seasonal lag of maximum and minimum temperatures. The revolution of the earth causes maximum and minimum solar energy (outside the Tropics) to occur at the time of the summer solstices in each hemisphere. Thus, June and December represent the time of maximum and minimum solar energy receipts in the northern hemisphere. The reverse holds true in the southern hemisphere. The months of maximum and minimum solar energy are not the warmest or coldest. Table 3.4 provides examples showing that there is a month or more lag between the time of maximum and minimum solar radiation and the warmest and coldest months.

Temperature extremes are those temperatures farther from normal and that occur least frequently. For extreme temperatures of either heat or cold to be reached, conditions for incoming and outgoing radiation must be optimum. For high temperatures, there must be high intensity solar radiation, long hours of sunlight, clear dry air, little surface vegetation, and relatively low wind velocities. These conditions are optimal near the Tropics of Cancer and Capricorn near the time of the solstice. Here the atmosphere is dominated by high pressure and clear skies. The highest recorded surface temperature was measured in El Aziz, Libya, in September 1922. Temperatures reached 58°C (136°F). In North America, the highest temperature recorded is 57°C (134°F) in Death Valley, California, in July 1913. At any location, the record high temperatures occur when the air is clear and dry.

Record lows occur under conditions that favor the loss of energy from the surface and lower atmosphere by direct radiation to space. These conditions occur with high-pressure systems near the poles. The coldest temperature ever recorded is −89°C (−128°F) on the Antarctic Continent at Vostock in 1983. The cold temperature is in part due to the high altitude of the ice mass. In the northern hemisphere, the lowest temperature recorded is −68°C(−90°F) at Verkoyansk, USSR. In North America, the coldest temperature on record is −62°C at Prospect Creek, Alaska. The records of both cold and heat occur away from the oceans. The ocean greatly modifies temperature extremes.

Table 3.4   Sample Data Illustrating Seasonal Temperature Lag

	Average Temperature at Solstice		Average Monthly Temperature	
	June	*Dec.*	*Warmest Month*	*Coldest Month*
Charleston, SC (33°N)	26°C (79°F)	11°C (51°F)	July 28°C (82°F)	Jan. 10°C (50°F)
Urbana, IL (40°N)	22°C (72°F)	0°C (32°F)	July 25°C (77°F)	Jan. −1°C (30°F)
Naples, Italy (41°N)	22°C (72°F)	11°C (51°F)	Aug. 25°C (77°F)	Jan. 9°C (48°F)
Moscow, Russia (43°N)	19°C (66°F)	−6°C (22°F)	July 21°C (70°F)	Jan. −8°C (17°F)
Edmonton, Canada (53°N)	14°C (57°F)	−8°C (18°F)	July 17°C (63°F)	Jan. −14°C (7°F)

FACTORS INFLUENCING THE VERTICAL DISTRIBUTION OF TEMPERATURE

In the standard atmosphere, the decrease of temperature with height in the troposphere (the **lapse rate** or **environmental lapse rate**) is given as an average of 6.5°C per kilometer (3.5°F per 1000 ft). This value reflects the difference in the average temperature of the surface and the temperature at 11 kilometers (15°C and −59°C, respectively). However, the measured lapse rate varies appreciably from the mean depending on the nature of the air mass and the surface over which the lapse rate is measured.

The nature of the underlying surface influences the vertical distribution of temperature. For example, temperature decreases most rapidly with altitude over continental areas in summer. Figure 3.4 shows a cross section of the atmosphere up to 23 kilometers (14 mi). The profile is along the 80th meridian (which passes through eastern Canada, the United States, Cuba, and Panama) from the North Pole to the equator in January and July. The heavy lines on the graphs show the tropopause. Note the following features:

1. The tropopause is lower over high latitudes than low latitudes. A well-marked break occurs in the middle latitudes.
2. The north–south temperature gradients are much steeper in winter.
3. The strongest horizontal gradients are in middle latitudes in both summer and winter. This is the region of maximum storm activity.
4. The coldest part of the troposphere occurs over the equator in the region of the tropopause.

The **diurnal range** of temperature at higher elevations is more than at sea level. Figure 3.5 provides a graphic model of this effect. The reason for the greater range is that the atmosphere is less dense at higher altitudes. Maximum daily temperatures are about the same or slightly less than at low altitudes. The main difference occurs at night when heat escapes much more readily at high elevations because of the lower density of gases.

Decrease of pressure with altitude changes our everyday interpretation of some values on temperature scales. The gas laws state the relationships between temperature

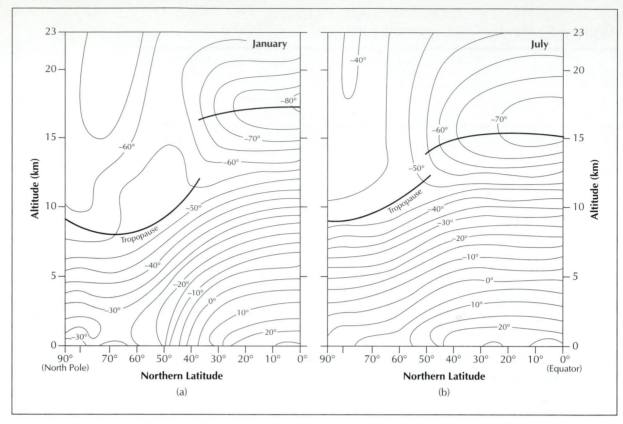

Figure 3.4
Mean vertical cross sections through the atmosphere in (a) January and (b) July. Both sections are along the meridian 80°W in the northern hemisphere.

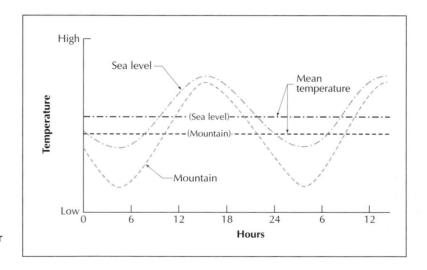

Figure 3.5
Daily range of temperature in a mountain and a lowland station at places of similar latitude.

and pressure. At higher elevations, pressure changes become important. Under reduced pressure, molecules of water vapor escape more easily from a water surface. Thus, at sea level, water boils at a temperature of 100°C (212°F). At Quito, Ecuador, at an elevation of 2824 meters (9350 ft), water will boil at 90.9°C (196°F). At the top of Mount Everest, the boiling point of water is only 71°C (160°F).

FACTORS INFLUENCING THE HORIZONTAL DISTRIBUTION OF TEMPERATURE

The temperature that occurs at any location is essentially a result of the net radiation available and the way that radiation is budgeted. It has already been pointed out that net radiation ($Q*$) depends on gains of solar and terrestrial energy; the available energy is then used for sensible heat transfer and evaporation. Box 3.1, the heat budget, expresses these exchanges in a little more detail using symbols; some

QUANTITATIVE EXPRESSION: BOX 3.1

The Heat Budget

Consider a column of the earth's surface extending down to where vertical heat exchange no longer occurs (Figure A).

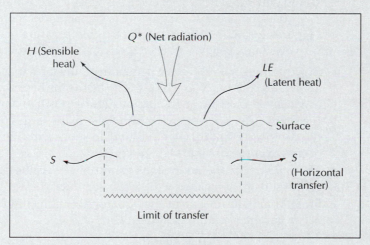

Figure A

The net rate (G) at which heat in this column changes depends on the following:

  Net radiation $(K\uparrow - K\downarrow) + (L\uparrow - L\downarrow)$,
  Latent heat transfer (LE),
  Sensible heat transfer (H), and
  Horizontal heat transfer (S).

In symbolic form:

  $$G = (K\uparrow - K\downarrow) + (L\uparrow - L\downarrow) - LE - H \pm S.$$

Since

  $$(K\uparrow - K\downarrow) + (L\uparrow - L\downarrow) = Q*,$$

then

  $$G = Q* - LE - H \pm S$$

in terms of $Q*$

  $$Q* = G + LE + H \pm S.$$

The column will not experience a net change in temperature over an annual period; that is, it is neither gaining nor losing heat over that time, so $G = 0$ and can be dropped from the equation.

  $$Q'* = LE + H \pm S.$$

This equation will apply to a mobile column, such as the oceans. On land where subsurface flow of heat is negligible, S will be unimportant. The land heat budget becomes

  $$Q* = LE + H.$$

The ratio between LE and H is given as the *Bowen Ratio*.

components of this representation are considered later in this chapter. The object here is to survey those physical components that influence the energy flows of the heat budget. They include latitude, surface properties, position with respect to warm or cool ocean currents, and elevation.

Latitude

The highest temperatures on Earth do not occur at the equator, but near the Tropics of Cancer and Capricorn. A partial explanation of this occurrence is the migration of the vertical rays of the sun between 23.5° N and 23.5° S. The vertical rays of the sun move quickly over the equator, but slowly as they progress north and south. Thus, between 6° north and south, the sun's rays are near vertical for 30 days around the equinoxes. Between 17.5° and 23.5° N and S, near vertical rays occur for 86 days around the solstice. The longer period of high sun and the concurrent longer days allow time for surface heat to accumulate. Thus, the zone of maximum heating and highest temperatures is near the Tropics of Cancer and Capricorn. The predominance of clear skies near the Tropics enhances heating even more compared with the very cloudy equatorial belt.

If the earth were a homogenous body without the land–ocean distribution, its temperature would change evenly from the equator to the poles. It is possible to compare actual temperatures with hypothetical data for a uniform surface and identify areas where temperature anomalies occur. Anomalies represent deviations from temperatures that would only result from solar energy. Isolines called **isanomals** are drawn through points of equal temperatures to produce isanomalous temperature maps of the world.

Figures 3.6 and 3.7 show isanomalous temperature maps for January and July, respectively. Some general observations can be made from studying the maps.

a. In winter in each respective hemisphere, the land masses have large negative anomalies of as much as −20°C (−4°F) below the hypothetical mean. By contrast, ocean areas show no anomalies or slightly positive ones.

b. In summer in each respective hemisphere, the largest anomalies are also over the continents, but they are positive. In some areas, a small negative anomaly occurs over the oceans.

c. The patterns of the anomalous conditions over the oceans are distinctive in shape and are associated with ocean currents.

d. Small anomalies occur in the equatorial realms, and the highest anomalies occur in upper middle latitudes.

The major factors that alter the zonal solar climate and hence the patterns of temperature over the globe are identified from these observations. The controls are of two types. First are those factors that are due primarily to geographical location on the earth's surface. This basically determines the amount of energy received from the sun. In contrast to these location factors are temperature characteristics that result from the transport of energy by the mobile atmosphere and ocean. This dynamic effect can substantially change the temperatures of a place.

Surface Properties

The disposition of solar energy striking a surface largely depends on the type of surface. Of particular note is **albedo**. Surfaces with high albedo absorb less incident radiation, so there is less total energy available. Polar icecaps are maintained because they reflect as much as 80 percent of the solar radiation falling on them.

Even if two surfaces have a similar albedo, the incident energy does not always result in similar temperatures since the heat capacities of the surfaces may differ. The heat capacity of a substance is the amount of heat required to raise its temperature. *Specific heat* is the amount of heat (number of calories) required to raise the temperature of 1 gram (.035 oz) of a substance through 1°C (1.8°F). As Table 3.5 shows, the specific heat of substances can vary appreciably.

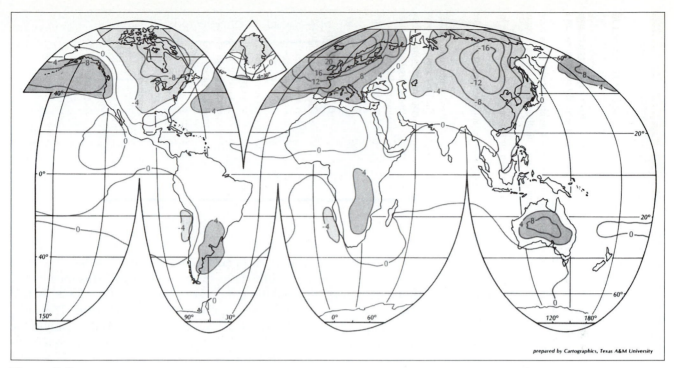

Figure 3.6
Isonomalies of temperatures (°C) in January.

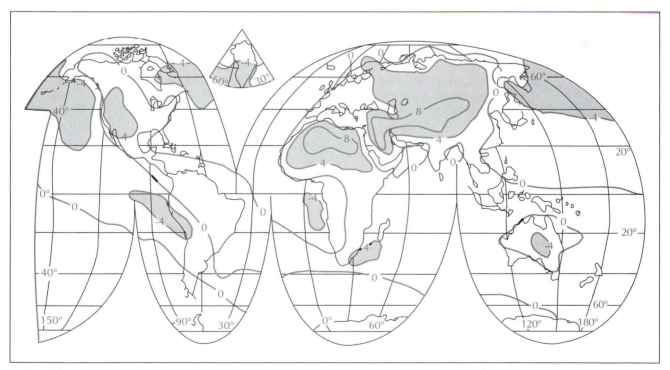

Figure 3.7
Isonomalies of temperatures (°C) in July.

Table 3.5 Specific Heat of Selected Substances

Substance	Specific Heat (cal/g/°C)
Water	1.0
Ice (at freezing)	0.5
Air	0.24
Aluminum	0.21
Granite	0.19
Sand	0.19
Iron	0.11

Of particular significance is that the **specific heat** of water is some five times greater than that of rock material and the land surface in general. This means that the amount of heat required to raise the temperature of water 1°C (1.8°F) is five times greater than required for the same temperature increase on land. The same amount of energy applied to a land surface and a water surface would result in the land becoming much hotter than the water. The difference increases due to the different heat conductivity of the earth materials and water (Table 3.6). Loose, dry soil is a very poor conductor of heat. Only a superficial layer will experience a rise in temperature from solar radiation. Water has only a fair conductivity, but its general mobility and transparency permit heat to circulate well below the surface. A natural undisturbed soil with a vegetation cover may have daily temperature changes to a depth of one meter. A quiet pool of water has daily temperature variations that can be measured to a depth of perhaps six meters (20 ft).

Land masses heat much more rapidly than oceans in summer. However, since this heat concentrates near the surface, it rapidly radiates to the atmosphere as winter approaches. Hence, land masses tend to experience extreme temperatures, whereas water bodies are more equable and show less change (Fig. 3.8).

As noted in an earlier chapter, absorbed energy raises the temperature of the absorbing surface, which then radiates according to its temperature. Net energy—that received by the surface but not radiated back to the atmosphere—is then either passed to the air in the form of sensible heat (H) or used to evaporate moisture from that surface and transferred as **latent heat** (LE) as shown in Box 3.1. The amount of sensible heat available to pass to the atmosphere depends in part on the fraction of net energy that

Table 3.6 Thermal Conductivity of Selected Substances

Substance	Heat Conductivity*
Air	0.000054
Snow	0.0011
Water	0.0015
Dry soil	0.0037
Earth's crust	0.004
Ice	0.005
Aluminum	0.49

*The number of calories passing through an area 1 cm^2 in a second when temperature gradient is 1°C/cm. It is expressed in cal/cm^2/sec/°C/cm.

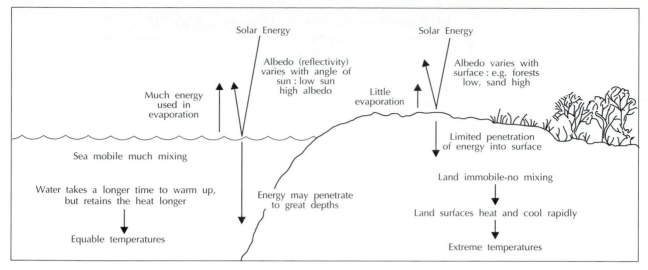

Figure 3.8
Solar energy striking land and sea surfaces divides in different ways. The water surfaces are conservative, warming and cooling slowly; the land masses experience large temperature changes.

goes to evaporate water. Over the oceans, evaporation is continuous. On land, it depends on the amount of water available at the surface. The significance of this process is determined by comparing the ratio of sensible heat to latent heat for different parts of Earth. The Bowen Ratio, which expressed this relationship, is H/LE. The larger the ratio, the more sensible heat, rather than latent heat passing to the atmosphere.

The relative division of LE and H in the energy budget equation varies at individual stations. Figure 3.9 provides examples of a tropical moist station (West Palm Beach, Florida), a desert station (Yuma, Arizona), and a station on the west coast (Astoria, Oregon). Note the large amount of net radiation that passes to LE in the moist coastal locations. By contrast, the desert example shows much of the net radiation becomes sensible heat. Since the pattern of temperature follows the sensible heat flow, temperatures of places located in moist environs will never experience the very high temperatures found in arid locations. Cities along a coast have smaller ranges in temperature than inland cities. Cities that have the wind blowing onto the land have a lower range than coastal cities, where the dominant wind direction is offshore.

Inland cities, particularly in the northern hemisphere, have the highest annual ranges in temperature. In North America, ranges are greatest in the Great Plains of northern United States and southern Canada. The most extreme annual range is found in the Soviet Union. At Yakutsk, the annual range is 62°C (112°F). This represents a change in mean temperature of more than 13°C (24°F) per month from January to July. So distinctive is this effect that meteorologists use the term **continentality**, a measure of the continental influence on weather and climate. Various indexes to measure continentality have been devised, some of which are considered in the regional setting of later chapters.

Aspect and Topography

The combined influences of the steepness and direction a slope faces determine its aspect. Differences that occur on north-facing and south-facing slopes in the northern hemisphere illustrate the importance of aspect. A north-facing slope may have snow on it, whereas a south-facing slope is bare. The north slope gets less intense radiation and, as the sun gets lower in the sky, will be in shadow long before the south-facing slope.

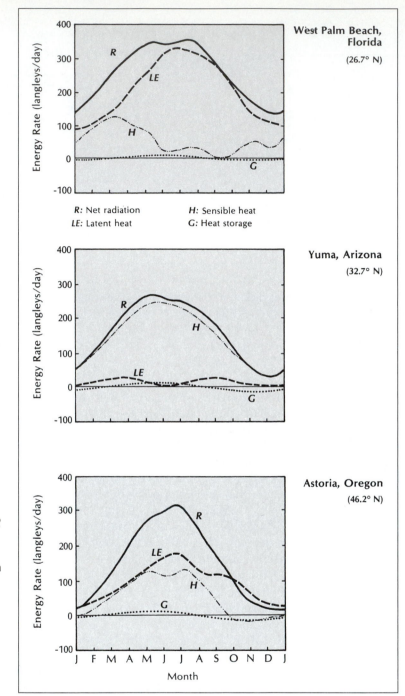

Figure 3.9
Average annual variation in the components of the surface energy budget at three different locations. *R* is net radiation, *LE* is latent heat, and *H* is sensible heat. Note that in the dry climate (Yuma), almost all net radiation goes to sensible heat (from Sellers W. D., *Physical Climatology*, © 1965, Chicago: The University of Chicago Press).

Slope aspect influences many natural phenomena. For example, the height of permanent snow and ice on mountains varies from one slope to another as does the tree line. Similarly, the depths of snow and frost differ on north- and south-facing slopes.

Topography also plays an important role in climates of some lowlands. On a continental scale, mountain ranges that run north–south have a very different effect from those that run east–west. The lack of any extensive east–west mountain barrier in the United States permits polar and tropical air to penetrate long distances into the continent. One result of this unobstructed flow of air is the high incidence of tornadoes in the United States.

Table 3.7   Theoretical Temperatures in Stationary Atmosphere and Actual
Temperatures by Latitude

	Equator	10°	20°	Latitude 30°	40°	50°	60°	70°	80°
Temperature in °C (°F)									
Northern Hemisphere									
Planetary temp.	33(91)	32(89)	28(83)	22(72)	14(57)	3(37)	−11(12)	−24(−11)	−32(−26)
Actual temp.	26(79)	26(80)	25(78)	20(69)	14(57)	5(42)	−1(30)	−10(13)	−18(−1)
Difference	−7(−12)	−6(−9)	−3(−5)	−2(−3)	0	+2(+5)	+10(+18)	+18(+24)	+14(+25)
Southern Hemisphere									
Planetary temp.	33(91)	32(89)	28(83)	22(72)	14(57)	3(37)	−11(12)	−24(−11)	−32(−26)
Actual temp.	26(79)	25(78)	22(73)	17(62)	11(53)	5(42)	−3(26)	−13(8)	−27(−17)
Difference	−7(−12)	−7(−11)	−6(−10)	−5(−10)	−3(−4)	+2(+5)	+8(+14)	+11(+19)	+5(+9)

Dynamic Factors

The largest imbalance of energy is between tropical regions and the poles. This
imbalance is partly alleviated by the transfer of latent heat (LE), sensible heat (H),
and heat stored in the water of the oceans (S). Table 3.7 provides the theoretical
planetary temperature for sea level and the actual mean annual temperature for
every 10° of latitude. The biggest differences are at the equator and for latitudes
above 60°. Tropical latitudes are cooler than the theoretical value, whereas high latitudes are warmer. The differences between the actual and theoretical values result
from the transport of energy over the earth by air and ocean currents. Every storm
system, circulation pattern, and evaporation/precipitation event aid the moving
of heat from tropical regions toward the poles.

TEMPERATURES OVER THE EARTH'S SURFACE

All of these factors play a role in determining the distribution of average temperatures over the earth's surface. Figures 3.10 and 3.11 show the distribution of mean
July and January temperatures, respectively. Figure 3.12 shows the annual average
range of temperature.

In each case, the general decline in temperatures from equator to poles is clear,
illustrating the basic influence of latitude. However, significant variations from a
simple zonal pattern exist, and these result from a combination of other temperature controls. Some of these variations are:

1. The maps show the temperature extremes that occur over continental land
   masses. The largest land mass, Asia, experiences average temperatures that
   range from 4°C (40°F) in January to more than 15°C (60°F) in summer. The effect, seen in Figure 3.12, clearly shows the effect of the size of the continent on
   the average annual range of temperature.
2. The oceans exhibit the results of heat transport by ocean currents. The displacement of isotherms toward the poles shows the warming effects of the
   North Atlantic and North Pacific Drifts. Cold ocean currents, such as the
   California and Canaries currents, cause temperatures to be lower.

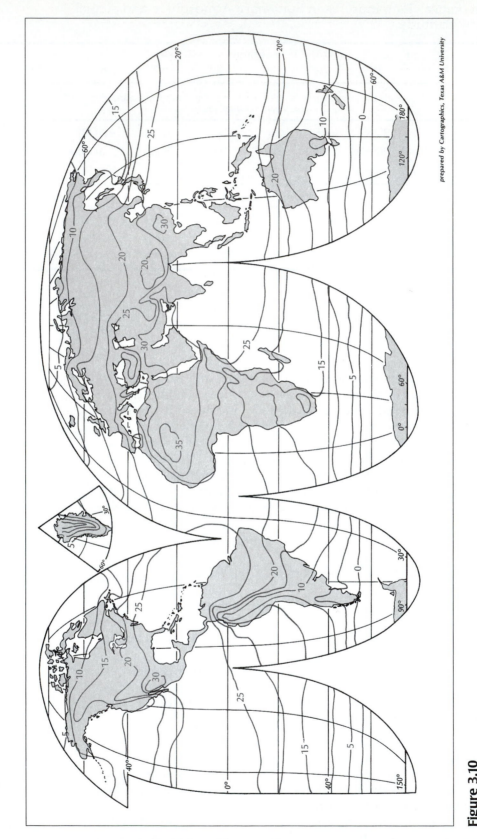

Figure 3.10
Distribution of average July temperatures over the earth (°C).

prepared by Cartographics, Texas A&M University

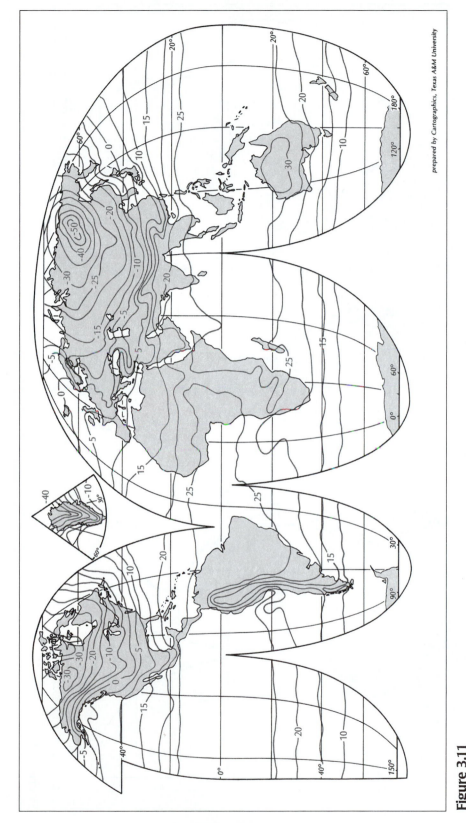

Figure 3.11
Distribution of average January temperatures over the earth (°C).

prepared by Cartographics, Texas A&M University

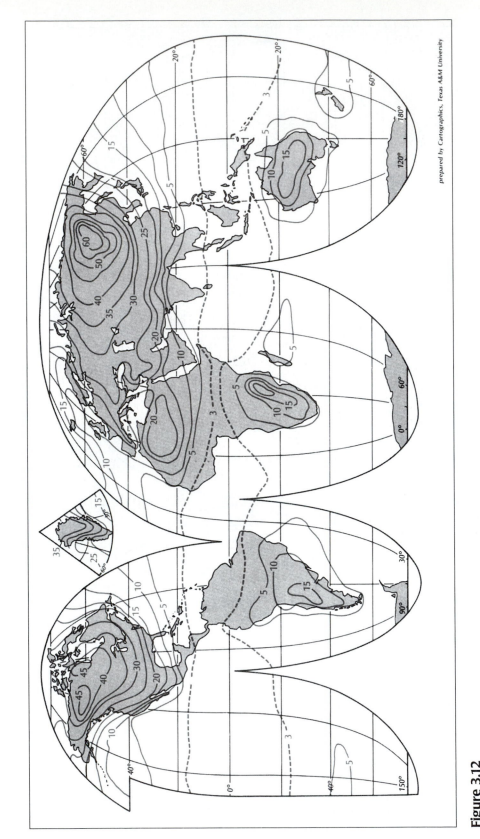

Figure 3.12
Distribution of mean annual range in temperature over the earth (°C).

3. The maps show actual temperatures that include the effect of altitude. Some world temperature maps have isotherms reduced to sea level, which eliminates the effect of elevation. Notice, for example, the isotherms over South America. The Andes show as a tongue of colder temperatures extending toward the equator.

4. The closest approximation to zonal temperatures occurs over the southern ocean and Antarctica. This area is the most extensive location where homogenous surfaces encircle the globe without interruption.

5. Compare Figures 3.10 and 3.11 and note that the hemispheric temperature gradient is steepest in winter. For example, in January, the northern hemisphere gradient is more than 60°C (110°F). In July, it is about 10°C (18°F). This pattern has highly significant results in atmospheric circulation.

SUMMARY

Temperature measures the amount of heat energy available over the earth's surface and is a function of the net radiation at a location. There is no direct relationship between maximum and minimum solar radiation and highest and lowest temperatures. A lag exists in both daily and seasonal temperatures. Care must be taken when using mean temperature data for means do not provide a full assessment of the temperature regime. Some measure of the variation of data about the mean is necessary for complete description. Similarly, the use of the mean for a given period obscures temperature changes within the time period.

The lapse rate represents the vertical distribution of temperature in the troposphere. In the Standard Atmosphere, the lapse rate is 6.5°C for each kilometer (3.5°F per thousand ft). The horizontal distribution of temperature over the earth's surface is a function of both location and the dynamic movement of the atmosphere. The former consists of such factors as latitude, surface properties, and aspect. The latter consists of energy transferred by the atmosphere and the oceans.

KEY TERMS

Albedo
Aphelion
Continentality
Diurnal range
Environmental lapse rate

Isanomals
Lapse rate
Latent heat
Perihelion
Plane of the ecliptic

Radiation intensity
Seasons
Solar equator
Specific heat

REVIEW QUESTIONS

1. Why does the earth have seasons?

2. What determines the intensity of solar radiation at a location at a given time?

3. What is the solar equator, and how does it differ from the geographic equator?

4. Assuming the same air mass is in place and calm conditions exist, outline the processes involved in temperature change over a 24-hour period.

5. Explain (a) the daily lag, and (b) the seasonal lag of temperature.

6. Explain isanomals and their significance.

7. What is specific heat, and what role does it play in the climate of a location?

8. Explain how temperatures in the center of a continent might differ from those at a coastal location in the same latitude.

9. Explain the Bowen Ratio and why it varies from location to location.

10. What are some of the temperature differences caused by aspect?

CHAPTER
4

Moisture
in the Atmosphere

CHAPTER OUTLINE

*T*he significance of water as an atmospheric variable is a result of its unique physical properties. Water is the only substance that exists as a gas, liquid, and solid at temperatures found at the earth's surface. This special property enables water to cycle over the earth's surface. While changing from one form to another, it acts as an important vehicle for the transfer of energy in the atmosphere.

CHANGES OF STATE

The chemical symbol of water, H_2O, is probably the best known of all chemical symbols. It tells us that the water molecule is made up of two atoms of hydrogen for one of oxygen. Water in all of its states has the same atomic content. The only difference is the arrangement of the molecules (Fig. 4.1). Little energy is available at low temperatures, and the bonds binding the water molecules are firm. The water molecules pack tightly in a fixed geometric pattern in the solid phase. As temperature increases, the available energy causes the bonds of the ice phase to weaken. Because they are not firmly set, bonds form, break, and form again. This permits flow to occur and represents the liquid phase of water. In this liquid stage, there is still bonding, but it is much less compact than in the ice phase. At higher temperatures and with more energy, the bonding of the water molecules breaks down and the molecules move in a disorganized manner, which is the gas phase. If the

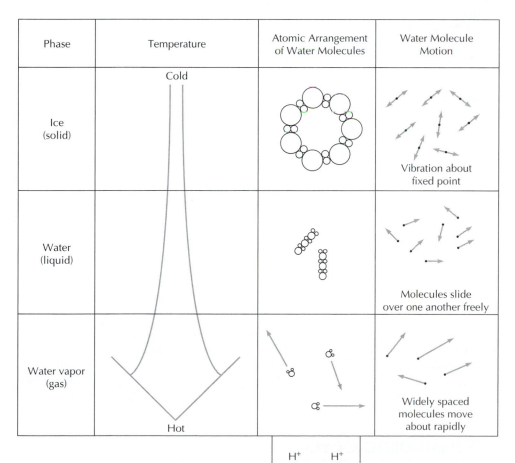

Figure 4.1
Schematic diagram of the arrangement and motion of water molecules in different phases.

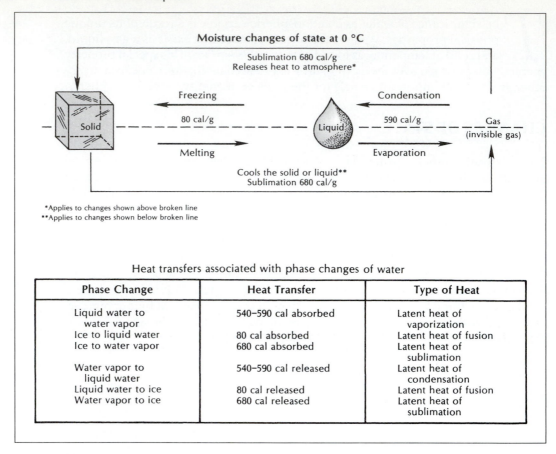

Figure 4.2
Changes of state of water.

temperature decreases, the molecules will revert to a less energetic phase and reverse the processes. Gas will change to liquid and liquid to solid.

Figure 4.2 illustrates these changes of phase. Note that the processes of melting, evaporation, and sublimation from the solid to liquid phase absorb energy. This added energy causes the molecules to change their bonding pattern. The amount of energy incorporated is large for the changes to the water vapor stage and much lower for the change from ice to water.

The energy absorbed is latent energy and goes back to the environment when the phase changes reverse. When water vapor changes to liquid, it releases the energy originally absorbed and retained as latent heat. The same is true when water freezes and water vapor sublimates to ice.

The significance of the release of latent heat shows in many ways. It provides energy to form thunderstorms, tornadoes, and hurricanes. It also plays a critical role in the redistribution of heat energy over the earth's surface. Because of the high evaporation in low latitudes, air transported to higher latitudes carries latent heat with it. This condenses and releases energy to warm the atmosphere in higher latitudes.

THE HYDROLOGIC CYCLE

The **hydrologic cycle** is a conceptual model of the exchange of water over the earth's surface. Figure 4.3 shows one model of the hydrologic cycle. It shows large-scale changes of state, with evaporation providing the moisture that

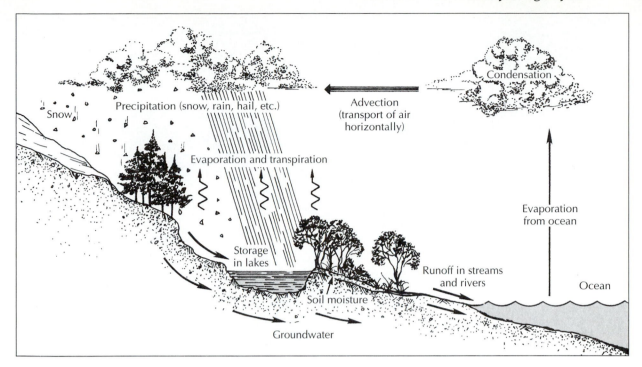

Figure 4.3
The hydrologic cycle.

condenses and becomes precipitation. Additionally, the diagram shows that a process of transpiration occurs. Transpiration is, in some ways, a special form of evaporation in that moisture does return to the air through evaporation but it is via vegetative processes. In dealing with evaporation and transpiration amounts, climatologists often combine them into a single parameter called **evapotranspiration**.

Some of the precipitation that falls to the surface passes to the soil to become soil moisture, which growing plants use; some passes deeper into the ground to become groundwater. Other precipitation runs off the surface and is collected in ponds, lakes, and reservoirs or flows as surface water in streams and rivers. Eventually water finds its way to the oceans and starts the cycle over again.

How is water distributed over the earth? The divided circle in Figure 4.4 shows that, of all the water available on Earth, 97 percent occurs in the oceans. The remaining three percent is mostly ice found in the large ice caps of the world. Almost all of the rest is in groundwater. Rivers, lakes, and soil moisture account for less than one percent of the total water. The atmosphere contains only 0.35 percent of all the water available. Yet this small amount is the reservoir that provides the moisture for clouds and precipitation that occurs over the earth's surface.

The actual exchanges of water provide some surprising data. Assume that the average precipitation of the world (the amount that would fall if every place on Earth got the same amount) is 85.7 centimeters (33.8 in) and let this amount equal 100 percent. Most water evaporates from the ocean, and the greater part of all precipitation falls over the ocean. As illustrated in Figure 4.5, the amount of water that falls on the land is appreciably less (23%). This is expected, however, for the oceans provide much of the available moisture and occupy a much greater area than the continents.

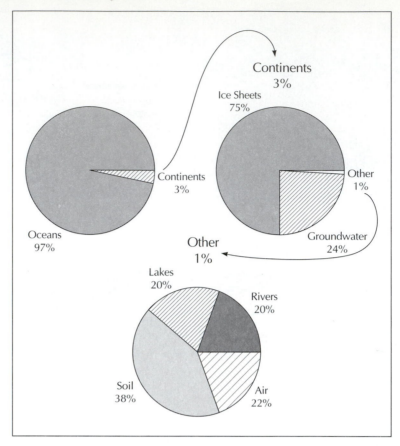

Figure 4.4
The proportional distribution of water over the earth's surface.

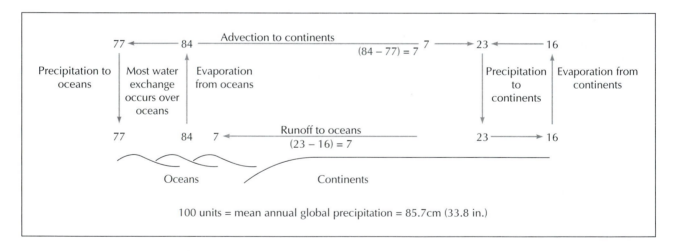

Figure 4.5
Estimates of water transfer within the hydrologic cycle. A value of 100 is used to denote average global precipitation.

RELATIVE HUMIDITY

One widely used measure of water vapor in the atmosphere is relative humidity. The calculation of relative humidity depends on the maximum amount of moisture that air can hold—the saturation level.

$$\text{Relative humidity} = \frac{\text{vapor pressure in the air}}{\text{saturation vapor pressure}}.$$

The relative humidity is stated as a percent so the fraction is multiplied by 100.

The maximum amount of water vapor that may be in the air is mainly a function of the temperature of the air. Warm air may contain more moisture than cold air. In fact, the maximum amount of moisture that may be in the air increases rapidly with increasing temperature. This is because evaporation rates increase rapidly with temperature. Figure 4.6 illustrates this feature.

Of special significance is the **dew point** or dew point temperature. This is the temperature to which a parcel of air must cool before condensation exceeds evaporation. The name provides a good guide to the meaning for dew provides visible evidence that the air has reached a critical temperature as far as the physical state of water in the atmosphere.

On a cool, calm night, the air near the ground loses heat by radiation to the air above. With little mixing of the air, a thin layer of air next to the ground becomes cold. If the temperature drops far enough, condensation will exceed evaporation and moisture condenses on the ground in the form of dew. The temperature at which this transition occurs is the dew point temperature. At temperatures below freezing, it is called the frost point rather than dew point. This is the temperature to which air must cool for the formation of ice crystals from water vapor. The formation of layers of ice on a car window after a cold night is a clear sign of cooling to the frost point. Often because of the way in which heat escapes from an object such as a car, ice appears on the windows and nowhere else.

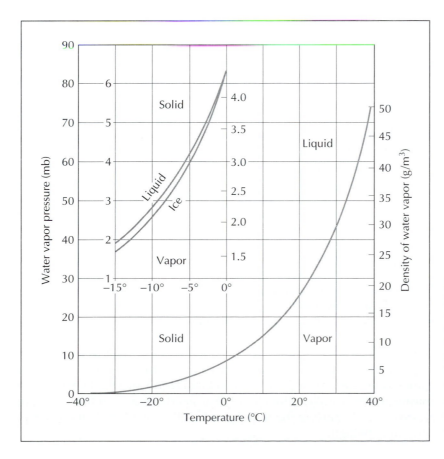

Figure 4.6
Saturation vapor pressure as a function of temperature. Values for below 0°C are shown in the inset (from Miller A., Thompson J. C., Peterson R. E., and Haragan D. R., *Elements of Meteorology*, 4e., © 1983, Prentice Hall, Upper Saddle River, N.J.).

EVAPORATION AND TRANSPIRATION

Evaporation is the process by which water changes from a liquid to a gaseous state. Water is sufficiently volatile in solid and liquid states to pass directly into the gaseous state at most environmental temperatures. Sublimation is the process of change from ice to water vapor. Since the source of water vapor is at the earth's surface, the amount of water vapor present in the atmosphere decreases with height. Most atmospheric moisture is found below 10,000 meters.

The amount of water that actually evaporates from a given water surface in a given period depends on the following factors:

1. Vapor pressure of the water surface. This factor depends on water temperature. The higher the water temperature, the greater the surface vapor pressure. When water temperature is higher than the air temperature, evaporation will always take place.
2. Vapor pressure of the air. The greater the vapor pressure of the air, the less evaporation there will be. The rate of evaporation varies directly with the difference between vapor pressure of the water surface and the vapor pressure of the air.
3. Wind. Air movement is usually turbulent, with moist air removed from near the water surface and replaced by dry air from above. Evaporation thus varies directly with the velocity of the wind. The higher the wind velocity, the more evaporation there is.

The analysis of terrestrial water balances is not quite as simple as that of water bodies. For land areas, the net atmospheric transfer of water vapor depends on the amount of water available and the processes of evaporation and transpiration. Transpiration refers to the loss of water from living plants. Transpiration cools the plant and keeps its temperature within the tolerable limits. Transpiration is difficult to measure in the field so climatologists use evapotranspiration. It is the combined loss from the surface through evaporation and transpiration. A wide variety of factors affect evapotranspiration. Among them are the following:

1. Radiation intensity
2. Atmospheric temperature
3. Atmospheric dew point
4. Length of day (photoperiod)
5. Wind velocity
6. Type of vegetation
7. Soil moisture conditions
8. Type of precipitation

Evapotranspiration from a mixed land-and-water surface is nearly always less than evaporation would be from a similar-sized water surface. On a global scale, evapotranspiration for the continents is some 470 millimeters per year. From the ocean, it is 1300 millimeters per year. Average evapotranspiration from the continents varies through time and space. The major variables are the amounts of water and energy available. In tropical areas, where there is ample water, evapotranspiration rates are very high. In the lower Amazon Valley and the central Congo River basins, rates of 1200 millimeters a year nearly approach the evaporation over the open ocean. In parts of the Atlantic and Gulf Plain of the United States, the amount is almost as high.

CONDENSATION NEAR THE GROUND: DEW, MIST, AND FOG

Water vapor in the atmosphere eventually condenses to form water droplets. It is the condensation process that leads to deposition of water at the earth's surface. The greater part of the condensation process occurs in the formation of clouds, but some does occur near the surface in the form of dew, mist, and fog.

Condensation occurs at the dew point of air at a given temperature and can lead to the deposition of dew or the formation of mist and fog. Recall that cooling to the condensation level is the main means by which saturation occurs. If condensation of water droplets occurs in the layers of air immediately above the ground, it forms mist or fog. Mist is the suspension of microscopic water droplets that reduces visibility at the earth's surface. It forms a fairly thin grayish veil that covers the landscape. In contrast, fog is the suspension of very small water droplets in the air that reduces visibility to less than 1 kilometer (5/8ths mi). In fog, the air feels raw and clammy; given the correct illumination, the fog droplets are often visible to the naked eye. Mist does not provide the same damp, raw feeling, and the individual droplets are too small to see.

While all fog looks the same, its causes are quite variable. The most common types other than those associated with frontal systems (see chapter on frontal systems) are air mass fogs. These form as either advection or radiation types.

Advection is merely the horizontal transport of air. Advection fogs can form either by the transport of warm air over a cold surface or the transport of cold air over a warm wet surface. In the former, the warm air in contact with the cold surface cools. If the dew point is reached, condensation will occur in the form of a fog. This can occur in several ways as illustrated in Figure 4.7. Steam fogs or arctic "sea smoke" forms when cold air blows over a warm sea surface, which results in rapid evaporation from the water. This saturates the cold air, resulting in steam fogs commonly seen in arctic regions.

Radiation fog is an extension of the way in which dew forms. It occurs under clear skies in relatively still air and forms when nocturnal cooling of the ground results in a chilling of the layers of air near the surface. If the temperature drops to the dew point, then a layer of fog forms. Compared with advection fog, radiation fog lasts but a short time and "burns off" in the early morning when the sunshine heats the layer of chilled air.

In some ways, fog is like stratus clouds at ground level. Another type of fog is that which forms when air rises up the side of a mountain. This **upslope** fog is due to adiabatic expansion.

FOGGY PLACES

The conditions favorable for the formation of fog occur in some areas more frequently than in others. The result is that the occurrence of fog is quite variable over both space and time. Figure 4.8 shows areas of the United States that frequently experience fog and relates the distribution to the way in which the fog forms.

Advection fogs are often widespread and persistent. The foggiest place in the United States is in the Libby Islands off the coast of Maine. This is part of the extensive fog belt associated with the Newfoundland coast and the offshore Grand Banks. Here, air flowing off the relatively warm waters of the North Atlantic Drift crosses the cold water of the Labrador Current. The warmer air cools below its dew point and fog results. A similar situation occurs off the Aleutian Islands in the North

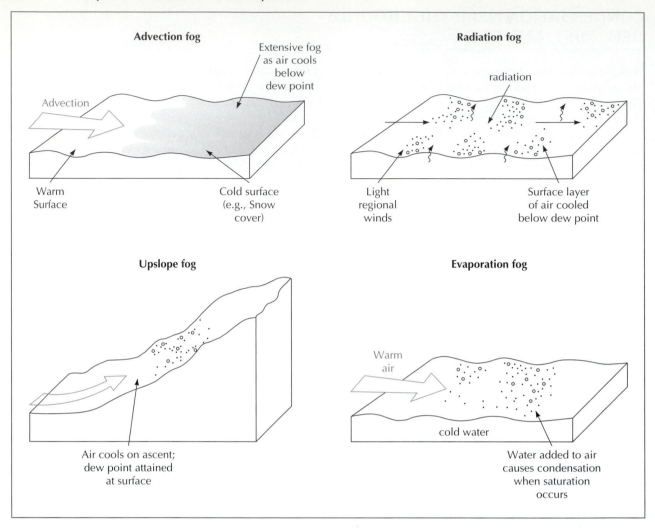

Figure 4.7
Schematic diagrams illustrating types of fog formation.

Pacific. Warm air from the North Pacific Drift comes in contact with the cooler waters of the Bering Current to give widespread advection fog. Such fogs are most frequent from March to September.

Fogs are common along the coastal areas of California. Cold **upwelling** water occurs close to the coast and air moving in from the warmer Pacific waters comes in contact with it. Often the fog starts here as low stratus clouds because wind action causes the air to mix. Air that has come in contact with the cold water cools further. When it rises, it results in fog forming above the sea surface. When this low layer of moisture reaches the hilly coast, people see it as fog at ground level. When fog actually forms at the sea level, it is usually very dense and results in minimal visibility.

Like the Newfoundland fogs, the fogs of California also occur during the summer months. During the dry summer, the moisture that condenses provides some moisture for the coastal redwood trees and, significantly, provides a humid environment that reduces loss of moisture through evaporation. In coastal Chile, South America, a similarly formed fog produces enough moisture to maintain green plants—the Garoua—in a desert region.

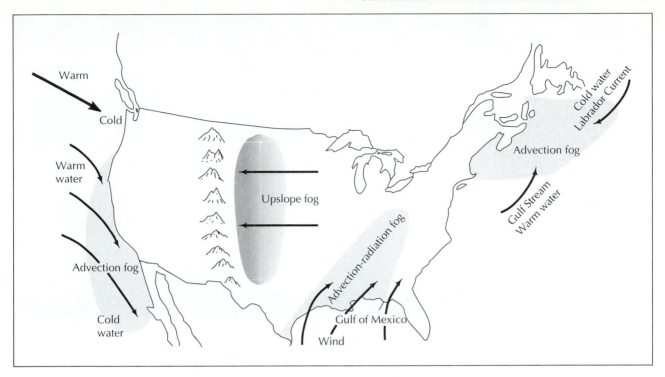

Figure 4.8
Winds associated with some of the common fogs (from Gedzelman S. D., *The Science and Wonders of the Atmosphere*, © 1980, John Wiley and Sons, N.Y.).

The upslope fog shown in Figure 4.8 is a result of adiabatic expansion. On the eastern side of the Rockies, the land slopes gradually upward from the Great Plains. When an east wind blows in the area, warm moist air slowly ascends the slope, cools, and condenses. Sometimes known as "Cheyenne fog," the right conditions can lead to a thick blanket of fog extending from Amarillo, Texas, to Cheyenne, Wyoming.

CONDENSATION ABOVE THE SURFACE: CLOUDS

Clouds provide the most readily visible weather phenomenon. Their beauty is not overlooked. Poems, songs, and paintings use clouds in highly expressive ways. Although atmospheric scientists also see beauty in clouds, they look to the form and appearance of clouds as important keys to both understanding and predicting atmospheric conditions.

All water droplets begin as microscopic particles. Some of the optical phenomenon in the atmosphere are a result of these particles. The spectacularly beautiful rays of the sun seen in Figure 4.9 occur when the sun is low in the sky and partially hidden by clouds. These rays, known as **crepuscular** rays, appear to fan out as a series of giant searchlight beams. The intensity of the rays is a function of how much light is scattered. This is an indication of the concentration of fine particles suspended in the air. Where air is free from debris, such as in polar areas, crepuscular rays do not occur. We can demonstrate this using a bright flashlight. In most places, it is possible to see the beam of a flashlight quite distinctly at night. In clean air, the rays are quite invisible.

Figure 4.9
Photograph of crepuscular ray (from A & J Verkaik/Skyart).

CLOUD FORMS AND CLASSIFICATION

Before 1800, clouds had no formal names and there was little knowledge of cloud mechanics. It remained for a young Englishman named Luke Howard (1772—1864) to provide a new perspective on clouds. In 1803, he presented a classification of clouds into main and secondary types and gave them Latin names. He distinguished three principle cloud forms:

Stratus (from Latin stratum = layer) cloud—lying in a level sheet

Cumulus (from Latin cumulus = pile) cloud—having flat bases and rounded tops, and being lumpy in appearance

Cirrus (from Latin = hair) cloud—having a fibrous or feathery appearance

The classification is so well designed that it remains the system in use.

However, as knowledge of clouds increased, a more comprehensive system was required. To gain worldwide uniformity, in 1895, an International Meteorological Committee published a system for naming and identifying clouds. This system was revised several times. The international standard is now under the auspices of the World Meteorological Organization (WMO), which publishes *The International Cloud Atlas*.

In the international classification system, the main types are genera (singular genus), which contain species and, in turn, several varieties.

Despite the infinite variety of clouds that occur, they group into 1 of 10 basic types or genera. The genera are in Table 4.1 and illustrated by the photographs in Figure 4.10.

A second part of cloud classification is the altitude at which clouds occur. Clouds of similar shapes occur at different levels in the troposphere. Table 4.2 shows clouds grouped as high, middle, and low clouds. The approximate height at which these occur varies with latitude. As already noted, the structure and thickness of the atmosphere varies from equator to pole. To some extent, the 10 genera of clouds relate to height. Those with the term *cirrus* (or prefix *cirro-*) are high clouds. Those with the prefix *alto* are middle clouds. Names for low clouds lack prefixes. An exception to this is the **nimbostratus** cloud, classified as both a middle and low cloud. The word *nimbus* (or prefix *nimbo-*) applies to a cloud from which rain is falling. It derives from the Latin for "violent rain."

Table 4.1   Cloud Genera (After WMO)

Cirrus (Ci). Detached clouds in the form of white, delicate filaments, or white or mostly white patches or narrow bands. These clouds have a fibrous (hairlike) appearance or a silky sheen or both.

Cirrocumulus (Cc). Thin, white patch, sheet, or layer of cloud without shading composed of very small elements in the form of grains, ripples, etc., merged or separate, and more or less regularly arranged; most of the elements have an apparent width of less than $1°$ (approximately the width of the little finger at arm's length).

Cirrostratus (Cs). Transparent, whitish cloud veil of fibrous or smooth appearance, totally or partially covering the sky, and generally producing halo phenomena.

Altocumulus (Ac). White or gray, or both white and gray, patch, sheet, or layer of cloud, generally with shading, composed of laminae, rounded masses, rolls, etc., sometimes partly fibrous or diffuse, and may or may not be merged; most of the regularly arranged small elements usually have an apparent width of between $1°$ and $5°$ (approximately the width of three fingers at arm's length).

Altostratus (As). Grayish or bluish cloud sheet or layer of striated, fibrous, or uniform appearance, totally or partly covering the sky, and having parts thin enough to reveal the sun at least vaguely, as through ground glass. Altostratus does not show halo phenomena.

Nimbostratus (Ns). Gray cloud layer, often dark, the appearance of which is rendered diffuse by more or less continually falling rain or snow, which in most cases reaches the ground. It is thick enough throughout to blot out the sun. Low, ragged clouds frequently occur below the layer with which they may or may not merge.

Stratocumulus (Sc). Gray or whitish, or both gray and whitish, patch, sheet, or layer of cloud that almost always has dark parts, composed of tessellations, rounded masses, rolls, etc., that are nonfibrous (except for virga) and may or may not be merged; most of the regularly arranged small elements have an apparent width of more than $5°$.

Stratus (St). Generally gray cloud layer with a fairly uniform base, which may give drizzle, ice prisms, or snow grains. When the sun is visible through the cloud, its outline is clearly discernible. Stratus does not produce halo phenomena (except possibly at very low temperatures). Sometimes stratus appears in the form of ragged patches.

Cumulus (Cu). Detached clouds, generally dense and with sharp outlines, developing vertically in the form of rising mounds, domes, or towers, of which the bulging upper part often resembles a cauliflower. The sunlit parts of these clouds are mostly brilliant white; their bases are relatively dark and nearly horizontal. Sometimes cumulus is ragged.

Cumulonimbus (Cb). Heavy and dense cloud, with a considerable vertical extent, in the form of a mountain or huge towers. At least part of its upper portion is usually smooth, or fibrous or striated, and nearly always flattened; this part often spreads out in the shape of an anvil or vast plume. Under the base of this cloud, which is often very dark, there are frequently low ragged clouds either merged with it or not, and precipitation sometimes in the form of virga.

Table 4.2   Approximate Height Range of Cloud Bases

Level	Ranges in Polar Regions (km)	Ranges in Temperate Regions (km)	Ranges in Tropical Regions (km)
High	3–8	5–13	5–18
Middle	2–4	2–7	2–8
Low		from surface to 2 km	

(a)

(b)

(c)

(d)

Figure 4.10
Selected photographs of cloud genera. (a) from Kent Wood © 1991; (b) Photo Researchers, Inc. © Roger Appleton; (c) Photo Researchers, Inc. © Van Bucher; (d) from Nationals Audubon Society © Robert H. Wright.

Figure 4.11 is a generalized diagram illustrating the classification of clouds according to genera and height. Note cumulonimbus, the thunder cloud, extends from low, through middle, to the high elevations of the height classification.

Differences in the structure and shape of clouds permit identification of cloud species. We add a species name to cloud genera to add further information about the cloud. A cloud is classified within only one genus. If it is a cumulus cloud, it cannot be a stratus or cirrus. However, the species name can apply to any genera. For example, the species *castellanus* are clouds that appear to have turrets like a castle. Such a shape applies to many clouds, including cirrus and stratocumulus. If there is no distinct structure or shape in the cloud forms, it is then unnecessary to provide the species name.

The full list of cloud species and their descriptions is in Table 4.3. The clouds listed here are secondary in importance to the 10 genera, and few persons commit them all to memory. The important factor is the potential for generating precipitation from the clouds.

The word *precipitation* is used to describe any of the various forms of water particles that fall from the atmosphere to reach the ground. It is a useful word for it describes water forms ranging from snowflakes and drizzle to hail and raindrops (Table 4.4). As discussed later in this chapter, the form that precipitation takes determines the nature of its impact on people and the environment. The precipitation process, the process that occurs in clouds to give rise to pre-

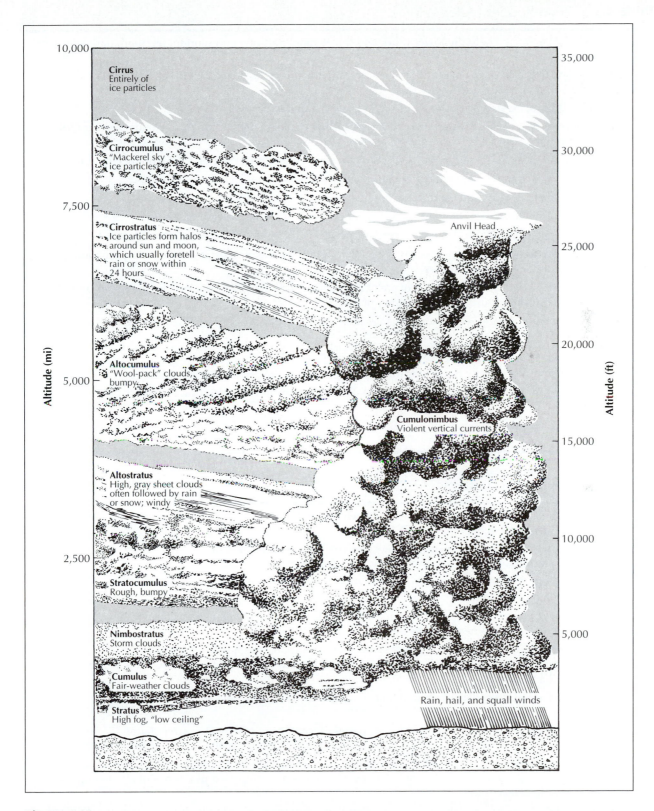

Figure 4.11
Cloud forms.

Table 4.3   Cloud Species (After WMO)

Arcus: shaped like an arc; refers to the lower (dark and threatening) part of a cumulonimbus cloud, particularly of the line-squall type.

Calvus: bald; refers to the absence of sprouting (cauliflower) structure as well as cirriform appendages in the upper part of a cumulonimbus cloud (see also *congestus* and *incus*).

Capillatus: hairy; fibrous or striated structure of upper part of cumulonimbus.

Castellanus: turreted; heap-shaped towers, resembling miniature cumulus protruding from clouds in the middle and upper troposphere, especially altocumulus.

Congestus: congested, heaped; sprouting, towering structures in the upper portion of a developing cumulus.

Fractus: fractured, torn; ragged fragments of clouds, notably stratus, cumulus, and nimbostratus.

Humilis: humble, small, flat; used to characterize nondeveloping cumulus clouds.

Incus: anvil-shaped; cirriform mass of cloud in the upper part of a developed cumulonimbus.

Intortus: twisted, entangled; used to describe a type of cirrus.

Mamma: shaped like udders; protuberances hanging down from the undersurface of a cloud; most pronounced in connection with thundery cumulonimbus.

Pileus: cap- or hood-shaped; accessory cloud of small horizontal extent above, or attached to, the upper part of a cumulus cloud, particularly during its developing phase.

Spissatus: spiss, compact; describes cirrus that is sufficiently dense to appear grayish when viewed in a direction toward the sun.

Tuba: tube-shaped; typical of clouds associated with tornadoes, water spouts, etc.

Uncinus: hooked; typical of streaky cirrus drifting in strong winds in the upper troposphere.

Velum: flap; an accessory cloud of considerable horizontal extent, sometimes connecting the upper parts of several cumulus clouds.

Virga: twig; trails or streaks of precipitation, hanging from the undersurface of a cloud but not reaching the ground.

Table 4.4   Types of Precipitation

Rain is precipitation of liquid water particles with diameters over 0.5 mm, although smaller drops are still called rain if they are widely scattered.

Drizzle is a fairly uniform precipitation composed exclusively of fine drops of water with diameters less than 0.5 mm. Only when droplets of this size are widely spaced are they called rain.

Freezing rain or *Freezing drizzle* is rain/drizzle that freezes on impact with the ground, with objects at Earth's surface, or with aircraft in flight.

Snow is precipitation of ice crystals most of which are branched. At temperatures higher than about −5°C (23°F), the crystals are generally agglomerated into snowflakes.

Snow pellets are composed of white and opaque grains of ice. The grains are mostly spherical and have a diameter of 2–5 mm. The grains are brittle and when falling on a hard surface bounce and break up. Snow pellets are also known as soft hail and graupel.

Snow grains are very small (less than 1 mm in diameter) grains of white, opaque ice. Snow grains are also called graupel.

Ice pellets are comprised of transparent or translucent pieces of ice that are spherical or irregular and have a diameter of 5 mm or less. They are composed of frozen raindrops or largely melted and refrozen snowflakes.

Hail is precipitation of small balls or pieces of ice (hailstones), with diameters ranging from 5 to 50 mm, falling either separately or agglomerated into irregular lumps. Hailstones are comprised of a series of alternating layers of transparent and translucent ice.

Ice prisms (diamond dust) are ice crystals often so tiny that they seem suspended in air. Such crystals may fall from a cloudy or cloudless sky. Mostly visible when they glitter in sunshine (hence diamond dust), they occur at very low temperatures.

Fog, ice fog, and *mist* are also considered forms of precipitation.

cipitation, is the end product of a whole set of events. The ascent of air, its cooling to form clouds, the various clouds that form, and their characteristics all occur as building blocks in the process by which precipitation is produced. Our concern is how and why air ascends, the behavior of water droplets (or ice crystals) in clouds, and the theories that explain the cause of precipitation. Once these are understood, the distribution of precipitation over the globe is more easily comprehended.

VERTICAL MOTION IN THE ATMOSPHERE

The condensation of water vapor to form mist and fog is largely the result of advection and radiation cooling. Condensation of water vapor to form clouds relies on another process—one that depends on vertical motion in the atmosphere. The significance of this process cannot be overstated for the presence or absence of clouds and the occurrence of precipitation ultimately depend on the upward motion of air. Here two essential questions are considered. First, what causes air to rise? Second, what happens to the air as it moves upward?

To help explain many of the processes that occur within the atmosphere, meteorologists often use the concept of a parcel of air. Such a parcel is considered as a volume of air that is small enough to have uniform properties (such as temperature and water vapor content) yet large enough to respond to the meteorological processes that influence it. While it can be any size, visualize it as being about one cubic meter in volume.

The Gas Laws, defined in chapter 1, provide a statement of the relationships among pressure, density, and temperature. One of the important facts resulting from the laws is that warm air is less dense than cold air. This density difference is demonstrated every time a hot air balloon rises. As shown in Figure 4.12, ballooning has become a popular pastime. The successful ascent of a balloon depends on inflating the balloon with hot air and, as long as the temperature inside the balloon is greater than that of the surrounding air, the balloon will continue to rise. In many ways, the hot air balloon can be considered to act like a parcel of air.

A parcel of warm air surrounded by cooler air will have a tendency to rise. Because pressure decreases with height above the surface, the rising air experiences a pressure decrease. Following the Gas Laws, the parcel will expand and cool. The rising of air, its expansion, and cooling form the basis for understanding the results of vertical motion in the atmosphere.

For air to move upward requires a mechanism that causes it to rise. One mechanism for lifting occurs when moving air encounters a physical barrier such as a mountain. Air cannot pass through the barrier and is forced to rise over it (Fig. 4.13a). Such lifting is termed **orographic** uplift, the term being derived from the word *orography*, which refers to the uneven shape of the earth's surface. Orographic lifting is a mechanical process. Lifting is also a result of interaction of the properties of air parcels. Such dynamic lifting occurs as a result of convection and frontal activity.

Convective lifting occurs in warm, moist air and is started by heating from the ground surface. When the surface is very warm, the air in contact with the ground is heated. It expands in response to the heating and, having a lower density than the surrounding air, it begins to rise (Fig. 4.13b).

Cyclonic lifting occurs along the boundary of air masses of different properties. The boundary between the air masses is termed a *front*, and uplift of air at these fronts is of major meteorological importance. Although dealt with in more detail in later chapters, the basic process involved at a cold front is illustrated in Figure 4.13c.

Figure 4.12
An ascending hot air balloon (from Ed Kosmicki/Photosource West.com/Snowmass Resort Association).

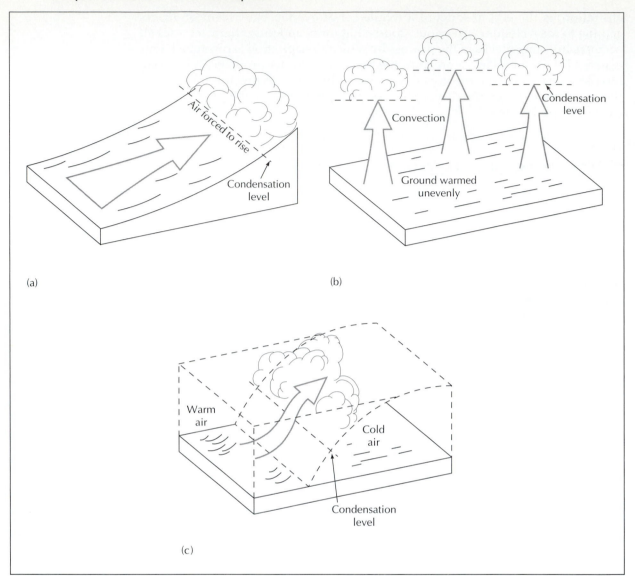

(a)

(b)

(c)

Figure 4.13
Processes responsible for the uplift of air.

Adiabatic Heating and Cooling

Once mechanical or dynamic lifting causes air to rise, it undergoes physical changes. Recall that atmospheric pressure decreases with altitude so that a rising parcel of air is subjected to decreasing pressure around it. The parcel of air expands. In the process of expansion, the distances among the individual gas molecules within the parcel are increased. Thus, during expansion of the air, its molecules have to do work (i.e., "push" other molecules out of the way). In doing this work energy is needed, and this energy comes from the gas molecules. Use of this energy reduces the kinetic energy of the molecules, and the temperature of the parcel decreases. Note that the decrease of temperature is an internal response to expansion of the air parcel; no heat is lost to or gained from the environment surrounding the parcel. Because of this, the process is called *adiabatic*—a process defined as the internal changes within a gas during expansion and contraction when no energy is removed from or added to the gas. Note that the adiabatic process applies to both expansion and contraction. This means that the adiabatic process applies to both rising and sinking air parcels.

The first law of thermodynamics states that the temperature of a gas may be changed by addition (or subtraction) of heat, a change in pressure, or a combination of both. In the adiabatic processes, the addition and subtraction of heat may be disregarded and expressed simply as

$$\text{Change in temperature} = \text{constant} \times \text{change in pressure.}$$

This adiabatic form of the first law of thermodynamics is of extreme importance in understanding many atmospheric processes, especially those concerning lapse rates.

Lapse Rates

In the standard atmosphere, temperature decreases with height in the troposphere at an average rate of 6.5°C per kilometer (3.5°F/1000 ft). This standard lapse rate is derived from the difference between the average surface temperature (15°C) and the average temperature at the tropopause (−59°C at 11 kilometers). At any given time, the lapse rate may vary substantially from this mean value. The standard lapse rate is a measured value. On days when a dense cold air mass occurs, the lapse rate measured will be quite different from that on another day when different conditions exist. This lapse rate may be measured when, for example, temperatures are taken at various elevations when ascending a mountain. Since it measures the temperature of the environment, it is called the **environmental lapse rate**.

When a parcel of air is forced to rise over a mountain, it cools at a rate independent of the environment. The cooling of the parcel will occur as a result of the adiabatic cooling process. Because the rate of cooling is based on a physical change within the parcel (a response to decreasing pressure), the rising parcel will cool at the adiabatic lapse rate of 10°C for every 1000 meters of ascent (5.5°F per 1000 ft). Unless condensation occurs within the rising air, this rate of change is a constant value and is called the **dry adiabatic lapse rate**.

REGIONAL VARIATION IN PRECIPITATION

The amount of precipitation received at the surface varies due to many factors. The basic element is the amount of water vapor in the air, which varies geographically and seasonally. The mean precipitable water content of the atmosphere at a given moment is 25 millimeters, with a maximum near the equator of 44 millimeters and a minimum in the polar regions of 2 to 8 millimeters depending on the season. In latitudes of 40 degrees to 50 degrees, it will range upward of 20 millimeters in the summer and drop to around 10 millimeters in the winter. The presence of water vapor is a necessary but not sufficient condition for precipitation. There is no direct relationship between the amount of atmospheric water vapor over an area and the resulting precipitation. To illustrate this, a comparison can be made between conditions over El Paso, Texas, and St. Paul, Minnesota. The average moisture content above these cities is about the same, and yet the mean annual precipitation is more than three times greater at St. Paul. Other factors must come into play to induce precipitation.

If the total amount of precipitation received over the surface of the earth were spread evenly, it would average 880 millimeters per year. It varies from near 0 to almost 12 millimeters. Figure 4.14 shows the mean precipitation over the globe. The total amount of precipitation received depends on several factors:

1. Whether air converges (to give uplift) or diverges (spreads out) in the area.
2. Air mass origin, an indication of the temperature and moisture conditions of the air.

76

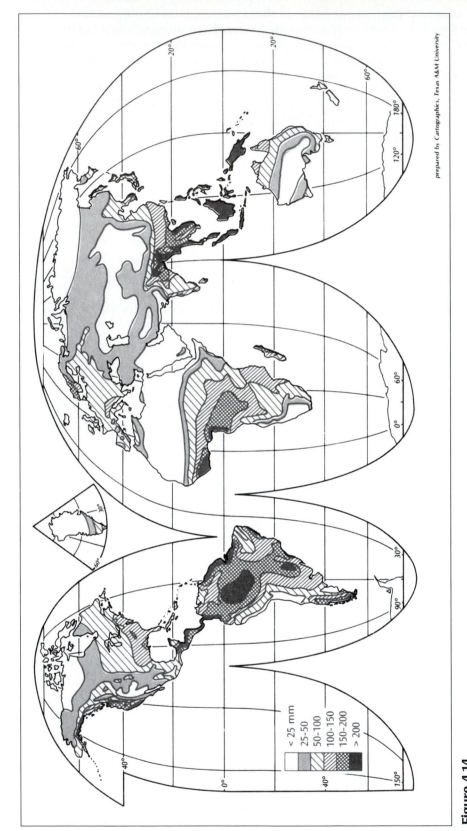

Figure 4.14
Mean precipitation over Earth (in mm).

prepared by Cartographics, Texas A&M University

< 25 mm
25–50
50–100
100–150
150–200
> 200

3. Topographic conditions.
4. Distance from the moisture source: The greater the distance from the source of the moisture, the less water vapor will be present in the air because of prior precipitation loss.

Combining these factors, the areas where precipitation is greatest are mountain areas of the Tropics, where there is frequent convergence of air from the ocean. Where such conditions exist, rainfall may reach 12 millimeters per year (Table 4.5).

Two general locations where precipitation totals tend to be above average for the earth are found relatively far apart. One is near the equator, where there is a zone experiencing convergence of moist tropical air most of the year. The equator lies beneath the low pressure convergence area caused by the **Intertropical Convergence (ITC)**. The trade winds moving toward the equator pick up moisture over the oceans and, when lifted in the ITC, yield abundant moisture. Precipitation is increased over coastal areas by orographic lifting and increased convection started from surface heating. Average precipitation ranges from 1.5 meters to 2.0 meters annually, but in some cases goes much higher.

The second situation that gives rise to above-average precipitation is found on the west side of the continents in mid-latitudes. Precipitation there is due to convergence of maritime air and orographic intensification. The zone is most

Table 4.5   Record Intense Rainstorms

Time	Amount		Place	Date
	mm	in.		
1 min	38	1.50	Barot, Guadeloupe, West Indies	Nov. 26, 1970
5 min	63	2.48	Panama	1911
8 min	126	4.96	Fussen, Bavaria, Germany	May 25, 1920
15 min	198	7.79	Plumb Point, Jamaica	May 12, 1916
20 min	206	8.10	Cuerta-de-'Arges, Romania	Date unknown
30 min	235	9.25	Guiana, VA	Aug. 24, 1906
42 min	305	12.00	Holt, MO	June 22, 1947
1 hr	401	15.78	Muduocaidong, Nei Monggol, China	Aug. 1, 1977
2 hr 10 min	483	19.02	Rockport, WV	July 18, 1889
2 hr 45 min	555	22.00	D'Hanis, TX	May 31, 1935
4 hr 30 min	782	30. 80	Smethport, PA	July 18, 1942
6 hr	840	33.07	Muduocaidong, Nei Monggol, China	Aug. 1, 1977
9 hr	1087	42.79	Belouve, La Reunion, Indian Ocean	Feb. 28, 1964
10 hr	1400	55.12	Muduocaidong, Nei Monggol, China	Aug. 1, 1977
24 hr	1870	73.62	Cilaos, Reunion Island	Mar. 15–16, 1952
5 days	3810	150	Cherrapunji, India	Aug. 1841
31 days	9300	366	Cherrapunji, India	July 1861
1 yr	26460	1041	Cherrapunji, India	Aug. 1860-July 1861

well defined in latitudes from 50 to 60 degrees. The totals run above 1.5 meters along the coasts. The amounts are not as high as in the Tropics because the moisture capacity of the air is much lower than that of the maritime tropical (mT) air of the Tropics.

The arid regions of the world occur in three great realms:

1. Extending in a discontinuous belt approximately between 20 to 30 degrees north and south of the equator. These areas constitute the great tropical deserts, which owe their aridity to large-scale atmospheric subsidence.

2. In the interiors of continents are found the continental deserts, which are arid as a result of their distance from the sea—the major source of the water vapor.

3. The polar deserts of the Arctic and Antarctic constitute the third great area of low precipitation. The low temperatures of these regions, along with subsidence of air from aloft, are probably the most important contributing factor to the low totals.

Mountain ranges play a significant role in the spatial distribution of precipitation. The windward slopes of mountains receive the greatest amount of precipitation. In the lee of the mountain ranges, the precipitation decreases markedly to give a rain-shadow effect. The predominant flow of air along the west coast of North America is from west to east. The mountains produce alternate zones of high precipitation and low precipitation—high on the mountain slopes and low in the intervening basins and valleys. In California, the Coast Ranges are the first barrier to the onshore winds. Precipitation is substantial and forests grow on the windward slopes. Precipitation increases with height to the crest at about 760 meters. In the Great Valley, precipitation is much lower, and a grassland environment exists. As elevation increases going eastward over the Sierra Nevada Mountains (2600 m), precipitation increases to two or three times that of the Great Valley (Fig. 4.15). Forests cover the slopes of the moun-

Figure 4.15
Precipitation totals across the Sierra Nevada Mountains illustrating orographic influence on precipitation and the rain shadow in the lee of the mountains. Rainfall totals range from over 125 inches on the windward side of the mountains to less than 10 inches in the valley to the east.

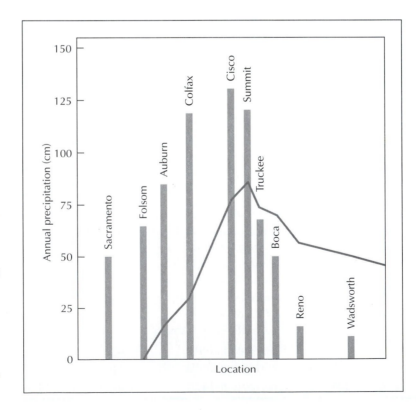

tains near the summit. The air crossing the crest descends slightly and warms adiabatically, and relative humidity drops. Precipitation declines rapidly, and the clouds begin to evaporate. Annual precipitation drops from around 1300 millimeters at the crest to about 150 millimeters at Reno, Nevada, giving Reno a desert climate. The arid zone extends from the Mexican border north into Canada between the Sierra Nevada-Cascade Range and Rockies because of the drying of the westerlies as they cross the mountains. The process is repeated as the currents continue to flow eastward over the Wasatch and Rocky Mountains. The Rockies, with a crest in excess of 3300 meters, cause still further drying of air from the Pacific. Little precipitation falls east of the Rockies from air originating over the Pacific Ocean. The chief moisture sources for the continent east of the Rocky Mountains are the Gulf of Mexico and the Atlantic Ocean.

VARIATION IN PRECIPITATION THROUGH TIME

Total annual precipitation alone is an insufficient measure of moisture availability for it does not take into account the manner in which the precipitation is distributed throughout the year. It is often the temporal distribution that is the dominant factor in determining use of precipitation. There are major discrepancies between the frequency of precipitation, measured in terms of the number of days per year on which measurable precipitation falls, and mean annual precipitation. In some parts of the world, precipitation falls nearly every day of the year. Bahia Felix, Chile, averages 325 days per year with measurable precipitation. Thus, there is an 89 percent chance that it will rain or snow on any given day of the year. Buitenzorg, Java, in the tropics, averages 322 days per year with thunderstorms. At the other end of the scale are those desert areas where rain is an oddity. Arica, Chile, averages only about one rain day annually; at Iquique, Chile, not far from Arica, 14 years passed with no measurable rainfall (1899–1913). It may seem strange that one of the areas with the least frequent rainfall is in the same country as one of the rainiest places in the world. The size and shape of the country, topography of the region, and general circulation of the atmosphere all play a part in the explanation.

Over most of the earth's surface, precipitation is of a seasonal nature, with a period of the year when precipitation has a high probability of occurring and a dry season when the probability is much lower (Fig. 4.16). Surprisingly perhaps, those sites that have the highest total annual rainfall are found in areas with a pronounced dry season. Cherrapunji, India, is an example. Although an average of more than 10 meters of rain falls each year, there are several months when rain is infrequent.

WATER BALANCE

To better understand the way in which water is allocated as part of the natural hydrologic cycle of an area, a water balance method proves invaluable. In this, by knowing the amounts that are used in various processes or are stored in the soil, the surplus and deficit can be calculated. The surplus is represented in surface water, and large surpluses lead to flooding: The deficit is a water shortage in relation to plant needs. For water planning and agriculture, it is essential to be aware of both of these values. A number of bookkeeping methods are available to assess the water balance. That introduced by the American climatologist C. W. Thornthwaite is outlined in Box 4.1.

The effect of high summer radiation is apparent in water balances of mid-latitude stations (Figure 4.17), where potential evapotranspiration peaks in summer.

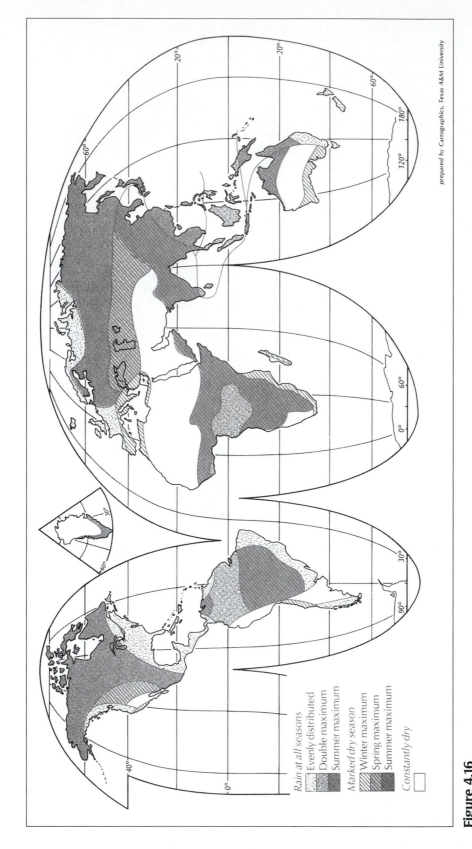

Rain at all seasons
  Evenly distributed
  Double maximum
  Summer maximum

Marked dry season
  Winter maximum
  Spring maximum
  Summer maximum

Constantly dry

prepared by Cartographics, Texas A&M University

Figure 4.16
Generalized map showing the seasonal distribution of precipitation. (*Source:* After A. A. Miller, *Climatology.* Metheun, 1965).

QUANTITATIVE EXPRESSION: BOX 4.1

The Water Balance: Calculation

At any location a balance between incoming and outgoing moisture is attained, and the balance will reflect the climatic regime that exists. This note outlines Thornthwaite's bookkeeping method using data shown in the table. Data are shown in centimeters.

	Jan	Feb	Mar	Apr	May	Jun	Jul	Aug	Sep	Oct	Nov	Dec	Year
1. PE	1	2	3	5	8	10	12	11	9	5	2	1	69
2. P	14	11	10	6	5	5	1	2	4	9	14	17	99
3. $P - PE$	13	9	7	1	−3	−5	−11	−9	−5	4	12	16	0
4. ST	0	0	0	0	−3	−5	−2	0	0	4	6	0	0
5. ΔST	10	10	10	10	7	2	0	0	0	4	10	10	0
6. AE	1	2	3	5	8	10	4	2	4	5	2	1	47
7. D	0	0	0	0	0	0	9	9	5	0	0	0	23
8. S	13	9	7	1	0	0	0	0	0	0	6	16	52

Row 1 is the adjusted potential evapotranspiration (PE) that is derived by substituting monthly temperature (°C) into an empiric formula derived by Thornthwaite.

Row 2 provides values for monthly precipitation (P).

Row 3 is the difference between precipitation and potential evapotranspiration ($P - PE$). It provides the amount of moisture available after evapotranspiration requirements have been satisfied. The excess above the difference will either go to soil water or occur as runoff. The amount of water stored by the soil is highly variable and will depend on the nature of the soil. In his initial work, Thornthwaite used a value of 4 inches (10 cm) as a general value to be applied to all water balance studies. Although this does enable comparison of the balance for different stations, it is not realistic in terms of precise water balance studies. Varying soil moisture-holding capacities have been introduced; however, for demonstration purposes, the 4 inches (10 cm) value is retained here.

Rows 4 and 5 give the amount of moisture stored in the soil (ST) and the change in storage (ΔST) since the previous month, using the 4 inches (10 cm) value

as a base. As long as precipitation is greater than potential evapotranspiration, the value remains 4 inches. However, as soon as potential evapotranspiration is greater than precipitation ($PE > P$), plants draw on the available soil moisture and the soil storage falls below capacity.

Row 6 shows the actual evaporation (AE). So long as $P > PE$, then $AE = PE$. The water in storage (ST) will then make up the deficit until $(P + ST) < PE$. That is, as soon as the cumulative excess of potential evapotranspiration over precipitation is greater than 4 inches (10 cm), soil moisture is assumed to be totally utilized. A deficit period then exists. At the end of the deficit period, when P is greater than potential evapotranspiration ($P > PE$), the 4 inches (10 cm) must be restored to the soil before any surplus occurs.

Rows 7 and 8 provide the balance of water in terms of deficit (D) and surplus (S). It is noted that the deficit does not occur as soon as potential evapotranspiration is greater than precipitation because of the period of soil moisture utilization.

Data can be shown graphically as in Figure 4.17. Notice that the graph is divided into four areas: surplus and deficit periods, period of utilization, and period of recharge.

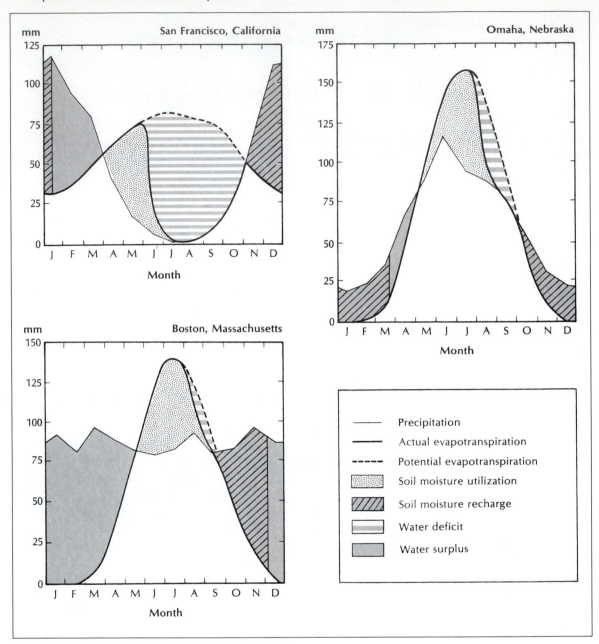

Figure 4.17
Selected water balance diagrams for North American stations.

In some cases, such as along the American west coast, this summer maximum of potential evaporation is out of phase with seasonal rainfall. Here precipitation occurs in the cool season when rainfall is more effective, with less water lost to evaporation.

In the interior areas of the mid-latitudes, the summer maximum rainfall coincides with the period of highest rainfall and the time when plants are actively growing. Irrigation is thus an integral part of agricultural practice. In the humid mid-latitude areas, precipitation may well exceed evapotranspiration for much of the year except during droughts.

An important aspect of the water balance in mid-latitudes is periodic freezing of streams, lakes, ponds, and soil moisture. This takes place where winter temperatures go well below freezing. Most parts of the mid-latitudes experience

Table 4.6 World Record Extremes of Precipitation

Longest period without rain	14 years	Iquique, Chile
Lowest average annual precipitation	0.5 mm	Arica, Chile
Greatest number of days per year with precipitation	325	Bahia Felix, Chile
Greatest number of days per year with thunderstorms	322	Buitenzorg, Java
Highest annual average precipitation	11.98 m	Mt. Waialeale, Kauai, HI
Greatest 24-hr snowfall (USA)	1.93 m	Silver Lake, CO, April 14–15, 1921
Highest one-season snowfall (USA)	28.5 m	Paradise Ranger Station, Mt. Rainier, WA 1971–1972

below freezing temperatures for varying periods during the winter. When such temperatures occur, soil moisture freezes and, depending on the severity of the cold spell, surface water freezes. This freezing stops the flow of runoff. It decreases the water supply available to plants, producing drought. The winter months are a time of year when moisture accumulates in the environment. Even though winter precipitation is generally less than summer precipitation over the Northern Hemisphere land masses, the demands on the water supply are also less. Evapotranspiration is reduced in winter because of lower temperatures and reduced plant growth. Over the land masses, some 70 percent of all precipitation goes directly back to the atmosphere by evaporation. In winter months, this may drop to 40 percent or less, thus allowing moisture to accumulate on and in the soil. Removing water from circulation by freezing shapes the annual pattern of stream flow. The U.S. Geological Survey uses a water year in all its calculations that is different from the calendar year. For most of North America, streams are at their low point in September or October. During the summer, evapotranspiration uses a big share of the water. As temperatures cool, the vegetation uses less water and moisture begins to collect. During winter, moisture accumulates either as soil moisture or snow. The spring thaw then causes streams to rise. Flooding may occur if a large amount of snow covers the land surface. Spring floods due to snow melt generally characterize poleward locations and mountain regions, but floods also affect other areas.

SUMMARY

Condensation occurs near the ground as dew, mist, or fog and at higher levels as clouds. Fog forms are a result of advection, radiation, and adiabatic processes. A high incidence of fog occurs in specific locations. Condensation nuclei play an important role in the condensation process.

The spatial distribution of precipitation shows that two global areas receive rainfall high above the global average while there are three great world realms that are classed as arid deserts. Total annual precipitation is an insufficient measure of moisture availability, and the way in which precipitation is distributed over time is highly significant. Precipitation seasonality forms an important component of the climate of a location.

KEY TERMS

Adiabatic lapse rate	Environmental lapse rate	Nimbostratus
Cirrostratus	Evapotranspiration	Orographic
Crepuscular rays	Hydrologic cycle	Upslope
Dew point	Intertropical convergence	Upwelling

REVIEW QUESTIONS

1. How much energy is absorbed and released when water changes from liquid to gas or gas to liquid at standard atmospheric conditions?

2. What is the relationship among evaporation, transpiration, and evapotranspiration?

3. What percentage of Earth's water is available for human use as groundwater and surface water?

4. What is the relationship between relative humidity and dew point?

5. What are the cloud conditions on most nights when radiation fog develops?

6. Why are salt particles in the air considered hygoscopic?

7. How is lifting condensation level related to dew point?

8. What are the environmental conditions at those places on Earth that have a large total annual precipitation?

9. What are the environmental conditions that favor the development of deserts?

10. What latitudinal zones on Earth's surface have above-average precipitation?

C H A P T E R
5

Motion in the Atmosphere

CHAPTER OUTLINE

*W*ind is the name given to air moving horizontally over the earth, and *air currents* refer to air moving vertically. Winds transfer heat over the earth's surface and carry water vapor from oceans to continents. Without wind systems there would be no life as we know it on the land masses. There are two main characteristics of wind: direction and velocity. The direction given to a wind is the direction from which it blows. They are named for the main points on the compass rose or assigned a bearing in degrees from north. A wind blowing from east to west is an east wind or a 90° wind. When the direction of the wind is changing, it is veering or backing. A veering wind is one changing in a clockwise direction, and a backing wind is one changing in a counterclockwise direction. This chapter deals mainly with wind for vertical motions. Air currents have already been considered while explaining cloud formation and precipitation mechanisms.

ATMOSPHERIC PRESSURE

Atmospheric pressure is the force per unit area or the weight exerted by the atmosphere on a surface. There are a number of widely used measures of atmospheric pressure. In the United States, pressure charts are shown using the millibar. This is derived from a bar—a force of one million dynes per square centimeter. (A dyne is the force needed to accelerate a mass of one gram one centimeter per second per second. A millibar is one thousandth of a bar.) In the standard atmosphere, pressure at sea level at 15°C is 1013.2 millibars. Commonly used elsewhere is the standard SI unit, the pascal. Derivation of the pascal and various pressure equivalents are shown in Table 5.1A.

In popular usage in the United States is the equivalent pressure in inches of mercury. A column of mercury about 30 inches in height just offsets the weight of a column of atmosphere of the same area. A column of mercury 1 inch on each side and 30 inches high weighs 14.7 pounds. This is the weight of a column of the standard atmosphere one square inch in area. Twenty-nine and ninety-two hundredths linear inches of mercury are equivalent to 14.7 pounds per square inch or 1013.2 millibars. In the real world, surface pressures vary routinely from about 950 millibars (28 in) to 1050 millibars (31 in). Table 5.1B provides some extremes of pressure.

The variation in atmospheric pressure and density that takes place with height in the standard atmosphere was presented in chapter 1. Like many other atmospheric variables, pressure changes with time and space. For example, at the two poles, it is higher than average. This is because of the cold temperatures and subsidence of air. There are other zones where it averages less than the global mean.

There are seasonal changes in pressure in some areas. Central Asia experiences marked differences from summer to winter. These changes are due to large changes in temperature with the seasons. In mid-latitudes, there are often day-to-day changes in pressure resulting from traveling storms. There are also regular daily variations. There are maxima around 10 a.m. and 10 p.m. and minima around 4 a.m. and 4 p.m. The exact cause of these daily changes is unkown, but they relate to the daily flood of solar energy that moves around the earth.

Areas below or above normal pressure develop in the atmosphere. Some specific terms refer to these areas. In all cases, the pressures are relative and do not have fixed values. A **low pressure** is an area in the atmosphere where pressure is less than the surrounding area while a **trough** is an elongated area of low pressure. A **high pressure** is an area of the atmosphere where barometric pressure is higher than the surrounding area and a **ridge** is an elongated area of high pressure.

Humans are normally not sensitive to small variations in pressure. We are most likely to be aware of them when going up or down in an airplane or driving up or down a long mountain road. In taking off and landing in a modern jet aircraft, the pressure changes are large. Cabins in commercial jets are now pressurized so the pressure normally stays at least 50 percent of that at sea level. We are most sensitive

Table 5.1   Pressure Units and Extreme Sea-Level Pressure Readings

A. Pressure Units

Pressure = Force per unit area

SI force unit is the Newton (N)

   (the force required to accelerate 1 kilogram 1 meter per second per second)

SI pressure unit = Pascal (Pa)

   (1 Pa is 1N per square meter)

Cgs pressure unit is the Bar

   (1 Bar is 10^6 dynes per sq. cm)

Standard Sea Level Atmosphere is

   101,325 Pa which is expressed as 1013.25 hPa

Or   1.01325 Bars which is expressed as 1013.25 mb

Or   Based upon height of column of mercury

   29.92 inches of mercury (76 cm) at sea level

B. Extremes of Sea-Level Pressure Readings

Pressure mb	inches	Event
1083.8	32.00	Highest recorded. Agata, Siberia, USSR. December 31, 1968
1078.2	31.85	Highest recorded in North America. Northway, Alaska. January 31, 1989
1063.9	31.42	Highest in the lower 48 states. Miles City, Montana. December 24, 1983
870	25.7	Lowest recorded. Typhoon Tip, Pacific Ocean near Guam. October 12, 1979

to pressure changes when we have a head cold and there is congestion in the eustachian tubes. Air cannot flow in or out of the inner ear to keep even with the outside pressure. When there is much difference, it may cause severe earache.

FACTORS INFLUENCING AIR MOTION

Laminar and Turbulent Flow

The notion of a standard atmosphere suggests an elusive uniformity. A fluid is any substance that has no rigidity. Gases and liquids are fluids, and hence the atmosphere is a fluid. It is like a river flowing around the earth. Since it is a fluid, it acts according to all the principles of fluids as well as the laws of gases. In fluids, there are two classes of flow. There may be movement as **laminar flow**. In this case, each particle in the fluid moves in essentially a smooth line without any rotational motion. The flow of air over the wing of an airplane is in part laminar. The flow of a thin sheet of water down the face of a spillway or dam is laminar. If small disturbances start to develop in laminar flow, the tendency is for the smooth flow to counteract the disturbance and keep smooth or streamlined flow.

   When fluids moving with a laminar flow are subjected to unequal pressure or heat, they begin to develop motion in three directions in the form of convection currents or eddies. The new flow is turbulent flow with disordered flow over a variety of scales. In a fast-moving river, eddies almost always exist imbedded in the overall downstream movement of the river. Around a bridge pier, swirls of water or standing eddies may remain in nearly the same position downstream from the pier. In paddling a canoe or rowing a boat, small eddies form around the end of the paddle or oar in response to the differential pressure of the blade. In **turbulent flow**, individual particles may move in any direction—sometimes upstream, the opposite direction of the stream's mean flow. Whenever the velocity in a moving fluid reaches a critical point, it becomes turbulent.

When small-scale disturbances develop in turbulent flow, they often grow into larger and larger disturbances. Since the atmosphere is a turbulent fluid, many small disturbances grow into large ones. The butterfly effect is the principle that a butterfly stirring the air in one place can cause a change in the weather at some later time and at a different place. Although this is an exaggeration, it illustrates that very small eddies in the circulation can grow to be major weather systems.

In response to differences in temperature and pressure, the troposphere is especially turbulent. There are very large-standing eddies and countercurrents that make up the general circulation. Some of these vortices may be thousands of kilometers across. Tropical cyclones are eddies or vortices measuring several hundred kilometers in diameter. Tornadoes are intense vortices usually no more than 2 or 3 kilometers (1.2–1.5 mi) across. Many of these storms grow from relatively small initial disturbances.

Newton's Laws of Motion

Newton's first **law of motion** deals with inertia. It states that a body will change its velocity of motion only if acted on by an unbalanced force. In effect, if something is in motion, it will keep going until a force modifies its motion. On Earth, a parcel of air seldom moves continuously and in a straight line. This is because, as Newton's second law states, the acceleration of any body—in this case, the parcel of air—is directly proportional to the magnitude of the net forces acting on it and inversely proportional to its mass. Note that these laws concern acceleration, which is change of velocity with time.

By identifying the forces that act on a parcel of air, it becomes possible to understand more fully the processes that lead to the acceleration (or deceleration) of air. If we consider a unit parcel of air ($m = 1$), then Newton's second law becomes

$$\text{Acceleration} = \text{Sum of Forces}$$

$$\text{or } F_a = \Sigma F.$$

The ΣF is made up of the atmospheric forces so that:

$$\text{Acceleration} = \text{Pressure Gradient Force} + \text{Coriolis Force} + \text{Frictional Forces} + \text{Rotational Forces}.$$

Each of these forces (or accelerations) may be dealt with separately.

Pressure Gradient

Differences in heating and internal motion in the atmosphere produce differences in atmospheric pressure. This difference in pressure over space is a **pressure gradient**, and air will move down this gradient from high to low pressure. If no other forces occurred, then air on Earth would move in a straight line from high to low pressure. The rate at which air moves depends on the steepness of the gradient. When small differences in pressure occur over large areas, a weak gradient gives rise to weak winds. A steep gradient causes rapid motion or high winds. Figure 5.1 illustrates this effect. Box 5.1 provides a quantitative expression of pressure gradient.

Atmospheric pressure changes slowly in a horizontal direction. The change may be as little as two millibars in 160 kilometers (100 mi). In the summer over North America, the temperature difference between the Gulf of Mexico and the

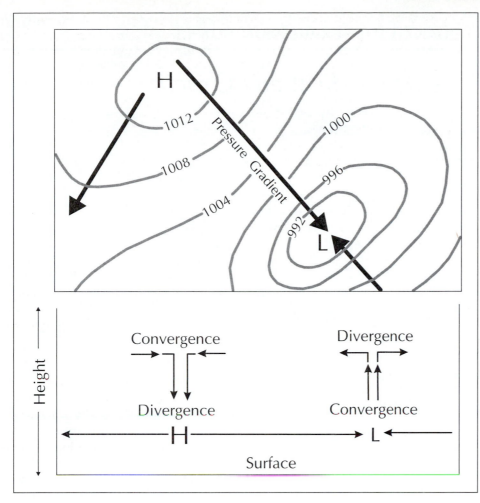

Figure 5.1
A map of a simple high- and low-pressure system. In a high-pressure system, pressure decreases outward from the center. In a low-pressure system, pressure decreases toward the center. Lower figure shows convergence and divergence associated with high- and low-pressure systems.

Canadian border is small. As a result, the density of the air is about the same and the north–south pressure gradient is weak. Since the pressure gradient is weak, mean wind velocities are low. In the winter, temperatures differ more. It is still warm over the Gulf of Mexico, but relatively cold over Canada. Density of the cold air is greater and there is a pressure gradient that slopes downward toward the south. This transports cold air south and wind velocities are higher.

The wind will not blow directly down the pressure gradient; once air is underway, other forces, as defined before, come into effect.

Coriolis Effect

The rotation of the earth results in a process called the **Coriolis effect** (also referred to as the Coriolis force, or Coriolis acceleration), which acts on any object moving free of the earth's surface. This includes aircraft, ballistic missiles, and long-range artillery (Table 5.2). The rotation of the earth causes winds in the northern hemisphere to turn in a clockwise direction or to the right. In the southern hemisphere, the winds turn counterclockwise or to the left (Fig. 5.2). Mariners have long known and recorded the process. The French mathematician G. G. Coriolis (1792–1843) first explained the process, hence the name.

The rate of curvature imparted to the moving air is a function of the wind's velocity and the latitude. At the equator, the Coriolis effect is zero. It is maximum at the two poles. Although the Coriolis effect is relatively small, it is significant because air streams travel long distances over the earth. Box 5.1, the Coriolis Effect, provides a quantitative look at Coriolis.

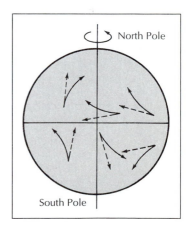

Figure 5.2
The Coriolis effect acts on the wind to deflect it to the right in the northern hemisphere and to the left in the southern hemisphere. Deflection in each hemisphere is always in the same direction regardless of wind direction.

QUANTITATIVE EXPRESSION: BOX 5.1

Pressure Gradient and Coriolis

Consider the conditions shown in the parcel of air in Figure B.

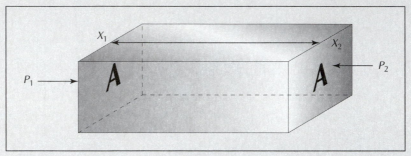

Figure B

The parcel has a cross-sectional area A and is influenced by forces producing pressure on *surface A* with the distance between the two acting forces given by $X_1 - X_2$. The pressure is expressed by area P_1 and P_2. If there is a difference between P_1 and P_2, $P_1 > P_2$, then a net force exists and air will accelerate from P_1 toward P_2. This is the pressure gradient and air will move from the high to the lower pressure. Given that the air has a density d, then the net forces may be shown as

$$\text{Net force} = \frac{\text{difference in pressure}}{\text{density of air}} \times \text{distance},$$

where the net force is expressed by the pressure gradient F_p

$$F_p = (P_1 - P_2)/d \times (X_1 - X_2).$$

In scientific notation, the difference between two values is given as delta (Δ), so the equation may be shown as

$$F_p = \Delta P/d \times \Delta X$$

or

$$F_p = \frac{1 \Delta P}{d \Delta X}$$

where F_p is the pressure gradient, ΔP is the pressure difference and ΔX is the distance between the two pressures used to derive ΔP. Air density is given by d.

The steeper the gradient, the faster the flow of air.

As soon as air is in motion, the Coriolis effect or force (F_c) will come into effect. The Coriolis effect is a result of Earth's rotation and is a function of the angular velocity of Earth (7.29×10^{-5} radians per second), which is represented by Ω, the latitude (ϕ), and the velocity of the air in motion (v). Since Coriolis is zero at the equator and attains its maximum at the poles, the sine of the latitude ($\sin \phi$) is used in the computation of F_c. Thus, the equation becomes

$$F_c = 2\Omega \sin \phi v.$$

Since F_c is an acceleration, the Coriolis effect may be expressed in, for example, cm/sec^2. The Coriolis acceleration of air moving at 10 m/s (36 mph) would be 0 at the equator, 0.09 cm/sec^2 at latitude 40°, and 0.12 cm/sec^2 at latitude 60°.

Friction

Like Coriolis, **friction** can only exert an influence once air is in motion. It acts in opposition to air motion and results from contact with the surface, which induces turbulent flow. Friction works in opposition to the pressure gradient force to reduce wind velocity and the Coriolis effect. With the rotational effect reduced, the pressure gradient dominates. Thus, at the surface, winds blow from high to low pressure at an angle across the isobars. Over the oceans, where friction is less than over land, the winds cross the isobars at angles of 10° to 20° at six meters above the surface. Oceanic

Table 5.2   Examples of Drift Due to the Coriolis

1. A person walking at a rate of 10 m per min (4 mph) toward a point in space drifts 40 m per km (250 ft/mi).

2. An auto traveling 100 kilometers per hr (60 mph) drifts 4.5 meters (15 ft) in 1.6 kilometers (1 mi). The tendency to drift is offset by friction with the road.

3. A bullet fired at 120 meters per sec (400 ft/sec) drifts 2.5 mm in 1 sec (0.1 in 400 ft).

4. An artillery shell fired at 750 meters per sec (2500 ft/sec) drifts 60 meters (200 ft) per 33 km (20 mi).

5. A missile launched from the North Pole toward New York City at a rate of 1.6 km per sec (1 mi/sec) lands near Chicago.

6. An airliner leaving Seattle for Washington, DC along a great circle route and not corrected passes over South America near Lima, Peru.

winds blow at or about 65 percent of the velocity dictated by the pressure gradient. Over the land masses, where the surface is rougher and friction greater, the wind crosses the isobars at an average angle of 30° (Fig. 5.3). Because of greater friction, winds blow with a velocity of about 40 percent of gradient velocity.

Rotational Forces

Rotating bodies are subject to the law of **conservation of momentum**. This states that if no external force acts on a system, the total momentum of the system stays unchanged. Angular momentum is a function of the mass of the rotating body, the angular velocity (degrees per unit time), and the radius of curvature. The product of these elements stays constant when there is no external interference. A tetherball serves as a good illustration. As the ball goes around the pole and the radius shortens, the angular velocity increases to keep the momentum the same as it was initially. Another example of the conservation of momentum is that of a figure skater going into a spin. The moving skater turns into a slow spin with arms and perhaps a leg extended. When the skater pulls in his or her limbs when stretching upward, it reduces the radius of the spin and he or she turns faster and faster.

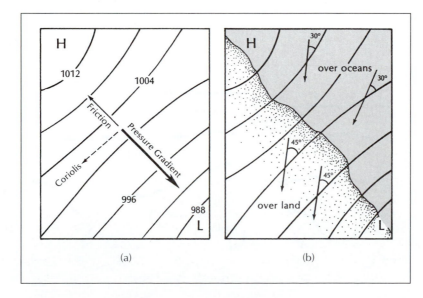

Figure 5.3
(a) The relative direction of the three forces that affect surface wind direction and velocity, (b) and the mean angle of wind direction relative to the pressure gradient over land and the sea.

Figure 5.4
In a converging system such as a mesocyclone, vertical stretching reduces the radius of the rotating air and leads to an increase in wind velocity. In this example, the mass of the rotating air (M) is set to 1 for the sake of simplicity. It is the same for the initial rotating mass and for the stretched mass.

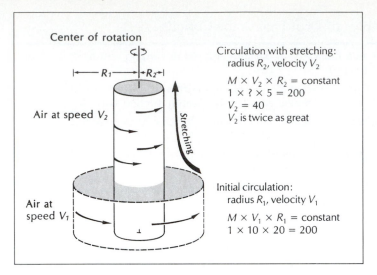

Center of rotation

Air at speed V_2

Stretching

Air at speed V_1

Circulation with stretching:
radius R_2, velocity V_2

$M \times V_2 \times R_2 =$ constant
$1 \times ? \times 5 = 200$
$V_2 = 40$
V_2 is twice as great

Initial circulation:
radius R_1, velocity V_1

$M \times V_1 \times R_1 =$ constant
$1 \times 10 \times 20 = 200$

Since air has mass, it also has momentum. As it moves north or south, the rotational velocity changes as its distance to the earth's axis changes. In actual practice, the increase in angular velocity is much less than it might be. Friction dissipates the energy as does diffusion in large mid-latitude eddies or cyclones. Conservation of momentum is a part of the acceleration process in mesocyclones and tornadoes. As the converging air travels a shorter and shorter path, the velocity increases accordingly. Figure 5.4 provides a schematic model of conservation of momentum. Box 5.2 examines quantitative aspects of vorticity and angular momentum.

THE RESULTING PATTERNS

The processes outlined previously produce the winds that blow at every scale, from the lightest hint to hurricane forces. As shown in Box 5.3, quantitative study of these processes is based upon fundamental equations relating to Newton's laws of motion. However, the forces do not act in the same way in all parts of the atmosphere for clear differences are found between surface and upper level winds. The reason for this is shown in Figure 5.5. The turbulent flow induced is a result of friction produced by moving air in contact with a surface. The vertical extent of the irregular flow will vary depending on the

Figure 5.5
Moving air in close to the surface is influenced by friction; above the friction layer, the lack of friction causes geostrophic flow. A friction layer and a geostrophic layer may be identified.

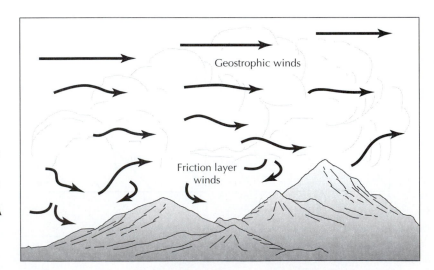

Geostrophic winds

Friction layer winds

QUANTITATIVE EXPRESSION: BOX 5.2

Vorticity and Angular Momentum

Figure C shows a cyclonic circulation. The rate of spin of an air particle is given by its vorticity (Q). In a constant circular motion, such as that shown in Figure C, vorticity is computed by the circulation path (C) divided by the area enclosed by that path, A. The value of C is the circumference of the circle ($C = 2\pi R$) multiplied by the rotational velocity V. Thus,

$$Q \text{ (Vorticity)} = \frac{C \text{ (Circumference)}}{A \text{ (Area)}} = 2\pi RV/\pi R^2,$$

which simplifies to $= 2V/R$.

The rate of spin depends on the radius of curvature of the circulating air and its velocity. The same variables, V and R, are used to derive the *angular momentum*, the amount of rotational spin. To obtain this, the mass of the rotating body is incorporated. Of significance in respect to the atmosphere is the fact that angular momentum is conserved when rotating air moves from one place to another. This *conservation of angular momentum* is given by:

$$\text{mass } (m) \times \text{velocity } (V) \times \text{radius of curvature } (R) = \text{constant}$$
$$\text{or } mVR = K.$$

This relationship shows that if a unit mass of air ($m = 1$) changes its radius of curvature, then its velocity must also change. A simplified example of this, using quite arbitrary and nondimensioned values, is given in Figure C.

A frequently seen example of the conservation of angular momentum at work occurs when an ice skater changes from a slow to a rapid whirl. This effect occurs because, after gaining momentum, the skater stretches upward to reduce the radius of spin. This stretching up motion enables air to increase its velocity as shown in Figure 5.4.

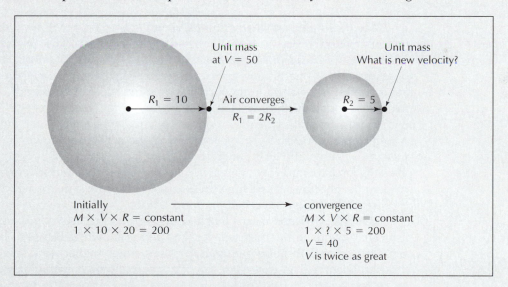

Figure C

roughness of the surface. However, for every surface and for wind at different speeds, it is possible to identify either geostrophic winds or friction layer winds.

Geostrophic winds occur above the friction layer. Without the influence of friction, the main forces acting on moving air are pressure gradient and the Coriolis effect. Consider the parcel of air in Figure 5.6a. It will move initially from high to low pressure and, in the northern hemisphere, be turned to the right by Coriolis. Eventually, these two forces are balanced, and the geostrophic wind then moves parallel to the isobars.

QUANTITATIVE EXPRESSION: BOX 5.3

The Equation of Motion

A major input into studies of the general circulation using computer models are the fundamental or primitive equations governing air motion. One of the equations relating to motion is an application of Newton's second law, which states that the acceleration experienced by a parcel of air depends on the net forces acting on it.

$$\text{Acceleration} = \text{sum of forces/mass}$$
$$= \Sigma F/m. \tag{1}$$

The equation of motion summarizes the forces (ΣF) and assumes a unit mass of air—1 kg, for example. Acceleration is the change in velocity (ΔV) between Times 1 and 2 (Δt) so that the equation may be written

$$\Delta V/\Delta t = \Sigma F. \tag{2}$$

The horizontal forces acting on a unit mass of air are pressure gradient (F_p), Coriolis effect (F_c), and friction (F_f). This may be written as

$$\Delta V/\Delta t = F_p + F_c + F_f. \tag{3}$$

Application of the equation of motion explains the difference between geostrophic and friction layer winds. Geostrophic winds are represented by

$$\Delta V/\Delta t = F_p + F_c, \tag{4}$$

whereas friction layer winds are those represented in Eq. 3.

Note that this equation provides acceleration and not air velocity. Even if the sum of the values is 0, there can still be air motion. Also, other forces, such as centrifugal force, may be added to the equation when applicable. The gradient wind equation would include the additional force.

When depicted on weather maps, as in Figure 5.6b, the flow of air parallel to the isobars is apparent. Note that in these upper air charts, pressure is depicted by an elevation; instead of showing actual pressure values, as on the surface map, these upper level charts show pressure surfaces. In this case, the 500 millibars surface is shown as the altitude at which this pressure occurs. For example, where high pressure is found 500 millibar level will be higher in the atmosphere (Fig. 5.7).

If pressure gradient and Coriolis are of equal magnitude, then the moving air should continue in a straight line for there are no net forces. However, around low pressure systems, the air follows an almost circular path. In such instances, the pressure force toward low is greater than the outward directed Coriolis force. A similar effect occurs around high pressures. The resulting flows, which are a form of geostrophic winds, are called **gradient winds**.

The lack of friction means that geostrophic and gradient wind can blow much faster than winds in the friction layer. The winds shown in Figure 5.6b are seen to commonly exceed 75 miles per hour. In places, much higher wind speeds are seen. These are identified as *jet streams*—corridors of high winds within flow of air above the friction layer.

Surface winds, because of friction, do not attain a balance between pressure gradient and Coriolis. As shown in Figure 5.8a, when air is moving, the initial state, a balance is eventually established among the pressure gradient, Coriolis, and friction. The resultant winds blow across isobars rather than parallel to them (Fig. 5.8b). It has been noted previously that the angle they cross the isobars varies depending on the surface.

Not only is surface direction different from winds found aloft, but the rate at which they blow is appreciably lower. The 75 miles per hour winds commonly found aloft are ranked as hurricane force winds at the surface. In fact, it is only under conditions where well below normal pressures are found that high winds occur at the surface. In these cases, such as hurricanes and tornadoes, the rapid circular motion of air is influenced by the rotation forces outlined earlier.

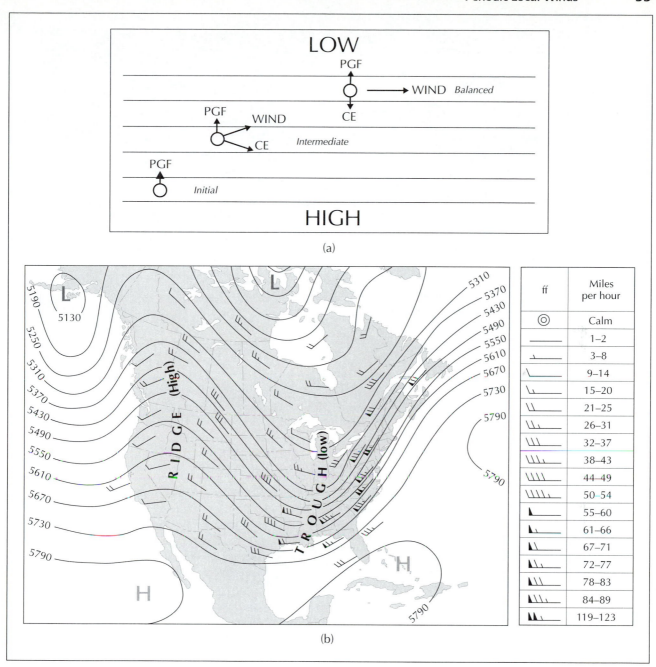

Figure 5.6
(a) The balance between pressure gradient force and Coriolis force result in a geostrophic wind.
(b) An upper air chart at the 500 mb level shows how winds generally blow parallel to the height contours. Note the high wind speeds (from Lutgens F. K. and Tarbuck E. J., *The Atmosphere*, 8e., © 2001, Prentice Hall, Upper Saddle River, N.J.).

PERIODIC LOCAL WINDS

Daily variations in atmospheric pressure develop in many parts of the world and lead to distinctive wind patterns. These variations in pressure result from moving air caused by changing surface temperatures through the day, and they result in

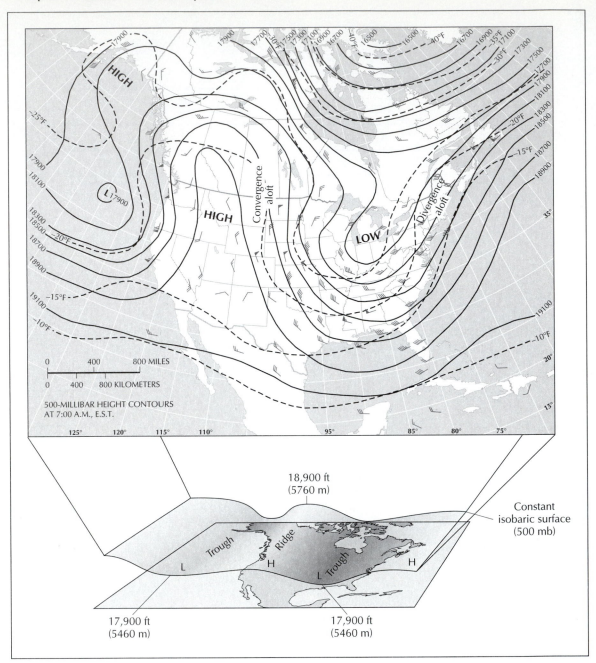

Figure 5.7
Pressure depiction on upper air charts is by the elevation of constant pressure levels. In this example, the height of a 500-mb surface is shown (from Christopherson R. W., *Geosystems*, 3e., © 1997, Prentice Hall, Upper Saddle River, N.J.).

daily changes in wind direction and velocity. There are many locally named winds (Table 5.3), and only selected examples are considered here.

Land and Sea Breezes

One of the daily winds associated with temperature differences of the ground surface is the land and sea breeze occurring along coastal areas and shorelines of large lakes. The land and sea breeze is a function of the change in temperature and pressure of the air over land in contrast to that over water.

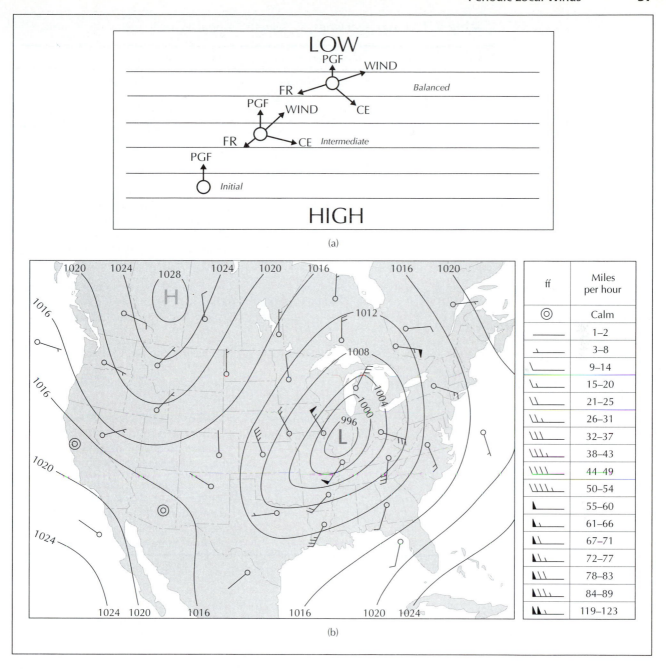

Figure 5.8
(a) The balance between pressure gradient force, Coriolis force, and friction result in friction-layer winds.
(b) A surface chart shows how winds relate to high- and low-pressure systems. Notice the low-wind speeds compared with those shown in Figure 5.6b (from Lutgens F. K. and Tarbuck E. J.. *The Atmosphere*, 8e., © 2001, Prentice Hall, Upper Saddle River, N.J.).

In the hours after sunrise, land temperatures rise rapidly while water temperatures rise slowly. When the overlying air above the shoreline heats, it expands and the density decreases. The air rises over the land surface and cold air flows in from the water to replace it. This flow of air landward is the sea breeze, which usually begins several hours after sunrise and peaks in the afternoon when temperatures are highest. This air will penetrate inland up to 30 kilometers or more along an ocean coast and to a lesser extent around lakes. The cell is quite a shallow one,

Table 5.3   Characteristics of Selected Local Winds

Name (origin)	Location	Characteristics
Bora (L, boreas, north)	Adriatic coast	Cold, gusty, northeasterly wind. Frequency at Trieste, 360 days in 10 yr. Mean winter speed 52 kilometers per hr (32 mph), summer 38 kilometers per hr (24 mph). May reach 100 kilometers per hr (62 mph) in winter. Adiabatically warmed wind.
Chinook (from Chinook Indian territory)	Eastern slope of Rockies	A warm wind that may, at times, result in sudden and drastic rise in temperature. May attain 60°–70°F (15°–21°C) in spring with relative humidity of 10%. Predominant in spring. Adiabatically warmed wind.
Etesian (Gr, etesiai, annual)	Eastern Mediterranean	Cool, dry, northeasterly wind that recurs annually. Summer and early autumn. Associated with regional low-pressure systems.
Fohn (German, possibly from Latin *favoniun* = growth, i.e., favoring wind)	Alpine lands	Similar to Chinook. Characterized by warmth and dryness. Most frequent in early spring. (Also Santa Ana, see text)
Haboob (Arabic)	Southern margins of Sahara (Sudan)	Hot, damp wind often containing sand. Of relatively short duration (3 hr). Average frequency of 24 per year. Early summer with the advance of the ITCZ.
Harmattan (Arabic)	West Africa	Hot, dry wind, characteristically dust-laden. All year, but most notable in the low-sun season. Associated with air flow from subtropical highs.

ranging up to several kilometers deep under the best conditions. The sea breeze is gusty and brings a decrease in temperature and an increase in humidity. The complementary land breeze is less well developed than the sea breeze. It occurs when the land surface cools at night and the water surface remains warm. The temperature and pressure relationships reverse with the higher pressure over the land. At night, however, the contrast between water and land is not as large as it is in the daylight hours. The water surface does not actually gain heat at night, but retains the heat acquired during the daylight period and thus remains relatively warm compared with the land. There is usually a lag in the development of the breeze due to the general inertia of the system. The sea breeze does not set in until well after the temperature differences develop.

Richard Henry Dana, Jr., described the effects of land and sea breezes on commercial shipping in 1840 in the days of sail in the following passage from *Two Years Before the Mast*:

> The brig Catalina came in from San Diego, and being bound up to windward, we both got under weigh at the same time, for a trial of speed up to Santa Barbara, a distance of about eighty miles. We hove up and got under sail about eleven o'clock

Table 5.3   *(continued)*

Name (origin)	Location	Characteristics
Khamsin (Arabic)	N. Africa and Arabia	Hot, dry, southeasterly wind. Regularly blows at a 50-day period (Khamsin = 50). Temperatures often 100°–120°F (38°–49°C). Same wind with adiabatic modifications includes Ghibli (Libya), Sirocco (Mediterranean), Leveche (Spain). Late winter, early spring. Regional low-pressure systems.
Levanter (from Levant, eastern Mediterranean)	Western Mediterranean	Strong easterly wind often felt in Straits of Gibraltar and Spain. Damp, moist, sometimes giving foggy weather for perhaps two days. Fall, early winter to late winter, spring. Regional low-pressure systems.
Mistral (maestrale of Italy = master wind)	Rhone Valley below Valence	Strong, cold wind channeled down Rhone Valley. May reach 100 kilometers/hr in north. Can cause sudden chilling in coastal regions. (Note also the Bise, and equivalent cold north wind in other parts of France.) Most frequent in winter. Regional low-pressure systems.
Norther	Texas, Gulf of Mexico to W. Caribbean	Cold, strong, northerly wind whose rapid onset may suddenly drastically lower temperatures (also Tehauantepecer of C. America). Winter. Related to circulation pattern over United States.
Pampero	Pampas of S. America	Southern Hemisphere equivalent of the Norther. Winter. Related to large-scale pressure patterns.
Zonda	Argentina	A warm, dry wind on lee of the Andes. Can attain 120 kilometers per hr (75 mph). Comparable to Chinook and Fohn. In dry weather, carries much dust. Winter.

at night, with a light land breeze, which died away toward morning, leaving us becalmed only a few miles from our anchoring-place. The Catalina, being a small vessel, of less than half our size, put out sweeps and got a boat ahead, and pulled out to sea, during the night, so that she had the sea-breeze earlier and stronger than we did, and we had the mortification of seeing her standing up the coast, with a fine breeze, the sea all ruffled around her, while we were becalmed, inshore. (p. 187)

Figure 5.9 shows the effect of a lake breeze on the temperature at Chicago, Illinois, compared with that of Joliet, Illinois. Up until about 3 p.m., temperatures were increasing in a similar manner in both cities. The lake breeze set in at Chicago at about 3 p.m., and there was a marked drop in temperature of some (3.8°C) 7°F in a little more than an hour. The temperature at Joliet, 56 kilometers (34 mi) southwest of Chicago, does not show the decline. The lake breeze was not strong enough to reach as far inland as Joliet. The wind shift producing the drop in temperature was nearly 180°, switching from west to east. Later in the evening when the lake breeze died, there was a rather sharp jump in the temperature in Chicago.

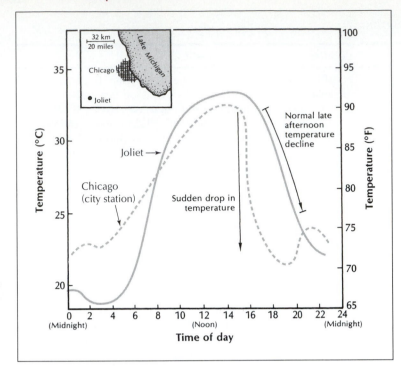

Figure 5.9
The effect of a lake breeze is shown by the difference in temperature over a 24-hour period between Chicago, on the lake, and Joliet, an inland location.

Mountain and Valley Breezes

Similar in formation to the land and sea breezes are the mountain and valley breezes characteristic of some highland areas (Fig. 5.10). In the daytime, the valley and slopes of mountains heat from the sun. The air near the surface heats, expands, and rises up the sides of the mountains. This breeze, called the *valley breeze* for the place of origin, is a warm wind and a daytime or late afternoon phenomenon. The clouds often seen forming over the hills in the afternoon are the result of condensation taking place as the air rises to cooler heights over the mountains. At night the valley walls cool. As the air at the surface cools, it flows down the slope due to greater density. This is the night breeze. In some mountain regions subject to frost, the valley slopes are the preferred places for orchards. The air moving up and down the valley slopes reduces the chance of the stagnant conditions conducive to frost formation.

Föhn Winds

As air descends, it is compressed and warmed. This property has given rise to a host of locally named winds that are collectively grouped using the European term **Föhn** winds. They sometimes result from synoptic patterns in concert with local topography, as in the case of the Santa Ana of California. This wind results from a high pressure located over the Rocky Mountain area leading to winds moving from high elevations toward the coast. As the wind descends, it is warmed and becomes increasingly dry. The wind can reach high velocities when

Figure 5.10
The mountain and valley breeze. The mountain breeze is a nighttime breeze; the valley breeze is a daytime event.

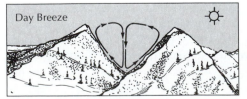

Figure 5.11
Cross-section of a mountain area showing the formation of a Foehn (Föhn) or Chinook wind.

channeled through valleys, such as the Santa Ana Valley. In the case of the hot, dry Santa Ana wind, dessication of vegetation can increase the propensity for fires and the wind is classed as a hazard. In the case of another compressional wind, the Chinook, the warm dry wind is considered an asset. This is considered in more detail in chapter 13.

The Chinook wind occurs on the eastern side of the Rockies when the synoptic situation causes winds to flow down the slope. As shown in Figure 5.11, the heating of the descending air can be considerable; in late winter, it can cause an early melting of snow in the high plains. This enables an earlier start to some agricultural activities.

STRENGTH AND DEPICTION OF WINDS

Today, numerous hand-held and automated instruments are available to measure wind speed and direction. Such was not always the case, particularly at a time when all ocean transport was by wind-reliant sailing ships. In 1805, British Admiral Sir Francis Beaufort provided a practical method for estimating the force of the wind. He devised a scale, from 0 to 12, that provided knowledgeable sailors, especially those aboard British "man-of-wars," with a list of common words to describe the force of the wind. Thus, for example, Force 11 was "that which would reduce a man-of-war to storm-stay sails."

Although the original **Beaufort Scale** made no mention of wind velocity, with the availability of instruments to measure the wind, its popularity caused it to be retained as a guide to wind velocity. It has been modified to provide wind speed limits to each Beaufort Number while, using appropriate descriptive terms, it has been applied to land area as well as oceans. The modern equivalent of the Beaufort Scale is shown in Table 5.4.

As indicated in Figures 5.6 and 5.8, for example, synoptic charts provide wind speed and direction by adding lines and pennants to a line showing the direction of the wind. To show average winds over a longer time period, **wind roses** are commonly used. They vary enormously from one depiction to another and may be shown as a single figure or located on a map. Most use lines perpendicular that are drawn proportional to the frequency of the wind from that direction. The examples shown in Figure 5.12 are but two of the ways in which wind roses may be constructed.

Table 5.4   The Beaufort Scale (after Christopherson)

Wind Speed					Beaufort Wind Scale	
kmph	mph	knots	Beaufort Number	Wind Description	Observed Effects at Sea	Observed Effects on Land
< 1	< 1	< 1	0	Calm	Glassy calm, like a mirror	Calm, no movement of leaves
1–5	1–3	1–3	1	Light air	Small ripples; wavelet scales; no foam on crests	Slight leaf movement; smoke drifts; wind vanes still
6–11	4–7	4–6	2	Light breeze	Small wavelets; glassy look to crests, which do not break	Leaves rustling; wind felt; wind vanes moving
12–19	8–12	7–10	3	Gentle breeze	Large wavelets; dispersed whitecaps as crests break	Leaves and twigs in motion; small flags and banners extended
20–29	13–18	11–16	4	Moderate breeze	Small longer waves; numerous whitecaps	Small branches moving; raising dust, paper, litter, and dry leaves
30–38	19–24	17–21	5	Fresh breeze	Moderate, pronounced waves; many whitecaps; some spray	Small trees and branches swaying; wavelets forming on inland waterways
39–49	25–31	22–27	6	Strong breeze	Large waves, white foam crests everywhere; some spray	Large branches swaying; overhead wires whistling; difficult to control an umbrella
50–61	32–38	28–33	7	Moderate (near) gale	Sea mounding up; foam and sea spray blown in streaks in the direction of the wind	Entire trees moving; difficult to walk into wind
62–74	39–46	34–40	8	Fresh gale (or gale)	Moderately high waves of greater length; breaking crests forming sea spray; well-marked foam streaks	Small branches breaking; difficult to walk; moving automobiles drifting and veering
75–87	47–54	41–47	9	Strong gale	High waves; wave crests tumbling and the sea beginning to roll; visibility reduced by blowing spray	Roof shingles blown away; slight damage to structures; broken branches littering the ground
88–101	55–63	48–55	10	Whole gale (or storm)	Very high waves and heavy, rolling seas; white appearance to foam-covered sea; overhanging waves; visibility reduced	Uprooted and broken trees; structural damage; considerable destruction; seldom occurring
102–116	64–73	56–63	11	Storm (or violent storm)	White foam covering a breaking sea of exceptionally high waves; small- and medium-sized ships lost from view in wave troughs; wave crests frothy	Widespread damage to structures and trees, a rare occurrence
> 117	> 74	> 64	12–17	Hurricane	Driving foam and spray filling the air; white sea; visibility poor to nonexistent	Severe to catastrophic damage; devastation to affected society

From Christopherson R. W., *Geosystems*, 3e., © Prentice-Hall, N.Y., p. 143.

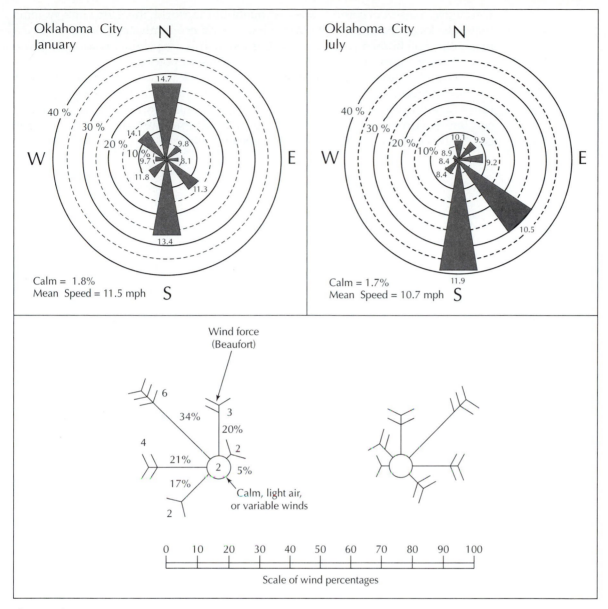

Figure 5.12
A wind rose may show the prevailing direction and speed of the wind for given periods. These upper figures shows seasonal wind roses for Oklahoma City, the lower figure the type of wind rose used on nautical charts.

SUMMARY

The atmosphere is never still. It moves whenever and wherever there are pressure differences in the atmosphere. Once air moves along a pressure gradient, several processes affect direction and velocity. The Coriolis effect deflects the wind to the right or left depending on hemisphere. Friction reduces the effects of the Coriolis force and determines how directly air will flow down the pressure gradient. Because different forces act at different levels of the atmosphere, friction layer and geostrophic winds are identified.

There are wind systems that blow regularly on a daily pattern. Examples include the land and sea breezes, the mountain and valley breezes, and winds warmed as a result of compression. Throughout the world, there are wind systems that have local importance. They influence many economic activities and have a major effect on human comfort. Some of these winds are peculiar to a single place, and others occur in similar circumstances at different places.

Since the atmosphere is continually changing, it is never exactly the same at any two points in time. Thus, weather over the earth is continually changing, sometimes changing faster than at other times and often in an unusual manner. The reason that weather is such a concern, especially in mid-latitudes, is because it is constantly changing and never exactly repeats itself.

KEY TERMS

Beaufort scale	Gradient wind	Ridge
Conservation of momentum	High pressure	Trough
Coriolis effect	Laminar flow	Turbulent flow
Föhn wind	Laws of motion	Wind rose
Friction	Low pressure	
Geostrophic wind	Pressure gradient	

REVIEW QUESTIONS

1. In what units is pressure measured?
2. Differentiate between laminar and turbulent flow.
3. Explain how Newton's second law of motion applies to air flow.
4. What is pressure gradient and how does it influence air motion?
5. What is the Coriolis effect?

6. How does the conservation of momentum apply to wind?
7. How and why do friction layer winds differ from geostrophic winds?
8. What causes a sea breeze?
9. Explain the cause of the Santa Ana wind.
10. How is a wind rose of value?

CHAPTER 6

Global Circulation of the Atmosphere

CHAPTER OUTLINE

*T*he general circulation of the atmosphere is extremely important to Earth. It is the general circulation that carries water from the ocean over the continents to provide precipitation and move heat energy from the tropical regions toward the poles, warming the high latitudes. The general circulation is not static, but changes constantly. Next to the seasons, changes in the general circulation are the most significant changes that take place in our weather. Earlier it was shown how a standard atmosphere provides a model of an average or mean atmosphere. Here we do the same with a simple model of the how the general circulation works. Then building on the processes that control the general circulation, we examine how it changes. In the final chapters of the book, we examine how changes in the general circulation bring about climatic change.

Between latitudes of about 35°N and 35°S, incoming solar radiation exceeds Earth radiation to space (Fig. 6.1). Toward the poles from 35°, there is a net loss of energy to space. For incoming solar radiation to exceed outgoing radiation in equatorial areas over a prolonged period, energy must be removed in some other manner than radiation. For outgoing radiant energy to exceed incoming radiation in the polar regions, energy must be transported in. The mechanism for transporting this energy is the general circulation of the atmosphere and oceans. It is a mechanism that tends to equalize the distribution of energy over the surface of the earth.

The atmosphere transfers about 60 percent of this energy and the ocean moves the rest. This complex circulation system consists of several semipermanent areas of convergence and divergence and the air flow in and between them. The energy moves primarily as sensible heat and latent heat of water vapor. Internal friction, or friction with the ground surface, converts the kinetic energy of the wind systems to sensible heat. There is an internal energy balance. The energy lost from the system by friction balances the rate of kinetic energy generation within the atmosphere.

Figure 6.1
(a) Latitudinal distribution of incoming and outgoing radiation. Between latitudes of about 30° north and south, incoming solar radiation exceeds outgoing Earth radiation. Toward the poles from 30° outgoing radiation exceeds incoming radiation.
(b) For the distribution shown in (a) to exist, there is an equator to pole movement of energy by the ocean and atmosphere. The total amount of energy transported is shown by the solid line. The dashed line shows the amount of heat transported by the atmosphere. The shaded area is the amount transported by ocean currents (modified from Gill, 1982).

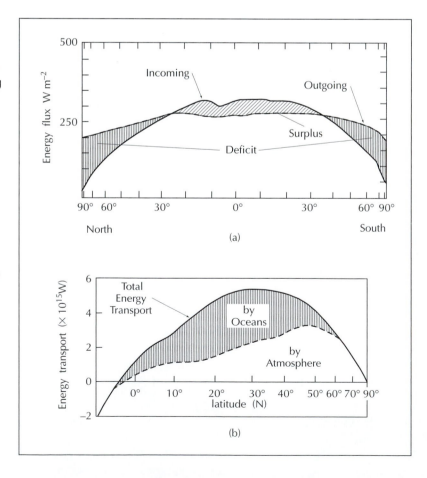

THE GENERAL CIRCULATION

In 1735, George Hadley proposed a cellular model to explain the primary circulation of the atmosphere (Fig. 6.2a). He based the model on the basic pressure differences brought about by uneven heating of the earth's surface. Hadley postulated that cold air descended at the poles and flowed along the earth's surface toward the warmer equator. This flow would be countered by rising warm air at the equator and a flow aloft toward the poles so that two large **Hadley cells** existed in each hemisphere.

Observation of global wind patterns and, later, theoretical work showing that the rotation of the earth prevents the development of a single convective cell led to a modified view. Hadley's model was gradually altered to include a three-cell structure between the poles and the equator. In the revised **three cell model**, there is still subsiding cold air at the poles and rising warm air at the equator (Fig. 6.2b). In a crude way, it illustrates the essentials of the primary circulation.

Meridional Circulation

A more detailed circulation pattern shown in Figure 6.3 is often used to show the general circulation. It actually represents a long-term average of the **meridional (north–south) circulation** over the globe. In reality, the circulation is much more complex than this model suggests. For example, the meridional transfer of air across the globe is dominant in low latitudes, but is some 60 percent weaker in mid-latitudes. Although the low-latitude cell exists, the circulation of middle latitudes consists essentially of a west-to-east (zonal) flow of upper air that determines the position and location of moving high- and low-pressure systems at the surface. It is the circulation patterns associated with the traveling high- and low-pressure systems that is responsible for energy transfer outside of the Tropics. Thus, although Figure 6.3 is commonly used to show the global circulation, it does not do justice to the actual processes that occur.

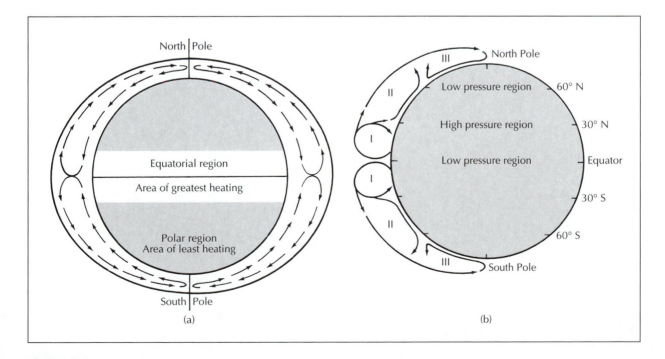

Figure 6.2
(a) The fundamental driving mechanisms of the general circulation.
(b) A three-cell model that results from differential heating, the earth's rotation, and the fluid dynamics of the atmosphere.

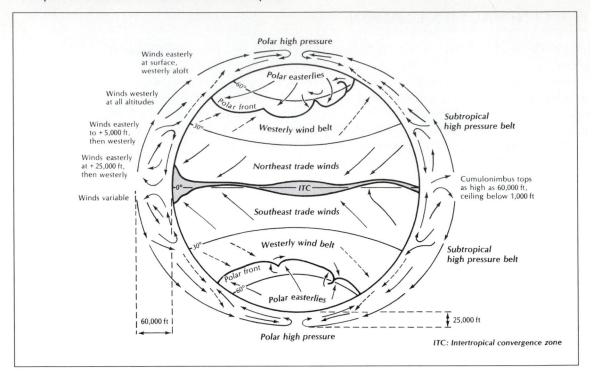

Figure 6.3
The vertical and longitudinal flows in the idealized general circulation.

Zonal Circulation

The meridional depiction need be augmented by one showing **zonal circulation**. Figure 6.4 provides a model of the primary circulation of the northern hemisphere that stresses the difference between tropical and mid-latitude circulation. The figure shows the Tropics dominated by Hadley cell and mid-latitudes by upper air

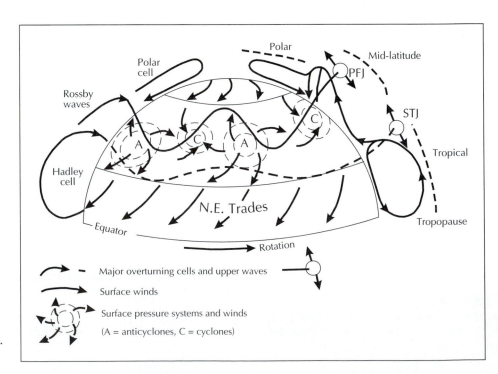

Figure 6.4
A general circulation model of the northern hemisphere showing the zonal flow dominant in middle and high latitudes. (After Hanwell, 1980.)

Rossby waves, so named for the meteorologist Carl-Gustav Rossby whose theoretical work provided the understanding of the mechanisms for their existence. The dominant air flow within the Rossby waves is west to east. Embedded in these upper air westerlies are corridors of rapidly moving air, or **jet streams**. Associated surface circulation, located in the old westerly wind belt, are the cyclones and anticyclones that provide the changeable weather of the region.

An understanding of global climates requires assessment of seasonal changes of climate. Neither of the models shown, whether emphasizing meridional or zonal flows, take seasons into account. Such is best considered using representations for summer and winter for both upper air westerlies and surface pressure maps. As shown later, such illustration provides clear evidence of the great changes that occur in the location of high- and low- pressure semipermanent systems.

Global Computer Models (GCMs)

Global Computer Models (**GCMs**) have become widely known as tools to forecast the effects of increased CO_2 on the energy balance of the earth. As their name indicates, GCMs are computer-generated maps; they use the basic equations governing air flow to assemble what conditions may occur under different scenarios. One of the problems with these models is that they all depend upon initial conditions established when the model is started. The conditions of the atmosphere at any point in time are sensitive to previous conditions. Every child has taken a dry dandelion blossom and blown the seeds from the stem. All the seeds begin their journey at nearly the same place and diverge farther and farther as distance from the stem increases. The seeds do not all take the same path because each begins at a slightly different place and hence becomes subject to different air currents as it moves away. The position of the seeds can be forecast much more accurately at a distance of 1 meter (3.3 ft) than at 10 meters (33 ft). Unfortunately, the initial conditions used in numerical weather forecasting are only approximate. All computer models of the atmosphere use sample points over the surface of Earth and at different heights in the atmosphere. The distance between sample points on the surface ranges upward from 100 kilometers (60 mi). A rectangle 10,000 kilometers^2 (3060 mi^2) can contain thunderstorms, hailstorms, and tornadoes within its boundaries. Sampling may completely miss these atmospheric storms. There may be differences in the thickness of clouds or amount of cloud cover. The ground surface may be all water or all land, or more likely part of each. The models assume that all of the atmospheric conditions within the area are the same. A single number represents all of the sample area. To predict the circulation accurately, we need to sample pressure, temperature, and moisture at every point in the atmosphere. Since the atmosphere contains an infinite number of points, the task becomes impossible.

Circulation of air follows an overall pattern, but it is never quite the same two times in succession, nor does it ever repeat itself exactly. If the initial conditions are even slightly different, the end product may be very different. Since initial conditions are never the same, the future state of the system can never be the same. The reasons that atmospheric forecasting is so difficult are because the initial conditions are never the same and it is not possible to know the initial conditions at every point in the atmosphere. Each set of slightly different initial conditions produces different weather.

For the same reasons, a weather forecast for 3 hours from now is more accurate than one for 3 days from now. Further, a current map of global circulation is more accurate than one 10 years in the future. Forecasting the future is similar to, but much more difficult than, predicting the path that a dandelion seed will take. Individual particles in a flow of water or air tend to take deviating paths. The more they deviate, the likelier they are to deviate still more. That chaotic nature makes predicting the future state nearly impossible.

Persistence, or the Joseph Effect

Forecasts may employ a property called **persistence**. Persistence involves the clustering periods of the same kind of conditions. It is the tendency for a particular set of conditions to continue or repeat itself. The simplest means of forecasting atmospheric conditions for an hour from now, or a day from now, is to forecast it to be the same as it is now. In fact, this method of forecasting is correct most of the time since all weather systems last for some amount of time. In the mid-latitudes, large weather systems exist for 5 to 7 days. Heat waves, cold waves, floods, and droughts are examples of persistent weather systems. When an area has more rain than usual, that area is likely to have more rainy weather. The same is true of dry weather. This persistence is known as the Joseph effect. The name comes from a chapter in the book of Genesis in the *Bible*:

> There came seven years of great plenty throughout the land of Egypt. And there shall arise after them seven years of famine . . . .

The 7 wet years followed by 7 dry years suggests a periodic phenomenon. We can forecast periodic phenomena with accuracy in amplitude and wavelength. Unfortunately, this is not the case with weather and climate because it is aperiodic or cyclical. For example, droughts repeat, but their intensity varies and the intervals between them vary. The reference to 7 wet years followed by 7 dry years indicates that persistence takes place over time intervals as short as minutes and hours to as long as years. Other evidence clearly shows that persistence operates over periods of decades and centuries. Geological evidence shows that it works over periods of thousands and millions of years. Add this set of ideas to producing a GCM and the problem becomes both more interesting and more difficult.

TROPICAL CIRCULATION

The circulation closest to the equator closely resembles that proposed in the three cell model and the Hadley cell. The cell extends from around 25° to 30° to near the equator (Fig. 6.5). Like Hadley's cell, it features surface flow toward the equator and counterflow aloft, with rising air at the equator and subsiding air near the Tropics of Cancer and Capricorn. This section of the primary circulation is called the Hadley cell since it operates effectively as Hadley outlined it in 1735. Associated with this tropical cell are two semipermanent belts of surface divergence and above-

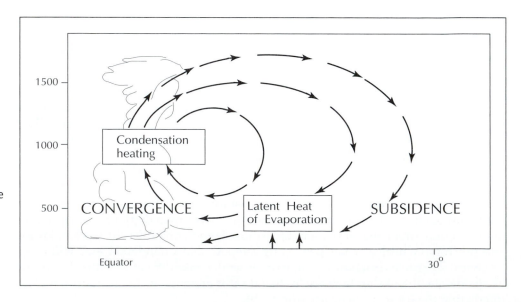

Figure 6.5
An idealized cross-section of the Hadley cell with the vertical scale greatly exaggerated (from Riehl H., *Introduction to the Atmosphere*, © 1972, McGraw Hill Book Co, N.Y.).

average sea level pressure located near the Tropics of Cancer and Capricorn. These **semipermanent high pressure** zones on either side of the equator act as a source region for air flowing both north and south at the surface. As the air subsides, it heats adiabatically and is quite warm. Since the temperature is increasing and no moisture is being added, the relative humidity is quite low. These zones have high amounts of sunshine, clear skies, and low frequency of precipitation. In fact, there are major deserts associated with these areas of subsidence: the Sahara and Sonora deserts of the northern hemisphere and the Great Australian, Atacama, and Kalahari deserts of the southern hemisphere. If the air that flows out of these areas of subsidence has a trajectory over a land mass, it remains warm and dry and becomes a continental tropical air mass. The high pressure is due to the subsidence, which gives rise to divergence at the surface. These zones of subsidence and divergence result more from the motion of the atmosphere and the Coriolis effect than from thermal factors. Over the oceans, the surface winds are light and variable. In the days of wind-driven sailing vessels, these zones were known as the **horse latitudes**. When ships were becalmed, horses were thrown overboard to lighten the load.

In addition to serving as a mover of heat, the primary circulation is a mover of moisture. The general circulation moves moisture from low to high latitudes and from oceans to land masses. The ocean yields much more water to the atmosphere through evaporation than does land. Evaporation from the ocean is highest where there are the biggest differences between the vapor pressure of the air and the water. Ocean temperatures are the highest at latitudes about 30° either side of the equator. This is a result of the high-intensity radiation and clear atmosphere associated with the divergence. The high temperature of the water and the low moisture content of the air create a large difference between the vapor pressure of the air and the water. Thus, the subtropical oceans near the divergence zones are the major source regions of the precipitation for land areas of low and mid-latitudes.

Trade Winds

Lying between the subtropical belts of subsidence and the equatorial convergence zone is a region of surface winds flowing toward the equator known as the **trade winds**. These wind systems are most frequent in belts 10° in width and centered 15° on either side of the equator. They are the most regular wind systems found at the surface of the earth. The winds, which average four to seven kilometers per hour (2–4 mph), are most persistent over the eastern half of the oceans and are more dependable there than elsewhere. Both north and south from 15° these winds are less distinct. Their origin is in the subsidence of the subtropical high-pressure zones, and they flow into the equatorial low-pressure belt. The Coriolis effect gives the winds a westward turn, so they become northeasterly winds in the northern hemisphere and southeasterly in the southern hemisphere. The trade winds often have layers, with a surface layer that becomes increasingly moist and unstable as the air moves toward the equator. Above there is an overlying layer that is dry and stable.

As soon as an air mass begins to move from the source region, alteration of the temperature and moisture characteristics begins. The air masses are changed by heating or cooling from the surface, either by conduction or radiation. Moisture content is altered by addition or removal of water by evaporation, condensation, and precipitation. At any given time, an air mass is the result of the nature of the source region and the changes that the air mass has undergone while moving over the surface.

As the air from the subtropical high moves toward the equator, it not only increases its moisture and instability, but adds latent heat. The following excerpts from *Mutiny on Board the HMS Bounty* illustrate the effects of the sea surface on the overlying air. The entries in the log were made as the *Bounty* sailed south in the North Atlantic toward the equator while in the trades.

The thermometer was at 82°F in the shade and 81 1/2° at the surface of the sea, so that the air and the water were within a half a degree of the same temperature.

Monday the 4th. Had very heavy rain; during which we nearly filled our empty water casks. So much wet weather, with the closeness of the air, covering everything with mildew. The ship was aired below with fires, and frequently sprinkled with vinegar; and every little interval of dry weather was taken advantage of to open all the hatchways, and clean the ship, and to have all the peoples wet things washed and dried.

Monday the 16th. The weather, after crossing the line, had been fine and clear, but the air so sultry as to occasion great faintness, the quicksilver in the thermometer, in the daytime, standing at between 81° and 83°F, and one time at 85°F. In our passage through the northern tropic, the air was temperate, the sun having then high south declination and the weather being generally fine till we lost the Northeast trade wind; but…such a thick haze surrounded the horizon, that no object could be seen, except at a very small distance. (pp. 30–31)

The trades are known for providing the route across the Atlantic followed by the sailing ships of the age of exploration. These same winds play a major role in the Pacific Ocean as well. It is the trade winds that carried Thor Heyerdahl and his crew across the Pacific Ocean on the first modern crossing by raft.

The wind did not become absolute still—we never experienced that throughout the voyage—and when it was feeble we hoisted every rag we had to collect what little there was. There was not one day on which we moved backward toward America, and our smallest distance in twenty-four hours was 9 sea miles, while our average run for the voyage as a whole was 42 1/2 sea miles in twenty-four hours. (Heyerdahl, 1950, p. 41)

The Intertropical Convergence Zone

Near the equator, the trade winds from both hemispheres converge to form a low-pressure trough where there is a gentle upward drift of air. This convergence zone is also called the **intertropical convergence zone** (abbreviated as either ITC or ITCZ). It is the area in which the trade winds from the north and south of the equator merge. The area of convergence is quite broad along the equator because of the extensive area of warm surface. Although this trough of low pressure is not very deep, it is quite consistent over the oceans and is deep enough to produce convergence most of the time. Where the converging trades have a trajectory over the ocean, the air contains large amounts of moisture, cloud cover is extensive, and precipitation is frequent.

There is less evaporation from the ocean along the equator than near the Tropics of Cancer and Capricorn. The reason is that equatorial air is quite humid and the vapor pressure at the water surface is lower. Sea temperatures are not as high near the equator as they are farther out near the Tropics of Cancer and Capricorn. The addition of fresh water from precipitation and from streams draining adjacent land masses cools the ocean. Less solar radiation reaches the surface as a result of the more extensive cloud cover, which further reduces water temperatures.

A major source of water vapor in the stratosphere is the warm, moist air rising at the ITCZ. The tropopause marks a transition in the vertical temperature distribution. Temperature above the tropopause in the stratosphere is higher than at the tropopause. This acts as a cap on rising air. The heated air in thunderstorms in the ITCZ routinely breaks through the tropopause into the stratosphere. Since the air rising through the tropopause cools at the adiabatic rate, it rapidly becomes cooler and more dense than the surrounding air. It then drifts back downward. Some of the air is carried away from the equator and subsides in the subtropical high. The rising air in the ITCZ gains a lot of heat from condensation. It radiates away much of it to space as it flows toward the subtropical high. Thus, the Hadley cells take heat from the ocean surface near the equator and moves it to the upper troposphere and toward the poles.

In the center of the ITCZ over much of the Atlantic and Pacific Oceans are the **tropical easterlies**. These are stable winds of low velocity moving from east to west.

Do not confuse them with the trade winds. These easterly winds are quite regular in direction. Occasionally, a change in circulation will bring quite unstable weather in the form of squalls and general rainstorms. The converging winds on occasion will rise some distance away from the center of the convergence zone. When this happens, quite stagnant conditions may result near the surface. Sailors of the 16th century called these stagnant conditions of low-velocity winds with ill-defined direction the **doldrums**. While traveling to the western hemisphere from Europe, sailing vessels became becalmed in this area just as farther north in the horse latitudes.

MID-LATITUDE CIRCULATION

The three cell model of the atmosphere has inadequacies, and the real world differs most from the model in the mid-latitudes. The model calls for a mean flow of air between the semipermanent high-pressure cells at 30° and the semipermanent low-pressure cells through near 60°. In reality, the mean meridional flow in these latitudes is very weak. Surface winds are quite variable in both direction and velocity. The greatest frequency and highest average velocity show a west-to-east flow. Maximum westerly velocities are at or about 35°N. Wind velocities at this latitude become more pronounced with height up to the tropopause. Thus, the primary flow in mid-latitudes is zonal (west to east).

It is in the latitudes of 35° to 40° north and south that the maximum transfer of energy takes place. With a relatively weak meridional flow, there must be some other means of transporting the energy through the zone. In this region, there is a mixing zone between the warm tropical air and the cold polar air known as the **planetary frontal zone**. It is here that the mid-latitude westerlies are strongest. The westerly flow increases up to about 12 kilometers (7.5 mi). In the upper part of the troposphere and the lower stratosphere, long meandering waves called Rossby waves develop in the planetary frontal zone (Fig. 6.6). These waves have an amplitude and wavelength that vary up to 6000 kilometers. They are always in a state of flux, sometimes being standing waves and sometimes moving with the westerlies.

The basic flow of the westerlies is from west to east at a mean latitude of about 45°. The westerlies and the subpolar front divide the cold and dry arctic air from the warmer and more moist tropical air. When the westerlies are blowing primarily west to east, it is called a *zonal flow*. When there is zonal flow over North America, the arctic and tropical air are kept separated, and there is cold air north of the polar front and warm weather south of it. Since the dominant wind direction is

Figure 6.6
Model of the development of waves in the westerlies: (a) When the westerlies are zonal (flowing from west to east), there is minimal mixing of tropical and polar air. When waves form with air flow becoming more north–south, a great deal of cold and warm air is exchanged. (b) A wave has begun to form. (c) A wave exhibits large oscillations. Maximum exchange of warm and cold air occurs at this time. (d) As the system begins to stabilize, masses of cold air become isolated toward lower latitudes.

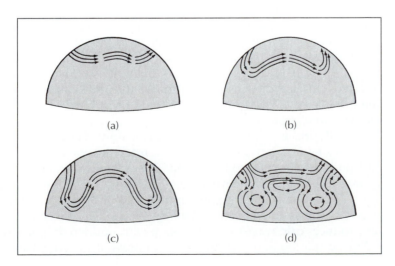

(a) (b)

(c) (d)

west to east with zonal flow, air from the Pacific Ocean flows over the Rocky Mountains and crosses the continent. In the process of crossing the mountain ranges, the air warms and dries. This brings periods of quite warm, dry, and stable weather to areas of the continent south of the jet stream.

When the westerlies begin to loop widely, there is a greater north–south or meridional flow, although the net direction of flow is from west to east. Meridional flow causes exchange of warm and cold air. Warm air, and hence heat, is carried northward, and cold air is carried southward. When there is high meridional flow, temperatures may fluctuate rapidly and widely on a week-to-week basis. In the first week of February 1991, temperatures went from record lows to record highs over much of northern United States and southern Canada.

Figure 6.6d illustrates a situation in which the meandering is large enough that cells of cold or warm air become isolated from the main air flow. Sometimes these large masses of air stagnate and control the weather of the area in which they are located for up to several weeks. These are called **blocking** systems. These masses of air can bring either very warm or very cold conditions. Blocking highs divert moisture, keeping storms away from the region, and enhance the probability of drought occurring.

The westerlies shift back and forth between a zonal pattern and a meridional pattern at irregular intervals, which makes long-range forecasting difficult. Near the surface, large eddies or vortices develop that are 1000 to 2000 kilometers across and have lifetimes of several days. These eddies are the cyclones and anticyclones of the mid-latitudes. They are rotational in character and thus carry cold air from the poles toward the equator and warm air from the subtropics toward the poles. These eddies carry energy toward the poles, but the net flow of air northward is very small. It is basically an exchange process of warm for cold air. Thus, there is a high-energy transfer with a low net flow of air toward the poles. It is also the high frequency of these cyclonic cells that forms the subpolar low-pressure zone in latitudes 50° to 60°. This convergence zone differs markedly from the equatorial convergence zone. The air flowing into it has very different temperature and moisture conditions.

In mid-latitudes, the primary circulation is very different from that of the Tropics. The subtropical high over the oceans and the polar high act as major source regions for air moving toward the convergence zone. These two source areas are very different. One is very warm and the other is quite cold. Although this difference alone is enough to produce different air streams in the mid-latitudes, moisture properties of the air streams magnify the differences. Thus, in the tropical convergence zone converging air streams have similar temperatures and moisture content, whereas in the mid-latitudes they are very different.

The **upper air westerly** flows dominate the circulation of the mid-latitudes. The temperature gradient between the equator and poles determines the strength and location of the subtropical jet stream. As polar regions warm in summer, the gradient is least and the strength of the westerlies diminishes. The jet migrates toward the pole. When a steep gradient exists, the westerly circulation is strong and the jet migrates closer to the Tropics. The limit of the mid-latitudes is along the axis of the semipermanent high-pressure cell located near 30°. From here to the equator, the Hadley circulation operates.

The surface circulation depends on the flow of upper level westerlies. The subtropical jet migration and the westerly circulation weakening produce very different conditions from season to season. The movement and relative dominance of air masses will vary, as will the frequency and paths of traveling low-pressure systems.

The movement of the polar front toward the equator results in the movement of polar air masses over much of the mid-latitudes. The invasion of extremely cold air produces cold waves. Cold waves are any sudden drop in temperatures within a 24-hour period such that measures to prevent extreme distress to humans and animals may be necessary. In the United States, winter cold waves claim more lives than any other weather-related event. Statistics compiled by the National Center for Health show an average of 355 deaths per year for the period 1949 to 1978. Although deaths directly attributable to cold take a high toll, evidence shows that cold strongly affects the death rate in indirect ways. This is particularly the case in subtropical areas where cold waves are infrequent. In Florida and Alabama, death rates nearly doubled during one severe cold wave in 1965.

Equally cold outbreaks of arctic occur in Europe and Asia. In 763 A.D., a cold wave froze both the Black Sea and the Straights of Dardanelles in Europe. In 1236 A.D., the Danube reportedly froze solid. In 1466, the winter in Flanders was so severe that it was necessary to cut wine rations issued to solders with hatchets and distributed in frozen chunks. In 1691, crazed by the lack of food in the frozen countryside, wolves invaded Vienna, Austria, killing humans for food.

POLAR CIRCULATION

Over the north and south polar regions, thermally induced high-pressure systems exist, although knowledge of them is limited. The high-pressure cells exist some distance from the geographic pole toward the continental **cold pole** (the point of coldest temperature). This displacement in the Antarctic is not large, but in the northern hemisphere it is significant. In the northern hemisphere winter, there is always a heat flow through the pack ice and from the open water areas. The cold pole is on the land mass. Since there are two major land masses in the northern hemisphere, there are often two cold poles—one over North America and one over Eurasia. Upper air divergence and temperature inversions are characteristic of the cold poles. There is a predominance of high-pressure systems and divergent air flow over polar areas most of the year. The size of the area affected changes, expanding during the winter months. Outward from the polar highs are belts of weak easterly winds that are more pronounced in the summer in the northern hemisphere.

Pressure gradients are weak in the tundra, so this is not a stormy region. In fact, it is one of the least stormy areas of Earth's surface. High-pressure systems dominate with cold, still, dry air prevailing, particularly in the winter. The North American Arctic is even less stormy than other areas. Because of the mountainous character of Alaska, the cyclones of the Pacific Ocean do not move over the Arctic coastline. Some of the storms penetrate far into the arctic, but few do compared with other areas. Spring and fall are the seasons with the most active weather. It is the same in most high-latitude locations. Winds are light in velocity and variable in direction. The average velocity is less than 15 kilometers/hour and less than 10 kilometers/hour in the Canadian tundra.

In the region surrounding the Antarctic continent, there is an unbroken sea extending completely around the earth. In the winter, there is a much greater frequency of storms with high winds blowing from a variety of directions. This is the part of the world ocean that was named the "roaring forties," "the furious fifties," and the "shrieking sixties" by the sailors of the 16th century. Sailors on board vessels trying to sail around Cape Horn at the tip of South America named the winds. The finding of the Straight of Magellan made the passage around South America considerably less trying but by no means easy.

Winds are a predominant factor in the weather of the **polar deserts**. Westerly winds are most common at the surface to around 65°, where they give way to low-level easterly winds that extend on to about 75°. Cyclonic storms develop over the oceans in the westerlies and move around the Antarctic from west to east. They normally do not penetrate far inland, and they account for much of the precipitation and weather along the coast. Winds of a high enough velocity to move snow and produce blizzard conditions occur at Byrd Station about 65 percent of the time.

They are of high enough velocity to produce zero visibility about 30 percent of the time. A slight increase in wind velocity brings a large increase in blowing snow. The amount of snow moved by the wind varies as the third power of the velocity. It is unfortunate for those working in the Antarctic, but the most accessible locations are also the windiest. Mawson's base at Commonwealth Bay experiences winds that are above gale force (44 kph or 28 mph) more than 340 days a year. At Cape Denison, the mean wind velocity is 19.3 meters/second (43 mph). During July of 1913, it averaged 24.5 meters/second (55 mph).

SEASONAL CHANGES IN THE GLOBAL PATTERN

Some major elements of the general circulation shift location from season to season. The tropical Hadley cell shifts north and south with the seasons. This is a result of the shift in the thermal equator. As described earlier, the latitude that receives the most sunshine during the year changes from about 30° north to about 30° south latitude. This change in the heat equator causes the Hadley cell to move north and south as well. Most areas of the earth between the equator and the thirtieth parallel experience seasonal changes in wind direction and precipitation amounts. This is especially true over the continents since the land surfaces heat and cool more rapidly than does the ocean.

Figure 6.7 shows the semipermanent highs and lows and the dominant winds that occur at the earth's surface. Note, however, that the actual system is not the same as the ideal shown in Figure 6.2. The earth is not a smooth surface, and the distribution of the land masses in the world ocean alters the winds.

The major semipermanent pressure zones tend to shift north and south with the seasons. Consider, for example, the location of the ITCZ. This convergence zone is mainly south of the equator in January (Fig. 6.7a), but is north of the equator in July (Fig. 6.7b). The subtropical highs and subpolar lows also shift location. This seasonal migration of the pressure belts and associated winds is a direct result of the changing distribution of temperature that occurs with the seasons.

Along the equator, the ITCZ moves north and south during the year. This results in the convergence of warm moist air streams from both sides of the equator, leading to frequent and abundant precipitation.

Monsoons

In addition to the north–south shift in the general circulation, there is also an east–west shift in the relative position of the subtropical highs and subpolar lows. In mid-latitudes, there is a large contrast in insolation between winter and summer. Because of the relatively low specific heat of Earth materials compared with water, the continents heat and cool more rapidly with the seasons. The heating of the land masses in summer weakens the subtropical highs, and in some cases breaks them down altogether and replaces them with a low.

During the winter half-year, the subtropical high and polar high expand and often meet over the land masses that have undergone major cooling. In midwinter, high-pressure ridges or cells exist over the northern hemisphere land masses. During the summer months, the subpolar low expands over the continental areas. Sometimes tropical and subpolar lows merge to form a single trough of low pressure extending from the Tropics to 60° to 70° of latitude.

The change in pressure systems brings about a marked reversal in wind and moisture conditions with the season. During the summer months, convergence predominates and high humidity and precipitation are general over the continents. In the winter months, when high-pressure systems predominate, divergence over the land is more frequent than convergence. As a result of the latter phenomenon, humidity and precipitation are both lower in winter than in summer. This seasonal shift of pressure, with all its ramifications, is a **monsoon**. The largest land areas in mid-latitudes are in the northern hemisphere. Here is where the monsoons are most pronounced. Large parts of Asia and Africa are subject to this seasonal shift. It need be noted that the monsoon climates are also subject to interannual variation that have extended effects on the region's inhabitants. This aspect is discussed further in chapter 7.

Asian Monsoon

The seasonal shift in pressure, wind, and humidity is greatest over Asia. During the winter, subsiding dry air covers much of the continent. This results in a dry season that is comparatively much drier than the North American winters. The subsidence produces offshore winds over much of coastal Asia. As these winds move out over the oceans, they evaporate large amounts of water from the sea surface and are sources of moisture for the offshore islands. During the summer months, the low-pressure system and convergence are strong, and the onshore flow of moisture is large (Fig. 6.8). Moisture-laden winds blow north from the Indian Ocean and the Arabian Sea. They converge over the northern plains of India and the slopes of the Himalayas. One of the rainiest regions of the world is found in the Assam hills. Cherrapunji, India, is an extreme example of seasonal rainfall. Here the annual precipitation has reached 25 meters (82.5 ft) and most comes in 4 months. Resulting precipitation on the coastal areas and oceanic sides of offshore islands is heavy.

In the spring and fall, there are intervening periods between the rainy and dry seasons when the weather is unsteady but not too unpleasant. The highest temperatures of the year often occur late in the spring or early in summer. This is before lots of moisture begins to move onto the continent. The clear skies before the development of the onshore winds allow more radiation than does the moist maritime air that follows. The rains and humidity, which reduce absolute temperatures, actually increase sensible temperatures.

The monsoon is the product of many factors working together. One is that the ITCZ migrates to between 25° and 30°. A second is that the land mass interior records a large change in pressure that enforces the migration of the ITCZ. It is also likely that the Himalaya Mountains are a very significant topographic barrier, which splits the circulation over Asia. In the winter, the jet stream is over the mountains. The polar high expands and covers much of north-central Asia. Divergence from this strong system north of the Himalayas produces extremely strong offshore surface winds. South of the Himalayas and the jet stream, the anticyclonic flow is much weaker, although offshore flow is most common. In the summer as the temperature differential over the continent diminishes, the jet stream breaks down. The subtropical high and ITCZ shift rapidly northward and bring the sudden shift of winds onshore over the Indian subcontinent. North of the Himalayas, a weaker convergent system develops with attendant onshore flow of air. The precipitation is primarily convectional rainfall randomly distributed within the flowing moist air. Some traveling cyclones associated with the ITCZ add further rainfall. The peak season of rainfall moves northward over the subcontinent as the zone of convergence moves northward toward the slopes of the Himalayas.

North American Monsoon

The monsoon applies to North America as it does to Asia. In summer, the subtropical high shifts north over the Pacific coast, and there are virtual desert conditions over much of southern California. Summer is the dry season for all of the United States west of the front ranges of the Rockies. On the eastern side of the continent, the low-pressure trough is more persistent, and convergence is frequent over

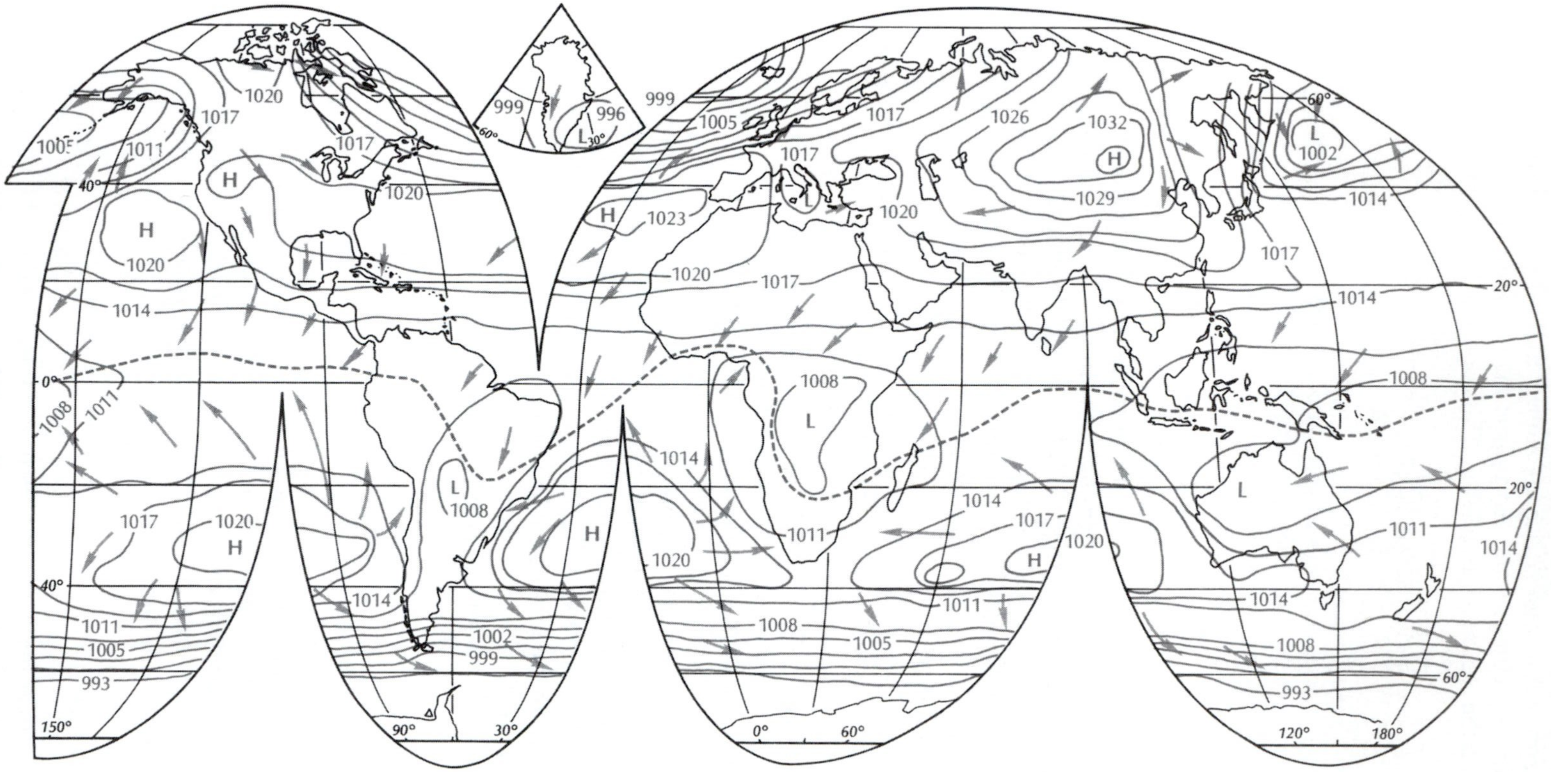

Figure 6.7
Mean sea-level atmospheric pressure in January (a) and in July (b). Pressure is in millibars, and arrows indicate prevailing air flow.

(continues on next page)

prepared by Cartographics, Texas A&M University

(a)

prepared by Cartographics, Texas A&M University

(b)

Figure 6.7
(continued)

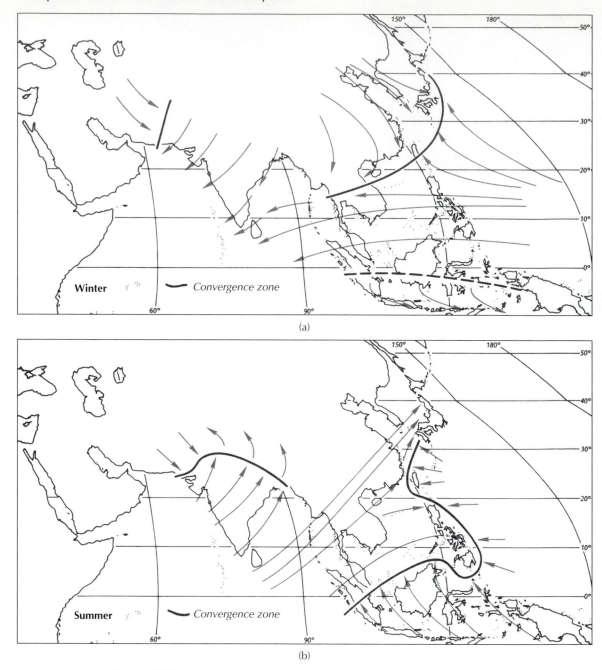

(a)

(b)

Figure 6.8
Summer and winter monsoons over Asia. The normal location of the convergence zone shifts substantially, so that most of the area has distinct rainy and dry seasons.

this part of the continent. Onshore winds form the Atlantic high-pressure system centered over the Azores Islands carry considerable moisture. The convergence and lifting over the land areas produce quite a bit of rainfall, and there is a summer maximum of precipitation over much of the eastern part of the country. In the winter, the subtropical high over the Pacific and the subpolar low move toward the equator as the solar equator moves southward. As the subpolar zone of convergence moves southward, precipitation increases along the West Coast. To the east of the mountains, winter high-pressure systems are more frequent, and the Great Plains experiences a season of low precipitation, although not absolute drought.

SUMMARY

The general circulation is important in transporting heat energy from the equator toward the poles. This heat transfer reduces the imbalance in energy between the Tropics and the polar regions. The general circulation also transports water from ocean to land. The winds follow a pattern resulting from the temperature difference over the earth, the Coriolis force, and the differential heating of land and water.

In equatorial regions, the winds are controlled by the subtropical highs and the intertropical convergence zone. In mid-latitudes the airflow is between the subtropical high and subpolar low. The convergence of air into the subpolar low is of sharply contrasting temperature and moisture. This results in the development of the mid-latitudes lows, which produce so much of the variable weather in mid-latitudes. Over the poles subsidence predominates producing a flow of cold dry air toward the equator.

The seasonal imbalance in energy between the northern and southern hemispheres causes the planetary circulation system to shift north and south. As the solar equator moves north toward the Tropic of Cancer, the general circulation moves northward. Because of the extensive amount of land in the northern hemisphere, there is a land-to-sea shift in relative position of the subtropical high and subpolar low. This produces a seasonal reversal of pressure and wind direction known as the monsoon.

KEY TERMS

Blocking	Jet streams	Semipermanent high pressure cells
Cold pole	Meridional circulation	Trade winds
Doldrums	Monsoons	Upper air westerlies
GCMs	Persistence	Zonal circulation
Hadley cells	Planetary frontal zone	
Horse latitudes	Rossby waves	

REVIEW QUESTIONS

1. What are Hadley cells?
2. What is a GCM? Give examples.
3. Why are the initial conditions used in a model so important?
4. Differentiate between zonal and meridional circulation.
5. Explain the location of the trade winds and why they were so named.
6. What are the major characteristics of the ITC?
7. What is the planetary frontal zone and why is it identified?
8. Explain the importance of shifts in the thermal equator.
9. Why is the tundra not a stormy area?
10. Explain the North American monsoon.

CHAPTER
7

Oceans and Interannual Variation in Climate

CHAPTER OUTLINE

*I*n the previous chapter, we discussed the general circulation of the atmosphere. The general circulation is not static, but changes over time. There are the regular changes that take place with the seasons. In some areas, the seasons are defined in terms of changing wind directions and seasonal precipitation, rather than by temperature as those of us living in mid-latitudes do. The general circulation changes from day to day, week to week, and month to month. No 2 years of weather are ever alike because of these changes. It is only necessary to examine the weather record for the past 50 years or so to find extreme short-term shifts in the weather. There are exceptionally cold or warm years or wet and dry years. There are clusters of cold or warm years or wet or dry years. The problem with these changes is that much of human activity is adjusted to the usual pattern, including the seasonal changes. Whenever the circulation changes very far from the normal, it brings about all kinds of problems for living organisms. It results in benefits for some areas, but widespread costs for others. Hazards such as floods, droughts, cold waves, and heat waves are examples of such deviations in the general circulation.

Many processes and events can cause the planetary circulation to change. Some of these events and processes are external to the planet, such as changing solar activity. Other processes that affect weather result from changes on the planet, such as changes in the dust content of the atmosphere or changes in ocean currents. Some events cause changes in the weather to occur rather quickly. The resulting weather changes may last for months or in some cases for several years. These changes in the weather between 1 year and the next do not herald a change in climate; they are part of the global climate.

OCEAN CURRENTS

The major winds systems that blow over Earth's surface exert friction on the surface of the ocean. This friction results in the movement of water in the general direction of the wind.

The result of the prevailing winds of the general circulation is a series of ocean currents that follows the same pattern as the general circulation. These currents have a great influence on climate. On a global scale, there are semipermanent high-pressure systems over each of the major ocean basins. As a result of the wind systems around these high-pressure systems, there are circular patterns of ocean currents. These currents move much slower than the wind systems that produce them, although they can move fast enough to be a hazard to small craft. The faster currents may move several kilometers per hour, whereas the average is more likely to be several kilometers per day. These are surface currents and do not go very deep into the oceans—typically no more than about 300 meters (Fig. 7.1).

The surface currents that most inhabitants of North America are familiar with are those off our east and west coasts. These currents are part of the larger circulation of the North Atlantic and North Pacific. A detailed examination of the currents in the North Atlantic Ocean serve as an illustration of where these currents flow and their effect on weather and climate. North of the equator in the Atlantic Ocean is a current that moves from east to west. Here the trade winds drag the surface water westward in a slow moving stream called the North Equatorial Current. The current splits on reaching South America, with most of the westward moving water flowing northwest along the coast into the Caribbean Sea and northward into the Atlantic off the North American coast (Fig. 7.2).

Tropical regions of Earth receive more than the average amount of solar radiation as learned in previous chapters. Polar regions are the major areas of Earth radiation to space. There is a continuous flow of heat from tropic to polar regions. This process cools equatorial regions and warms polar regions. This greatly reduces the imbalance in radiation between the two areas. Between the island of Cuba and

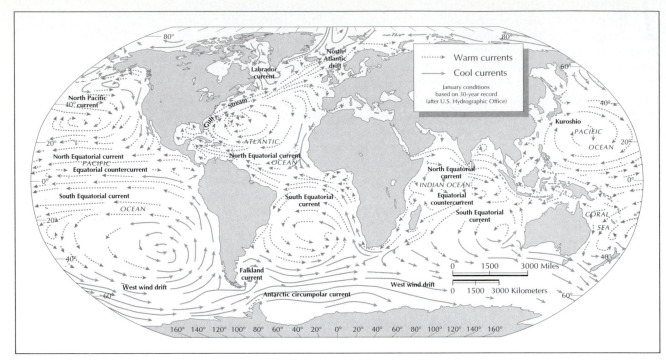

Figure 7.1

Primary circulation of ocean currents. Currents flowing toward the poles are relatively warm and carry large amounts of energy into mid-latitude and polar regions. Currents flowing toward the equator are relatively cool. *Source:* Christopherson R. W., *Introduction to Physical Geography*, 3e., Prentice Hall, Upper Saddle River, N.J., (from Kump L. R., Kasting J. F. and Crane R. G., *The Earth System*, © 1999, Prentice Hall, Upper Saddle River, N.J.).

Figure 7.2

Model atmospheric and oceanic circulation in the Atlantic Ocean. The surface ocean currents follow the same pattern as the atmospheric circulation. Note the desert conditions along the west coast of Africa due to the stabilizing influence of cool water offshore. *Source:* Christopherson R. W., *Introduction to Physical Geography*, 3e., Prentice Hall, Upper Saddle River, N.J., (from Christopherson R. W., *Geosystems*, 4e., © 1999, Prentice Hall, Upper Saddle River, N.J.).

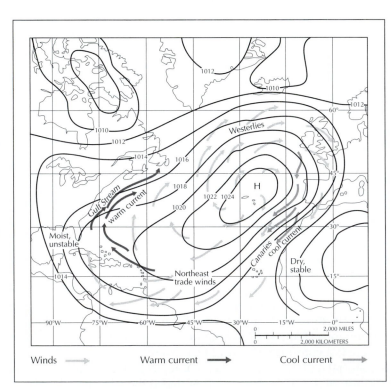

Florida, there is a strong current of warm water flowing out of the Gulf of Mexico and into the Atlantic Ocean. This warm water from the Gulf of Mexico merges with the northward flowing water from the Caribbean Sea near the West Indies and forms the Gulf Stream. The Gulf Stream, along with the other poleward flowing currents, transports a tremendous amount of heat from tropical regions toward the poles. Although some 60 percent of the poleward movement of heat is accomplished by the winds of the general circulation, about 40 percent of the heat is moved by ocean currents. The Gulf Stream is an extremely large current transporting warm water from the tropics, and hence heat, northward off the east coast of the United States. This current is a major factor in weather along the east coast. The current provides a lot of latent heat and water to the atmosphere through evaporation that plays a part in storm activity. Hurricanes often track northward along the Gulf Stream due to the heat available to maintain the storms. Because of the size of these storms, even if they remain offshore over the current, they can have major impact on coastal areas of the Carolinas. In the winter, the warm water often provides enough heat for mid-latitude lows to form, which then move north along the coast. Some of these lows become well-developed Nor'easters, which inflict major damage to beaches and coastal property. How far northward the warm water flows and how close to shore the current flows varies with the season and from year to year.

The Gulf Stream is one of a number of currents that flow along the western side of ocean basins. In the North Pacific Ocean is the Kuroshio Current. Its origin is near the equator and flows northward toward the islands that make up Japan; it then turns eastward into the Pacific Ocean. These warm currents are not limited to the northern hemisphere. They are found in the Indian and South Atlantic Oceans. These currents flow away from the equator as do the Gulf Stream and Kuroshio Currents of the northern hemisphere. The Agulhas Current flows southward along the east coast of Africa. All of these currents flow fairly fast and move a very large volume of water. They all affect the regional climate.

In the general vicinity of New England, the westerly winds direct the Gulf Stream away from North America. It spreads out as it goes eastward and is known as the North Atlantic Drift. The current remains warm as it flows toward northern Europe, which makes winter conditions there unusually mild for those latitudes. Some of the warm water flows north between Iceland and the British Isles and along the Scandinavian Peninsula (Fig. 7.3).

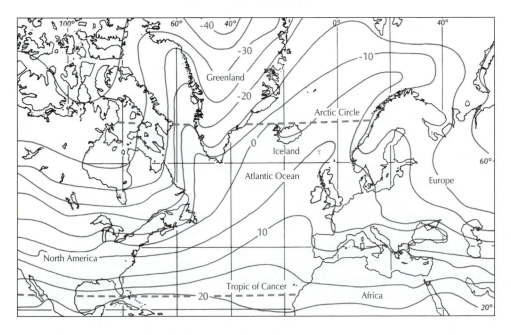

Figure 7.3
The Gulf Stream produces much warmer temperatures around the North Atlantic Ocean in winter than would be expected for the latitude. Note that the heat carried by the current produces warming beyond the Arctic Circle. Also note that the cool current off western Europe does not appreciably affect temperatures over the region. The isotherms are in degrees Centigrade for the month of January.

Part of the North Atlantic Drift splits and flows southward along the west coast of Spain and Portugal as the Canary Current, named for the Canary Islands. The water in this current is relatively cool as it flows south after having crossed the North Atlantic. It is cool only relative to the water temperature in mid-ocean. The equivalent current off North America is the California Current, which flows southward along the west coast.

The California Current flows southward roughly parallel to the west coast of North America. Since the water making up this current comes from the North Pacific, it is quite cool. All along the west coast of the United States, the near shore water is quite cool As the current flows south from Washington to California, it is to be expected the water would warm as it receives more solar radiation. However, upwelling cold water keeps the surface water rather cold all the way south past the Mexican border. This upwelling provides a cold surface over which the onshore winds cross before reaching land. The cold surface tends to stabilize the onshore moving air increasing the aridity along some coasts. This is particularly the case along southern California and Baja California. This cool current and upwelling keeps summer water temperatures off the California beaches fairly cool and much cooler than the water at the same latitudes off the east coast. Temperatures in the California Current are typically below 21°C (70°F), as surfers and bathers are only too aware. Another region that is exceptionally dry as a result of cool stabilizing water offshore is along the coast of Chile. Here precipitation seldom falls except during times of El Niño.

Changes within the planetary circulation systems of the atmosphere and ocean cause some changes in weather patterns. Both the atmosphere and ocean are fluids, moving freely in response to internal and external influences. Neither stays in the same state for very long. They are each a dynamic system and together form a very complex, ever-shifting system. There is no normal condition. By far the most important process at work in the combined system is the heat exchange between the ocean and atmosphere. Seventy-one percent of the earth's surface is water. Hence, it is the ocean surface that controls atmospheric circulation. The ocean absorbs far more solar radiation than does the atmosphere. The ocean has a much higher capacity to store heat than does the atmosphere. The ocean is the major source for heat in the atmosphere, and so small changes in the temperature of the ocean can cause major changes in atmospheric heating. This in turn alters the general circulation of the atmosphere.

THE WALKER CIRCULATION OF THE EQUATORIAL PACIFIC OCEAN

In the Pacific Ocean, the Hadley cells operate similarly to the general circulation model. The trade winds blow from the subtropical highs toward the equator. They blow from the northeast in the northern hemisphere and from the southeast in the southern hemisphere. This results in a westward drift along the equator. In the upper troposphere, there is a counterflow of air from west to east.

Over the Pacific Ocean, there is also a zonal or east–west shift in the heart of the ITCZ. This seasonal shift is due to a shift in the area of ocean surface, which has a temperature of more than 27.5°C (81°F). There is also an upper limit to the temperature of the sea, which is a little above 30° (86°F). It does not get much above this because evaporation cools the surface; the warmer the water gets, the more evaporation there is. The southeast trade winds blowing westward across the Pacific Ocean move the surface water westward. This westward movement of surface water results in sea level being a meter or more higher on the western side of the Pacific Ocean than on the eastern side. The water also warms as it moves westward.

Along the coast of South America, cold nutrient-rich water from below the surface replaces the westward drift of warm water at the surface. The temperature of the cooler water from below may be as low as 20°C (68°F). As this cooler water moves westward, it warms from solar radiation; by the time it reaches Micronesia, it has warmed to 24°C (75°F) or more.

The layer of warm water at the top of the Pacific Ocean is fairly thin, usually less than 100 meters deep. At the bottom of this layer of warm water is a fairly sharp boundary called the *thermocline*. Below the thermocline the water is a lot colder. The westward drift of warm water alters the depth of the thermocline south of the equator. In the western Pacific, the thermocline drops to a depth of 200 meters. At the eastern edge of the Pacific Ocean, the thermocline rises almost to the surface. To counterbalance the westward drift of water at the surface, there is an eastward flow of water along the thermocline. This is the Cromwell Current or the Equatorial Undercurrent. This is a strong current that in mid-ocean reaches a velocity of 1.1 m/s. Since the winds along the west coast of South America are weak and the cold water offshore stabilizes the lower atmosphere, there is fairly low precipitation along the coast. On the western side of the Pacific Ocean, there is much more precipitation since there is onshore flow of air and the air is more moist having traveled over the ocean.

In 1904, Sir Gilbert Walker became director general of observatories in India. He studied the monsoon system of Asia partly as a result of the severe famines of 1877 and 1899. By the 1930s, he showed there was a cyclical, interannual variation in the atmosphere over the southwest Pacific Ocean, which he called the Southern Oscillation. This oscillation brings major changes in pressure, winds, and precipitation over the southwest Pacific Ocean and the Indian Ocean. By the 1960s, enough data and information were available for scientists to conclude that the Southern Oscillation extends across the Pacific Ocean. This circulation system is the Walker Circulation. It includes the event called El Niño and the *anos de abundancia*.

El Niño

Off the west coast of South America, an event takes place that the local people call El Niño. It appears as a warm current of water that replaces the normally cold rising water off the coast of South America. Fishermen of Spanish descendents named this event. It implies "the Little One" after the Christ Child, as the event appears most often in late December. El Niño has occurred frequently since first reported in 1541 (Table 7.1).

Every few times, it is stronger than usual. It goes farther south and is exceptionally warm. It produces rain over the coastal desert. The rains bring a period of profound growth. These years are *anos de abundancia* (years of abundance). The desert provides abundant grazing for herds of sheep and goats.

In its greater context, El Niño is the collapse of the Walker Circulation. A temporary circulation known as El Niño-Southern Oscillation (ENSO) is replacing it. Interactions between the ocean surface and the atmosphere are responsible for the shifts in the Walker Circulation. Changes in sea-surface temperatures produce changes in atmospheric pressure and winds. Changes in atmospheric circulation change sea-surface temperatures. Once the process starts, the ENSO is self-perpetuating.

In low latitudes, surface temperatures of the ocean average around 27°C (80°F). Marked changes in sea surface temperatures in the Pacific Ocean accompany the Southern Oscillation. These surface temperature changes take place mostly within 7.5° of the equator. Temperature increases of more than +3°C (+5.4°F) take place in the warm phase or ENSO. There are large shifts in the location of the warm pool. During El Niño, the area of warm water expands eastward and the area of convergence and precipitation drifts eastward (Fig. 7.4). Rainfall increases over the eastern Pacific Ocean and along the west coast of South America. Pressure differences decrease and the trade winds die down. Pressure differences between Darwin, Australia, and the island of Tahiti show the Southern Oscillation. During El Niño, pressure increases at Darwin and rainfall decreases. Pressure decreases at Tahiti and rainfall increases eastward toward the international dateline. Sea-surface temperatures also increase toward the dateline. In the western Pacific, rainfall decreases and pressure increases (Fig. 7.4). The ENSO weakens when the supply of warm water moves away from the equator, and so the oscillation lasts anywhere from 14 to 22 months.

Table 7.1 El Niño and La Niña Events from 1950 to 2000. The Events Are Defined as Those Where Temperatures in the Eastern Pacific Ocean Were Warmer or Cooler by +/− 0.4°C.

El Niño Events	La Niña Events
1951–1952	1950–1951
1953	1954–1956
1957–1958	1956
1963–1964	1964–1965
1965–1966	1970–1972
1969–1970	1973–1974
1972–1973	1974–1976
1976–1977	1984–1985
1977–1978	1988–1989
1979–1980	1995–1996
1982–1983	
1986–1988	
1991–1992	
1993	
1994–1995	
1997–1998	

El Niño Episode of 1982–1983

Atmospheric pressures began to rise over the Indian Ocean in the summer of 1981. Sometimes the first sign of El Niño is in the early spring when there is a slight warming of the sea surface along the northwest coast of South America. In the spring and early summer of 1982, water temperatures stayed near normal. Across the ocean, the usually heavy rains of summer did not develop in Australia and Micronesia. Shortly after the summer solstice, the wind system reversed itself over the western Pacific. Still water temperatures off South America remained normal. In early September, the United Nations sponsored a conference on forecasting El Niño. The conference did not forecast the coming El Niño. With more uniform sea-surface temperatures across the Pacific, the trade winds weaken or even reverse. When the trade winds reverse and blow toward the east, it produces a surge of warm water known as the Kelvin wave. There is normally a warm current moving southward along the northern Pacific Coast. El Niño develops when this current goes much farther south than usual reaching Ecuador and Peru. On September 25, 1982, sea-surface temperatures near the village of Paita, Peru, rose 4°C (7.2°F) in one day. By the first of December, the rise had reached 6.5°C (11.7°F). In some parts of the Pacific Ocean, surface temperatures rose as high as 31°C (88°F) and were as much as 8°C above normal. Eventually, there was a swath of unusually warm water 13,000 kilometers long straddling the equator. By the end of December 1982, the thermocline off the coast of South America pushed downward 150 meters. One result of this was that the eastward flowing Cromwell Current almost came to a halt.

The El Niño Event of 1997–1998

The southern Pacific circulation shifted again in the fall of 1997 and heralded the arrival of another event. By October 1997, sea-surface temperatures had risen to some

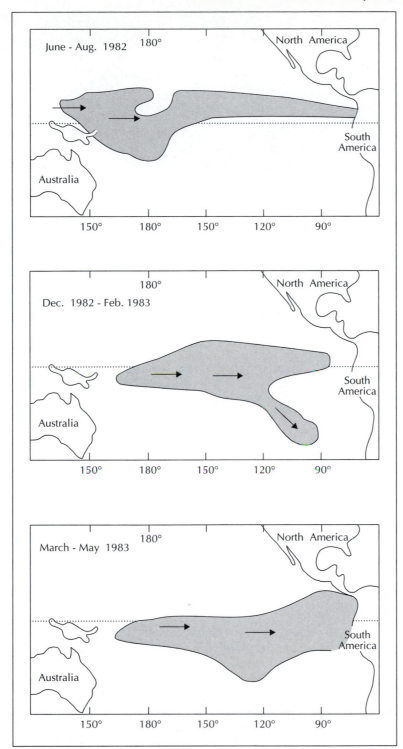

Figure 7.4
Shift in the location of the surface current and warm water pool during the El Niño of 1982–1983.

5°C (9°F) warmer than average in some areas of the eastern Pacific Ocean. This was as warm as the water temperature during the strong El Niño event of 1982–1983. Weather in the United States was far from normal during the winter. A large high-pressure system developed over the midwest of the United States. The high-pressure system blocked out the usual cold streams of air from the Arctic. Bismark, North Dakota, experienced a record warm December with temperatures averaging −2°C (28.5°F), well above the normal of −9°C (15.3°F). (Fig. 7.5).

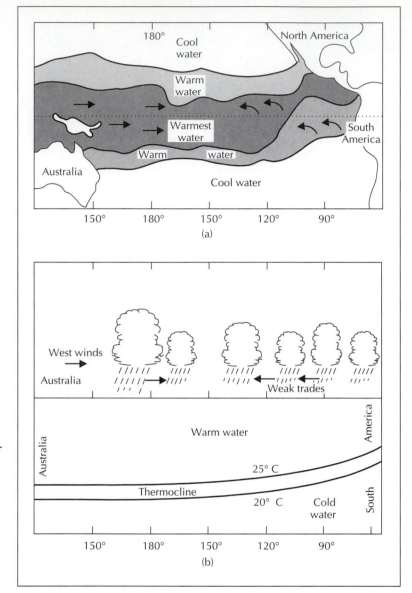

Figure 7.5
(a) Circulation during an El Niño. (b) The thermocline moves downward in the eastern Pacific Ocean. Cloud cover and convectional rainfall spread east across the Pacific. The warm pool of water shifts location. The surface winds slow or reverse direction, changing the surface temperature structure of the equatorial Pacific.

Winter storms were severe. In February 1998, the winter jet stream was crossing the United States from southern California to Florida. The result was storm after storm crossing the southern states beneath the jet. In the winter rainy season, California averaged double the normal amount of precipitation. Los Angeles set an all-time record for total rainfall in February. It received 34.8 cm (13.7 inches) of rainfall. In California, floods and mudslides caused a number of fatalities and did extensive property damage. Cities all across the south experienced above normal rainfall. Tampa, Florida, received 12.9 inches more than normal. Baton Rouge, Louisiana, doubled its long-term average. Florida also experienced a rash of tornadoes.

El Niño and Northern Hemisphere Weather

The Southern Oscillation directly affects atmospheric temperatures. In the immediate area of the Southern Oscillation, annual temperatures change by as much as 0.5° C. The ENSO affects atmospheric temperatures over the land masses within 20° of the equator. Temperature anomalies reach a maximum late in the calendar year and extend into the spring. For this reason, calendar year data tend to obliterate the effects of El Niño. The effect of the Southern Oscillation shows up much better in winter (October–March) data. Using winter data, when the Southern Oscillation is strongest, interannual variations of 0.3°C (0.5°F) appear.

Some studies show that the Southern Oscillation is connected to temperatures in the northern hemisphere. The Southern Oscillation affects temperatures in the northern hemisphere during the winter half year by as much as 0.2°C (0.36°F). The affected land areas are northwestern North America and southeastern United States. Large areas of the North Pacific Ocean developed a one to two degree increase. There was a large area of warmer weather over northwestern North America and a small area of cooler than normal temperature around the Gulf of Mexico.

Counterpoint—La Niña

In the summer of 1987, the temperature in the Pacific Ocean along the equator dropped 4°C (7.2°F). This brought unusually cool weather to the eastern Pacific. They call this sudden cooling of the equatorial water La Niña, Spanish for "the girl." The name was applied to the phenomenon for the first time in 1986. La Niña exaggerates the normal Southern Oscillation. During La Niña, the trade winds are stronger, the water off western South America is colder, and water in the western Pacific near the equator is warmer than normal. In the western Pacific, surface pressures are lower and heavy rainfall occurs (Fig. 7.6).

The strengthening of the Walker Circulation causes the coastal deserts of Peru and Chile to become even drier if that is possible. On the western edge of the circulation, southeast Asia gets even more summer precipitation than usual. In Bangladesh, more than 1000 persons perished in floods.

THE IMPACTS OF EL NIÑO

The event of 1982–1983 was of unusual severity. It was the most destructive event in the last 100 years and may well have been the most extreme case yet documented. Ecuador and Peru sustained the highest loss of life and economic damage.

The El Niño phase of the Southern Oscillation does bring *anos de abundancia* to the Peruvian portion of the Atacama Desert. In Peru, the benefits to some areas are more than offset by high costs in other parts of the nation. The costs of El Niño far exceed the benefits. The changes in the general circulation cause widespread social, economic, and environmental disruption. The event of 1982–1983 claimed some 2000 lives and incurred damage of over $13 billion. The same heavy rains that cause the coastal desert to bloom cause massive landslides and floods in the Andes Mountains back from the desert. In May 1983, Guayaquil, Peru, received almost 20 times the average rainfall for that month. The flow in some rivers increased to 1000 times the mean flow. The winds caused $400 million of damage in Ecuador alone.

In the western Pacific Ocean, the effects of the shift in the ocean–atmosphere system was equally destructive. Six typhoons hit the island of Tahiti during the season. This was after decades without a single occurrence. Eastern Australia suffered severe drought. The southern hemisphere subtropical high-pressure system dominates the weather in Australia. In the summer, this high usually weakens to permit low-pressure systems to move onshore, bringing rain. In the 1982–1983 season, few low-pressure systems reached the continent.

There were severe weather events in many parts of the world, and many are attributed to El Niño. Table 7.2 lists some of the events. Because they occurred at the same time as El Niño is not enough reason to attribute them all to the phenomenon. For example, the unusually warm weather of the eastern United States during the

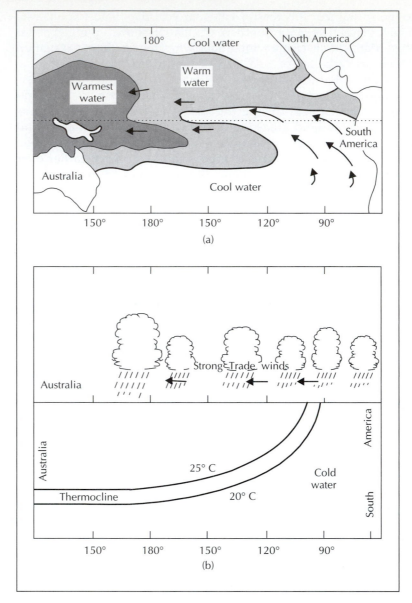

Figure 7.6
(a) Ocean temperatures and wind directions during La Niña. (b) The thermocline reaches the surface far west of South America. The warm pool of water is limited to the western Pacific Ocean. This restricts cloud cover and precipitation to the far western Pacific Ocean.

winter of 1983–1984 is attributed to El Niño. However, the last time El Niño occurred prior to 1983–1984 was in 1976–1977. The winter of 1976–1977 was one of extreme cold in the eastern United States.

Agricultural Losses

The greatest costs associated with El Niño are those to agriculture. El Niño has frequently produced drought over widespread areas. In 1982–1983 and again in 1997–1998, drought occurred in Australia, Indonesia, Mexico, the Phillippines, and South Africa. Damage estimates were $1 billion in South Africa, $600 million in Mexico, $500 million in Indonesia, and $450 million in the Phillippines.

Disease Outbreaks

Both floods and drought that occur tend to result in the proliferation and spread of disease organisms. Stagnant pools of water form under both sets of conditions. These stagnant pools of water are an excellent habitat for mosquitos that carry malaria,

Table 7.2 Worldwide Events Attributed to El Niño from 1982 to 1983

- Above-normal temperatures in Alaska and northwestern Canada.
- The warmest winter in 25 years in eastern United States.
- Drought in S.E. Africa.
- Drought in Australia and New Zealand.
- Drought in Southeast Asia, particularly Sri Lanka, India and Indonesia, Phillipines.
- Drought in Mexico.
- Floods in Louisiana and Cuba.
- Excessive beach erosion in California.
- A 200-mm (8-in) rise in sea level off California.
- Death of coral reefs in the Pacific Ocean.
- Slowing of the rate of Earth's rotation in late January.
- Reduction in the salmon harvest off western North America.
- Encephalitis outbreaks in eastern United States.
- Increased incidence of bubonic plague in southwestern United States.
- High mortality of seabirds on the Farallon Islands.

dengue fever, yellow fever, and encephalitis. Malaria outbreaks occurred in Columbia, Peru, India, and Sri Lanka during the 1982–1983 event. In the 1992–1993 event, El Niño may have produced a local outbreak of Hantavirus in the American Southwest. The heavy rainfall led to a burst of plant growth in the arid region. This in turn led to a rapid increase in deer mice that transmit the disease to humans. In areas of excessive rain, water-borne diseases such as dysentery and cholera erupt.

Wildlife Losses

El Niño brings abundance to the desert, whereas it creates havoc in the marine industries. Off the coast of Peru and Ecuador, there is a major anchovy fishing ground. The anchovy flourish in the cold and upwelling water, which is rich in nutrients. The nutrients provide food for the plankton, which in turn provide food for the anchovies. During El Niño, the upwelling continues but is restricted to the warm upper layer some 125 to 150 meters deep. There are fewer nutrients in the warm water, the supply of plankton drops, and the anchovies either leave the area or die in the alien environment. There is then no longer a steady supply of guano to support the fertilizer industry. Thus, both the fishing and fertilizer industries collapse.

Warm water and storms are extremely hard on marine life including coral reefs. The 1982–1983 event did extremely severe damage to reefs off Columbia, Costa Rica, Panama, and the Galapagos Islands. Reefs off the Galapagos Islands suffered losses from 50 percent to 97 percent. In the 1997–1998 event, coral reefs around the Caribbean Sea were badly damaged from Mexico south through Belize, Panama, and Costa Rica.

FORECASTING EL NIÑO

El Niño occurs every few years like many cyclical phenomena. The frequency appears to be on the increase. Since they have been well documented, they have averaged one about every 7 years. However, in the past few decades, they have occurred more frequently. There is one every 4 to 5 years, but they have varied from every other year to 10 years apart. In September 1982, a committee sponsored by the United Nations and meeting in Peru to discuss El Niño could not forecast the event that began later in the month. Since the event of 1982–1983, considerable progress has been made especially in terms of early detection.

There are several problems in forecasting changes in the Southern Oscillation. First, it is not a periodic phenomenon. It does not have a fixed interval or a fixed amplitude. Each event begins and develops differently. Some events first appear off South America, with the appearance of warm water, and then the system develops westward. Other events, including that of 1982–1983, appeared first in the western part of the tropical Pacific Ocean. It began with changes in the winds. This led to the eastward expansion of the pool of warm water and precipitation. This resulted in still further breakdown in the trade winds. Then warm water progressed still farther eastward and the system went on to completion.

Part of the reason that forecasting the event is difficult is because it is not yet known what causes the system to form. There are currently several hypotheses. One suggests that it is the result of huge amounts of heat released on the sea floor as a result of magma pouring out onto the sea floor. Another suggestion is that ENSO is a result of high snowfall over Asia the previous winter. That is, when there is a lot of snow on the Eurasian land mass in a given winter, there will be much more snow melt. A higher volume of meltwater reduces the normal summer heating of the land mass. Unfortunately, it will probably take at least several more ENSO events before any hypothesis can be adequately tested.

INTERANNUAL VARIATION IN THE MONSOONS

Much of Asia, the Middle East, and equatorial Africa are subject to the effects of an atmospheric circulation system known as the monsoon. Figure 7.7 delineates the region of Africa and Asia affected by the monsoon. The monsoon is a very pronounced seasonal shift in atmospheric pressure, which in turn brings about a marked change in wind direction, atmospheric humidity, and precipitation.

During the summer months, convergence predominates over Asia and high humidity and precipitation are the general conditions over the continent. In summer, the low-pressure system and convergence of air are strong. The summer monsoon winds blow across the subcontinent of India from southwest to northeast from April to October. The winds reaching the west coast of India have traveled over the ocean and have a very high moisture content. As the winds move onshore, they produce a lot of precipitation. At Bombay, the rains last from June until the end of September and total over 2 meters. The winds pass over the Western Ghats across the plains of Bengal toward the Himalaya Mountains. As a result of the direction of the winds and the topography of the Indian subcontinent, the annual rainfall varies tremendously. The Western Ghats divide India into two distinct climatic zones as they run nearly perpendicular to the direction of the summer winds. There is a narrow, very wet climatic zone on the western side of the Ghats and a much wider and drier zone on the eastern side of the mountains.

During the winter months when high pressure dominates the Asian land mass, divergence is more frequent than convergence. Divergence from this strong system north of the Himalayas produces extremely strong offshore surface winds. South of the Himalayas and south of the jet stream, the anticyclonic flow is much weaker, although offshore flow dominates. When divergence is present, subsiding dry air covers much of the continent, resulting in lower humidity and less precipitation. The subsidence produces offshore winds along much of coastal Asia. As the air streams out over the ocean, it evaporates large amounts of water from the sea surface and thus becomes a source of moisture for the offshore islands and along some coastal areas. There is a secondary rainy season in parts of India with the offshore flow of air and a second season harvest in March and April.

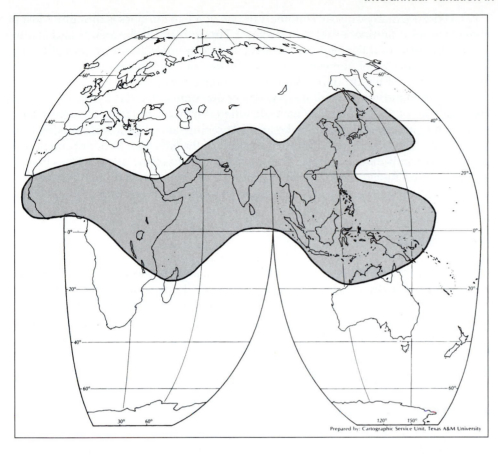

Figure 7.7
The monsoon region of Africa and Asia. Within this extensive area, seasonal reversals of wind direction are common and pronounced.

The largest share of the rainfall in northern India occurs during the summer onshore monsoon. The time of year when the Asian monsoon begins and ends is quite variable from year to year. The duration and intensity of the rainy season are also quite variable from year to year. The total amount of rainfall varies as a consequence. The agricultural economies of northern India and Pakistan are cruelly subject to the whims of the system.

The Asian monsoon is quite variable from year to year in the time of year when the onshore and offshore flow begins and ends and in the duration and intensity of the rainy season. The total amount of rainfall varies as a result. The agricultural economy of northern India and Pakistan are cruelly subject to the whims of this variable system. It is this summer monsoon that is so important for most of India. The primary growing season for rain-fed crops is during the summer rainy season, with the main harvest in December.

The demand and supply of food have been in a delicate balance for the human species throughout history. When the food supply has increased, there has been a gain in population; when food has been scarce, there has been some sort of trauma inflicted on the populace. Starvation results from too little food intake. During the long period of the hunting and gathering societies, starvation was probably near at hand for individuals, family groups, and tribes. The development of agriculture allowed the world population to expand rapidly. The basis for the supply of food—namely, agriculture—became more directly dependent on the weather. Famine is a phenomenon that affects whole populations of people over a broad area. It did not become a part of human experience until after agriculture began. However, as agriculture expanded, so did the frequency of famines. The number of times that famine has spread over the land is large. All histories of peoples and nations record famines. There are repeated references to famine in the *Holy Bible* and other ancient histories.

There are many causes of famine, but one of the major ones is drought. Most of the catastrophic famines result from drought. Drought affects the quality and quantity of crop yields and the food supply for domestic animals. In severe drought, there may be a major loss of domestic animals. The loss of the milk products or meat furthers the effect of the drought. Major famines have occurred throughout the Asian continent from the time agriculture spread over the continent. India, China, Russia, and the countries of the Middle East have all suffered many times from drought-related famine. A famine occurred during the life of Abraham (about 2247 B.C.). Another massive famine occurred in Egypt before the exodus of the Israelites. Drought and famine are common in India and China. The oldest record of famine in India goes back to 400 B.C. and in China to 108 B.C. From the time of the earliest noted droughts, there have been nearly continuous episodes of drought and famine in some part of Asia.

SUMMARY

Weather changes from day to day, week to week, and year to year. There are many factors that cause the weather to change over time. Changes in weather from year to year often bring disasters or economic stress. There is increasing knowledge that atmospheric circulation is very dependent on changes in the ocean surface, including changes in ocean currents. There is a global pattern of surface ocean currents that is similar to that of the general circulation of the atmosphere. In each of the major ocean basins, there is a circular series of currents that approximates the circulation around the semipermanent high-pressure systems. Such a system of currents exists in the North Atlantic Ocean. Part of that circulation along the east coast of the United States is a warm current known as the Gulf Stream. This stream of warm water affects the climate of both the eastern United States and western Europe.

Changes in ocean currents may bring substantial changes in weather and climate in large areas of Earth's surface. Perhaps the greatest impact of weather changes from year to year is when precipitation over land areas falls below normal. In subsistence or near-subsistence economies, the result is sometimes famine. Famine has plagued the human species from time immemorial. As the human population grows ever greater, numbers of people are affected. In the 1990s, there was famine almost every year in some part of the world.

KEY TERMS

Drought	Famine	Southern Oscillation
El Niño	La Niña	Walker Circulation
ENSO	Nor'easters	

REVIEW QUESTIONS

1. Where does El Niño traditionally appear first and have the most drastic effect on climate?

2. In what part of the world is the Walker Circulation found and to what does it refer?

3. How does La Niña differ from El Niño?

4. How does El Niño affect North America?

5. How does La Niña affect North America?

6. The monsoons of Asia are a dominant climatic element. How does variation in the monsoons affect the region?

7. What sector of the world economy suffers the greatest amount of dollar damage from El Niño?

8. How does the irregular shift in the Walker Circulation affect the climate of western South America?

9. What is the relationship between the large loss of birds in the eastern Pacific Ocean and El Niño?

10. In what way does El Niño affect North American weather?

C H A P T E R
8

Air Mass and Synoptic Climatology

CHAPTER OUTLINE

*T*he study of air masses is integrally related to synoptic climatology and both are examined in this chapter. *Synoptic climatology* may be defined as the study of climates in relation to atmospheric circulations; it emphasizes the relationships among circulation, weather types, and climatic regional differences. The term originated with military operations in the 1940s, although some of its methods have been used since the beginning of the century. The approaches used in synoptic climatology depend on scale, ranging from the global through continental and regional to local scales.

This chapter begins by looking at one of the basic causes of regional climatic differences—the role played by air masses. Thereafter, the circulation patterns associated with frontal systems are examined, and some special studies of synoptic climatology are then described. Included in this chapter is a brief discussion of satellites. Since the first weather satellite was launched in April 1960, images from space have been of high significance in meteorological studies. For satellite climatology to develop, enough time had to pass to have enough imagery for the longer term study. This has been achieved, and the availability of remotely sensed images has allowed the development of satellite climatology.

AIR MASSES

A significant step in understanding the workings of the atmosphere was the introduction of the idea of **air masses**. An air mass is a large, horizontal, homogeneous body of air that may cover thousands of square kilometers and extend upward for thousands of meters. Its uniformity is principally one of temperature and humidity. Air masses derive their properties from the surface over which they originate and are thus identified and classified by the area of the earth over which they form—the source regions.

Two basic categories of air masses are recognized on the basis of temperature and two on moisture properties. Air masses are classified as tropical (T) if the source is in low latitudes and polar (P) if the source is in high latitudes. Air masses originating over land, and therefore relatively dry, are labeled continental (c); those originating over the oceans, and hence moist, are called maritime (m). This classification identifies four individual kinds of air masses: maritime tropical (mT), maritime polar (mP), continental tropical (cT), and continental polar (cP). Two additional categories are sometimes used in reference to the extremes of continental polar air masses and maritime tropical air masses. Continental Arctic (cA) indicates exceptionally cold and dry air; equatorial (E) indicates very warm moist air. The basic classification of air masses is given in Table 8.1, and a map of source regions shown in Figure 8.1.

The general circulation of the atmosphere requires that energy exchanges occur over the globe. As a result, air mass movement from the source region acts as a mechanism for energy transfer in the atmosphere. The basic properties of the air mass will be modified the farther it moves from the source region. To indicate the manner in which air masses are modified, a third letter is sometimes added to the identification symbols. If the air mass is warmer than the land over which it is moving, then the letter *w* is added. Thus, an air mass designated mTw would indicate air that has its origin over the subtropical ocean and that is warmer than the surface over which it is moving. It may well be a warm, moist air mass moving onto land in the winter. If an air mass is colder than the surface it passes over, the letter *k* is added. Thus, a cPk air mass might be one passing from a Canadian source region and moving over warmer areas in the United States.

One of the best examples of air mass modification occurs when cP air masses pass over the Great Lakes in winter. Although cold, the lake water is warm relative to the air, and evaporation supplies moisture to the air mass. Once the air leaves the

Table 8.1   Classification of Air Masses

Name of Mass	Place of Origin	Properties	Symbol
Polar continental	Subpolar continental areas	Low temperatures (increasing with southward movement), low humidity remaining constant	cP
Polar maritime	Subpolar and arctic oceanic areas	Low temperatures, increasing with movement, higher humidity	mP
Tropical continental	Subtropical high-pressure land areas	High temperatures, low moisture content	cT
Tropical maritime	Southern borders of oceanic sub-tropical, high-pressure areas	Moderately high temperatures, high relative and specific humidity	mT
Equatorial	Equatorial and tropical seas	High temperature and humidity	E

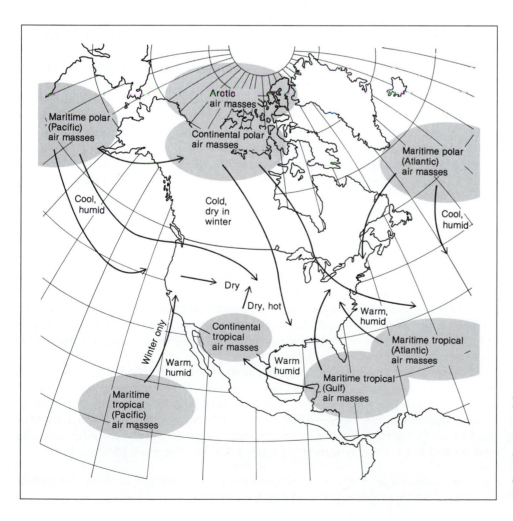

Figure 8.1
Air mass source regions for North America (from Moran, J. M. and Morgan M. D., *Meteorology: The Atmosphere and the Science of Weather*, 5e., © 1997, Prentice Hall, Upper Saddle River, N.J.).

lakes to pass onto the warmer land on the eastern or southern shores, it becomes unstable and produces flurries. As Figure 8.2a illustrates, greater snowfall may occur over the higher ground of the northern Appalachians. This lee-of-the-lake effect causes a remarkable gradient in snowfall amounts. Figure 8.2b provides snowfall distribution for the area.

In moving, the mass of air is not passing into a vacuum, but rather replaces an existing air mass. The leading edge of an air mass will thus be a site of conflict between air of different properties. This is where **fronts** form.

FRONTS AND MIDDLE LATITUDE CYCLONES

During World War I (1914–1918), remarkable advances were made in weather research in Norway. Cut off by the war from weather information of other countries, extensive station networks were established in the Scandinavian countries. This network of data enabled the meteorologists at Bergen in Norway to study the storm systems frequently encountered in northwest Europe. From their observations came models detailing the structure of **middle latitude cyclones**.

The vocabulary selected by the scientists to describe their model reflects the war background of the time. The Norwegians developed the Polar Frontal Model, with the term *front* used to identify a zone of transition between air of different properties; the analogy was to a front dividing the fighting armies. Various types of fronts were identified, as illustrated in Figure 8.3.

A warm front occurs when a warm air mass replaces colder air. Clearly, because warm air is less dense than cold, the warm air rises above the cold along a front that, in cross-section, has a slope of about 1:100 (1 unit of length vertically for every 100 of horizontal distance). This means that high clouds forming along the front may occur some 500 to 700 miles (800 km–1300 km) ahead of the location of the warm front at the surface. Generally, stable air ascends along the front, resulting in the formation of stratiform clouds. As the cross-section shows, clouds may range from cirrus (many miles ahead of the front) to nimbostratus (where the front intersects the surface). Widespread rainfall often occurs during the passage of warm fronts. The majority of warm fronts contain stable air. However, if the air in the warm air mass is highly unstable, then cumuliform clouds will occur along the front. These may be embedded in stratiform clouds, creating problems for small aircraft pilots.

Cold fronts occur when a cold air mass overtakes and replaces warmer air. Cold fronts are much better defined and generally move twice as fast as warm fronts. The front is much steeper, sometimes having a slope of 1:50. The rapid ascent of warm air along the front results in cumuliform clouds with the possibility of severe weather. In the northern hemisphere, strong cold fronts are usually oriented in a northeast to southwest direction and move east or southeast.

At times, the forces are such that the frontal boundary neither advances nor retreats. This stationary front creates conditions similar to those along a warm front and may persist for a number of days, causing low-ceiling weather over wide areas.

The analysis of these fronts, together with an understanding of traveling high- and low-pressure systems, led to the identification of the middle latitude cyclone. A cyclone is a low-pressure area with an organized circulation pattern of winds. In middle latitudes, cyclones develop at air mass boundaries and are characterized at the surface by the formation of fronts. The surface history of the development of a middle latitude cyclone is shown in Figure 8.4. Four stages are identified:

a. The initial stage in which a v-shaped pattern occurs and initiates the formation of the central low pressure and the two fronts.

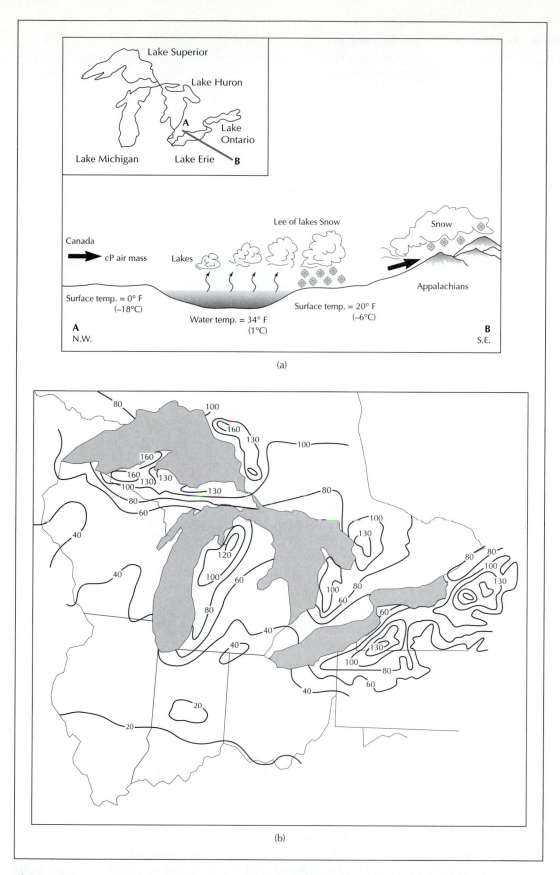

Figure 8.2
Snowfall in the lee of the Great Lakes. (a) Sample readings when winter cP air masses pass over Lake Erie, (b) Mean seasonal snowfall (inches) in vicinity of Lakes Superior and Michigan (from Morgan M. D. and Moran J. M., *Weather and People*, © 1997, Prentice Hall, Upper Saddle River, N.J.).

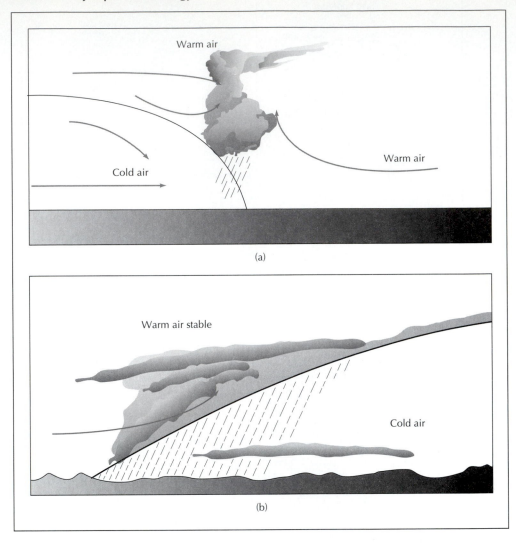

Figure 8.3
Schematic diagrams of (a) a fast-moving cold front with unstable warm air ahead of the front, and (b) a warm front of warm stable air. After *Aviation Weather*, Department of Commerce, Federal Aviation Agency (Washington, DC: Government Printing Office, 1965), 78 and 82.

b. The open stage in which the cold and warm fronts have formed a distinctive warm sector.

c. The occluded stage where one front overrides another.

d. The dissolving stage at the end of the life cycle.

Note that this classical representation has been modified over the years particularly in relation to the flow of air in the system. A recent view considers the conveyer belt model in which warm, cold, and dry air streams are conveyed into identified portions of the cyclone. Similarly, the idea of an occlusion resulting from one front overtaking the other is still accepted, but it has been found that there are alternatives to this. In some cases, elongation of the low-pressure area at the junction of the fronts can cause the occlusion. In other cases, the cold front may actually slide along the warm to cause a similar effect.

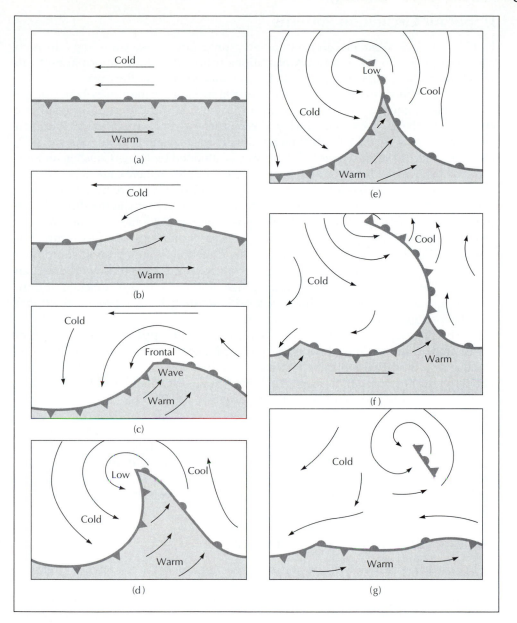

Figure 8.4
Schematic depiction of the life cycle of a frontal cyclone. See text for discussion. After *Aviation Weather*, Department of Commerce, Federal Aviation Agency (Washington, DC: Government Printing Office, 1965), 84.

CYCLOGENESIS

The identification of fronts and the concept of the structure of the middle latitude cyclone were of major significance in the development of the atmospheric sciences. However, the explanation of why and where, (e.g., a v-shape interrupted the isobaric pattern along a front) simply could not be explained with only surface knowledge. It remained for the acquisition and interpretation of upper air data to provide the explanation for the formation of cyclones—or **cyclogenesis**.

Recall that aspects of the upper air circulation were dealt with in the discussion of the general circulation of the atmosphere. At this point, emphasis is on the relationship between surface and upper air circulation patterns.

Upper Air Circulation Patterns

In 1941, the U.S. Weather Service initiated a program to take soundings through the atmosphere. Today, a network of stations routinely obtains information from which upper air charts are constructed. Such was not always the case.

The conditions existing through the vertical structure of the atmosphere is largely derived by measurements and sampling. The first indication of atmospheric changes with height were gained when thermometers were carried first by kites and later by balloons. In 1804, the first manned balloons measured conditions; by 1862, an altitude record of 9000 meters (29,000 ft) was attained by James Glaisher, an English scientist. A major breakthrough in monitoring the atmosphere came when recording instruments were developed in the early 1900s. Carried upward by a balloon that eventually burst, the instruments were parachuted back to Earth. In the 1920s, the radiosonde was introduced. This instrument package carried a radio transmitter that sent back to Earth continuous measurements, called *soundings*, of temperature, humidity, and pressure. A radiosonde tracked by direction finding ground equipment was introduced in World War II; named the *rawinsonde*, it measures wind direction and speed at various levels of the atmosphere. Radiosondes are launched at hundreds of ground stations throughout the world on a daily basis to provide the upper air information required for synoptic analysis.

The development of radar and, more recently, **Doppler radar** has been of major importance to monitoring the atmosphere. Of particular importance is the use of a modified Doppler system to produce a wind profiler. By simultaneously emitting three radar beams at different angles, changes in the backscatter beams can identify wind speed and direction at 72 levels up to a height of 10 kilometers (6 mi).

As previously described, upper air conditions are shown on constant pressure charts that indicate the height contour at which a given pressure is found. Three standard pressure charts commonly in use are the 850 millibars, 700 millibars, and 500 millibars. The average heights at which these levels occur are 5000 feet, 10,000 feet, and 18,000 feet (1525 m, 3050 m, and 5490 m).

Cyclones and Upper Air Circulations

Using the knowledge of winds presented earlier, the balance between pressure gradient force and Coriolis means that much upper air flow is geostrophic, moving approximately parallel to the contours. Observations of winds aloft show that the circulation is wavelike and that within the waves, troughs and ridges can be identified. These waves, with their ridges and troughs, migrate eastward across the United States in a complex of many interacting waves of variable lengths and amplitude.

Figure 8.5 illustrates how a wave may develop over the United States in the period of perhaps 4 days. The pattern passes from an west–east (**zonal**) flow to deep amplitude (**meridional**) waves and the eventual formation of a deepening trough, which may become cut off from the main west–east flow. The cutoff eventually disappears and a low-amplitude wave pattern is again established.

Winds in the upper air are strongest where the contours are closest together. The strongest bands within the westerly flow are jet streams, the cores of which are above regions of strong horizontal temperature gradients. The relationship between temperature and pressure aloft is shown in a cross-section through the atmosphere from warm to cooler latitudes (Fig. 8.6a). Initially, an equal surface pressure exists. Yet, because of the density differences between warm and cold air, constant pressure surfaces are higher above warmer latitudes than cold. Using the geostrophic relationship, winds aloft will move from warmer to cooler areas and be deflected by Coriolis. Westerly winds will occur.

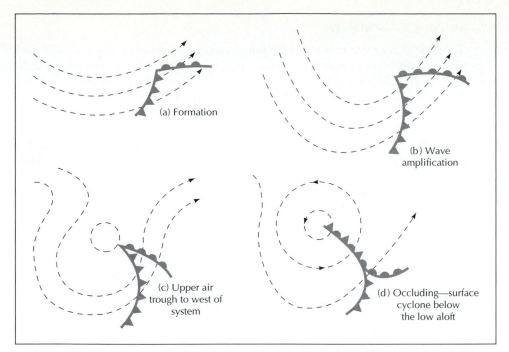

Figure 8.5
Development of a mid-latitude cyclone in relation to an upper air trough. System develops below a zone of divergence in the flow aloft (a). The cyclone deepens [(b) and (c)]. As a low forms in the upper air flow, the system occludes (d) and dissolves.

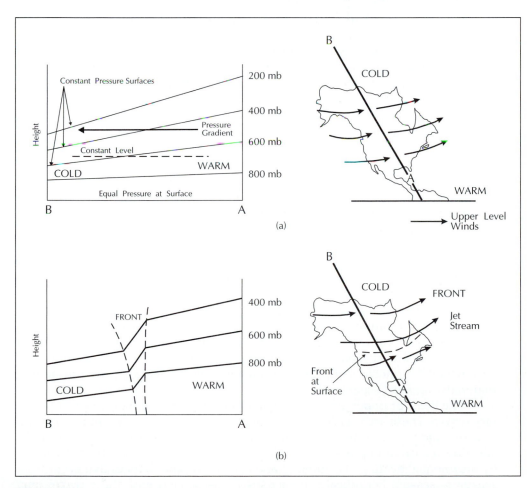

Figure 8.6
(a) A cross-section through the atmosphere from warm to cold latitudes shows that constant pressure surfaces are higher in warmer than cooler areas. This creates a pressure gradient toward the cooler air and, after Coriolis deflection, westerly winds. (b) If there is a strong temperature contrast (a front), then winds will blow faster to form a jet stream.

145

Suppose that a front exists and the temperature/pressure gradient is modified (Fig. 8.6b). At that junction, both gradients will be greatest and geostrophic winds above the front will be much faster. A jet stream will be identified with a frontal boundary situation. The creation of a specific surface low-pressure area along a front (and the eventual formation of a middle latitude cyclone) is also explained by the relationship between the surface and upper air flows.

In Figure 8.7, air moving within the ridge–trough pattern must undergo a number of changes caused by converging and diverging air. As the air converges on the western portion of the trough, air will slow down and create high pressure at the surface. On the eastern limb, air spreads out and speeds up to case divergence. This **divergence** aloft is accompanied by **convergence** at the surface, with the resulting formation of low pressure. Thus, the favored position for the formation of a low-pressure area along a frontal boundary can be identified.

The relationships between upper air and surface conditions clearly rely on the location of the upper air waves and jet streams. Such relationships have been dealt with at length in many studies in synoptic climatology, and these studies have been aided by the development of weather satellites.

SYNOPTIC CLIMATIC STUDIES

Synoptic climatology is the study of local and regional climates in terms of properties and motion of the overlying atmosphere. The term was first used in the 1940s in reference to analysis of past weather situations to assess the frequency of sets of conditions. The term **synoptic** is used by meteorologists to denote instantaneous weather conditions shown on a synoptic weather chart. By extension, synoptic charts are a major source for the study of synoptic climatology. To this source is added satellite imagery.

The basic procedures involved in synoptic climatic studies are the (a) determination of circulation types, and (b) statistical assessment of conditions in relation to the identified patterns. Clearly, an important step in an analysis is the classification of atmospheric properties and processes on the basis of synoptic patterns. The basic division is in terms of scale, with subsets relating to whether the methods employed are subjective or objective. Subjective studies may use numerous daily weather map sequences to identify a synoptic pattern associated with a distinctive climatic event and hence classify the types that occur. For example, to identify the type of circulation regime that gives rise to stagnant air masses, the researcher might need to classify conditions associated with blocking systems. Objective classification takes advantage of the availability of digital data banks and the use of computers and statistical packages to use correlation analysis, factor analysis, and various clustering techniques. A classification of conditions relating air masses to pollution levels is illustrative of this approach. A good idea of the nature of synoptic studies may be derived by considering the approaches used at the global, continental, regional, and local scales.

The heart of upper air global circulation is based on the presence of planetary waves. These are very long waves that appear to wrap around the earth sinuously. At any given time, one, two, or three waves may exist with their troughs and ridges located in preferred positions. These positions are partly determined by the unequal heating of continents and oceans and the changes that occur from season to season. Periodically, the jet stream associated with these waves may split to form meridional flow north and south of a high-pressure zone. This is a blocking high—a weather system that leads to extreme weather conditions, ranging from floods to drought, in large parts of North America. Note that a blocking high is not the same as the subtropical high, the latter occurring too far equatorward to split the flow of a jet stream.

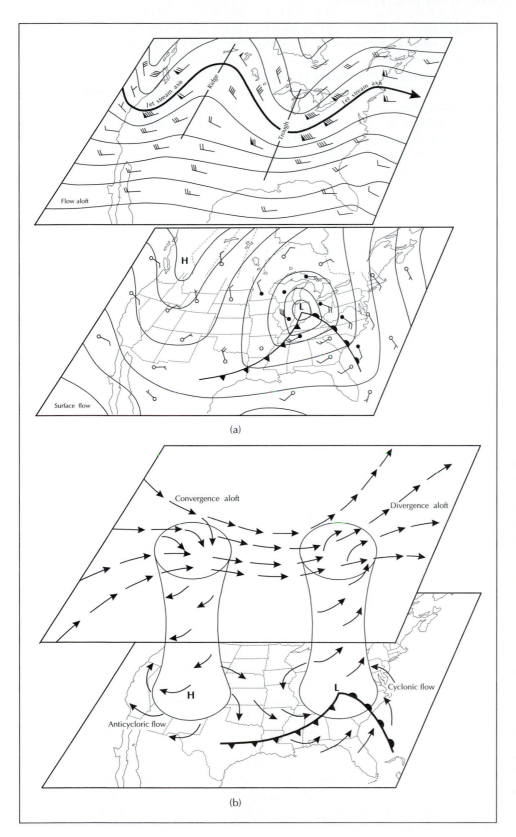

Figure 8.7
Relationship between upper air flows and surface pressure systems.
(a) Troughs and ridges at the jet stream level produce convergenece and divergence of air. Convergence, to the west of the trough, induces descending air and divergence at the surface; divergence aloft, to the east of the trough, induces rising air and convergence at the surface. (b) The resulting patterns of convergence and divergence result in high- and low-pressure systems at the surface (from Lutgens F. K. and Tarbuck E. J., *The Atmosphere*, 5e., © 1992, Prentice Hall, Upper Saddle River, N.J.).

The Hemispheric Scale

At the global or hemispheric scale, research often focuses on the patterns of circulation regimes. This may concern seasonal changes of pressure fields or analysis of circulation regimes. Thus, the role of zonal and meridional flow or the nature of blocking systems may be related to climates of large areas. Some objective schemes may calculate indexes relating to typical and atypical circulations. Such analysis is often related to a specific area to become a continental study, as in Figure 8.8, which provides a classic example of a bitter cold winter that occurred in the United States. The normal circulation of upper air flows associated with the Rossby regime is shown in Figure 8.8a. Compare these normal conditions with those showing an extreme winter season (Fig. 8.8b). Remarkable differences are seen and can be assessed by comparing:

1. the amplitude of the wave pattern,
2. the location of the high-pressure ridge over the Pacific Ocean,
3. the southerly extent of the low-pressure ridge normally off the northern Canadian islands,
4. the intense high pressure in the polar area, and
5. the deepening of the low-pressure system off the Aleutians.

In all, the pattern of the frigid year is quite different from the normal; the result is that winters in many places in the United States varied appreciably from what is normally expected. A comparison of the two figures permits an estimate to be made of the ways in which various locations felt the effects of different circulation patterns. Of immediate note are:

1. The high-pressure ridge that extends over the eastern Pacific Ocean causes the jet stream path to be far north of its normal position. This means that the West Coast of the United States does not fall in the path of frontal systems that normally bring rain from the Pacific. The area is dry.
2. The same ridging means that the warm, moist air would extend to Alaska. Usually dominated by cold air at this time, the regions would experience exceptionally mild conditions.
3. The exaggerated jet stream axis would be maintained by the movement around the low pressure over Labrador, and cold Arctic air would stream across the United States.

These conditions, hypothesized from the observed circulation patterns, actually occurred. For example, the temperature departure during January 1977 was as much as 10°C (18°F) below normal temperatures in the Midwest. As anticipated, large parts of the west received rainfall less than 50 percent of normal.

The synoptic climates that produced the unusual January conditions can be analyzed, but there is an important question of why such conditions occurred. In recent years, much research has been undertaken to understand which patterns and anomalies occur, and many of them are rooted in **teleconnections**— the influence of local climate conditions by events happening in other world areas. As already noted, U.S. climates are greatly influenced by ENSO events occurring in the Pacific Ocean—quite remote from the continental area. Other teleconnection correlations being studied include the Pacific Decadal Oscillation (PDO), which deals with sea surface temperature patterns over periods of 20 to 30 years, and the North Atlantic Oscillation (NAO), which deals with cool and warm phases in the North Atlantic Ocean. This is sometimes called the Arctic Oscillation (AO). Although these and other teleconnections are being investi-

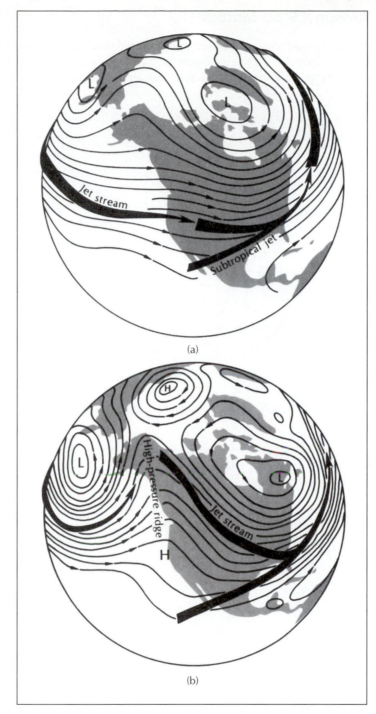

(a)

(b)

Figure 8.8
Schematic diagrams illustrating upper air circulations in winter: (a) the normal pattern; (b) circulation that results in a severe winter over much of the United States.

gated, none thus far appears to have as much influence as ENSO events on the North American climate.

The approaches to synoptic studies are many and variable, but all use either subjective or objective methods. Subjective studies use map sequences to identify a synoptic pattern associated with a distinctive climatic event and hence classify the types that occur. Objective schemes take advantage of the availability of digital banks and data availability and use various correlation and clustering techniques to classify synoptic types.

Regional Scale Studies

An early example of a subjective continental study was presented in the classic book *Compendium of Meteorology*. Figure 8.9 shows schematic diagrams of what are termed *weather types*, but that in effect demonstrate a synoptic classification. The author iden-

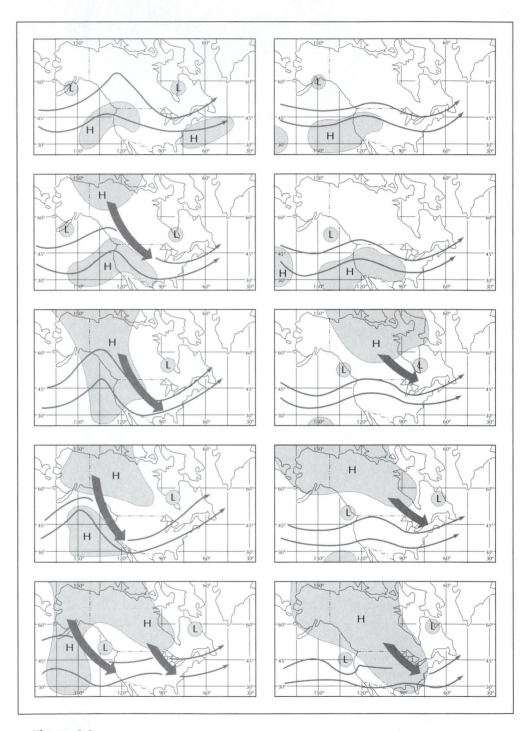

Figure 8.9

Principal circulation types for North America. Heavy lines indicate upper level mean flow. Shaded areas indicate quasistationary low-pressure centers (L) and persistent surface anticyclones (H). Arrows show polar air outbreaks (from Eliot R. D., "Extended range forecasting by weather types" in Malone T., ed. *Compendium of Meterology*, © 1951, American Meteorological Society, Boston.).

tifies eight situations to typify winter circulation types for North America. The four diagrams in the left column represent meridional circulation and those on the right zonal. Meridional flow types are characterized by upper level troughs and ridges of large amplitude that tend to steer circulation systems toward the north and south (meridional). Zonal flow types have upper air flows that are characteristically east–west without large amplitude waves. Each of the patterns is identified by a code and described in terms of the positions of the troughs and ridges in terms of meridional flows and by cyclone tracks for the zonal patterns. Thereafter, remarks about resulting surface conditions associated with each pattern can be made. Clearly, although there is more detail given in the original study, the basic methodology used provides a good example of the subjective classification method.

Many regional synoptic catalogs have been derived. Perhaps the best known are the *Grosswetterlage* (large-scale weather pattern), which involves the classified synoptic patterns over a large region. A daily catalog of Grosswetter is published in Germany to show conditions in Central Europe. Some regional scale studies provide air mass and frontal classifications and cyclone tracks that influence an identified region and relate them to upper air flows. The formation and movement of middle latitude cyclones act as mechanisms for the transfer of heat and moisture over the globe. The movement of the storms in the United States, especially those that influence the central and eastern United States, tend to follow four main tracks. Figure 8.10 shows the cyclogenesis area by name and common tracks of the storms.

Both the Alberta and the Colorado lows form in the lee of the Rockies. The Alberta system brings light snow and rain to the northern tier states—a contrast to the Colorado low. This storm system is often regenerated in southern Oklahoma when moisture from the Gulf is drawn into the circulation. In winter, these storms can produce blizzards, whereas they often produce severe thunderstorm activity in the Midwest. The Gulf low originates in the Gulf of Mexico and moves northeasterward to bring copious rain to the East Coast from Virginia to New England. The Cape Hatteras low greatly influences weather in the Mid-Atlantic states, bringing abundant rain and snow; these are referred to as *nor'easters* by New Englanders.

Local scale synoptic studies may define weather types according to some locally observed weather elements and group them accordingly. Frequently, locally

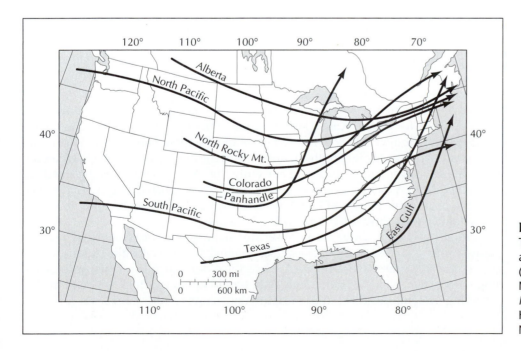

Figure 8.10
The main storm tracks across the United States (from Morgan M. D. and Moran J. M., *Weather and People*, © 1997, Prentice Hall, Upper Saddle River, N.J.).

observed weather events are intercorrelated and statistically condensed. This enables large amounts of data to be related to an identified type of weather or to the occurrence of weather-related impacts on people and the environment. Local snowstorms, dust storms, and high pollution events are typical topics of study.

SATELLITE CLIMATOLOGY

World War II spurred the use of rocketry; by 1947, a modified V-2 rocket took the first successful photographs of Earth's clouds from an altitude of about 100 miles (160 km). Thereafter, with more rocket use, it became evident to atmospheric scientists that viewing weather systems over wide areas from above could play a significant part in understanding the atmosphere. The use of the rockets to place a satellite in orbit became a significant goal and was achieved when the USSR launched Sputnik I in 1957.

Since the launching of the first specialized weather satellite, Tiros 1 in 1960, images of Earth and its cloud cover have become a common feature of TV weather programs. The changes that have occurred since the launching of the early satellites concern the orbit and sophistication of the sensors aboard the satellites. Thus, the types of data obtained from weather satellites are largely governed by their orbit and sensing instruments. Two fundamental orbits are used: the high-altitude, geostationary orbit and the low-level, polar orbit. In discussing these, it is possible to provide endless streams of acronyms, and the following accounts makes a conscious effort not to do so.

Orbits and Sensors

Geosynchronous satellites are placed in orbit such that the satellite circles Earth at the same rate as the equatorial spin of Earth so that it stays in the same point above Earth. To maintain the same rotational period as Earth, a satellite is in a fixed position at 35,786 kilometers (22,236 mi) above Earth's equator. An orbit permits a full disk image of the globe to be derived every half hour or so. Because of its high altitude, the satellite scan views a huge area (the viewing area of a satellite is called a *footprint*). An advantage of the geosynchronous orbit is that it requires no tracking; that is, just like your home TV satellite signal, it is not necessary to constantly adjust the receiver to the satellite position.

The first series of geosynchronous weather satellites, the ATS series, were in equatorial orbit over the equator south of Hawaii. A new series, the GOES satellite systems, first launched in 1975, were the first operational geostationary satellites to provide high-resolution picture transmission (HRPT).

Polar orbiting satellites are placed in orbit some 800 kilometers to 1500 kilometers (4968 mi–9315 mi) above the surface. The TIROS (for TV Infrared Observing satellite) series was the first meteorological polar orbiting satellites. The NOAA class satellites have orbits that pass very close to the poles on each revolution of the earth. The NOAA 15, for example, is in synchronous orbit at an elevation of 833 kilometers (516.5 mi). Both visible and infrared radiation are sensed and sent to Earth during the day, whereas at night both channels are in the infrared. Great improvements over time have been introduced, and currently the United States uses the satellites designated NOAA followed by a letter of the alphabet, which changes to a number after successful launch (e.g., NOAA-K became NOAA 15). Figure 8.11 provides various satellite views from different orbiting systems.

Satellites carry a wide variety of sensing instruments, but essentially may be classed as imaging and sounding systems. Imaging systems are either (a) those that give a series of instantaneous images such as the older vidicon systems, or (b) scanning sensors (i.e., scanning radiometers) that build up images line by line in tracks. Sounding systems are those that measure emissions of radiation from different

(a)

"The Perfect Storm" imagery close up from 1991.10.27 at 1824Z.

(b)

Figure 8.11
Example of images derived from (a) Geostationary satellite (NASA) and (b) polar orbiting satellite (NOAA).

Table 8.2   Spectral Bands on Advanced Very High Resolution Radiometer (AVHRR)

Band	Wavelength (μm)	Description	Examples of Use
1	0.58–0.68	Visible (green to red)	Cultural features; cities/farms
2	0.725–1.10	Near infrared	Land, water differences; penetrates haze
3	3.55–3.93	Infrared	Temperature differences; cloud heights
4	10.30–11.30	Infrared	Thermal mapping
5	11.50–12.50	Far infrared	Thermal mapping; water vapor correction

levels of the atmosphere to allow the construction of atmospheric profiles. The camera used on the TIROS series is the Advanced Very High Resolution Radiometer (AVHRR), which measures radiation brightness in selected wavelength bands. Each of these bands has a special purpose.

The GOES imager is designed to sense radiant and solar reflected energy from Earth. By imaging different wavelengths, as shown in Table 8.2, a variety of phenomena can be sensed. The GOES sounder takes readings of temperature, moisture (including clouds), and ozone through a column of the atmosphere using a radiometer that monitors energy from the visible wavelengths to the long wave, 15 microns. (See chapter 2 for a reminder of details of the electromagnetic spectrum.) The NIMBUS series of satellites has been used as the basic research and development vehicle for these sensors.

Satellite Studies in Climatology

Although important work is being completed in satellite climatology, its development is not as rapid as might be expected for a number of reasons. Climatology, by its very nature, demands long runs of homogeneous data. Weather satellite data only recently reached the 30 to 35 years required for the confident evaluation of climatic norms. Frequent changes of satellite orbital patterns and sensors result in nonhomogeneous data sets. Methods of data processing have changed greatly since 1960, and there have been few coordinated international programs. This has resulted in extremely variable archived data. Despite this, satellite-based climatic studies have produced significant results.

Consider some of the sensors that have been used:

ERB (Earth Radiation Budget) for monitoring energy budgets

SAMS (Stratospheric and Mesospheric Sounder) for temperature profiles

SBUV (Solar Backscatter, Ultraviolet Energy) for ozone profile retrieval and evaluation of ozone, terrestrial, and solar irradiances

TOMS (Total Ozone Mapping Spectrometer) for ozone monitoring

Each of these has contributed toward understanding specialized components of the atmosphere and has been especially useful in climatology. Many of the findings have been used in other sections of this book, and the following lists but a few examples.

Sea surface temperature related to ENSO

Evaluation of the solar constant for energy budget studies

Global temperatures for comparison with surface temperatures and global warming/cooling

The extent of ozone and its seasonal variations

Properties and location of sea ice and polar ice caps

Assessment of the variable gases in the atmosphere

As can be seen, satellite observations and measurement aid significantly in the understanding of the atmosphere.

SUMMARY

The introduction of air masses, fronts, and middle latitude cyclones is a major advance in the development of the atmospheric sciences. Although the surface patterns of fronts and frontal systems were well known, the understanding of the formation and life cycle of the middle latitude cyclone could not be fully explained until upper air data became available.

Studies in synoptic climatology use both surface and upper air data. The scale of analysis varies from planetary to local, whereas the fairly recent availability of large computing capacity has allowed the rapid development of objective studies that complement the older subjective methods. Explanation of many synoptic events may be related to planetary wave analysis and teleconnections. ENSO events have been widely studied and publicized.

Satellite imagery results from both polar-orbiting and geosynchronous platforms. The types of sensors aboard the satellites determine the types of imagery produced, with a wide range of electromagnetic radiation bands being used. Satellite climatology is becoming increasingly important as digital archives are growing.

KEY TERMS

Air masses	Fronts	Synoptic
Convergence	Geosynchronous	Teleconnections
Cyclogenesis	Meridional flow	Zonal flow
Divergence	Middle latitude cyclones	
Doppler radar	Polar orbiting	

REVIEW QUESTIONS

1. How are air masses classified?

2. Explain lee-of-the-lake snowfall.

3. Compare the major features of cold and warm fronts.

4. Describe the surface evolution of a middle latitude cyclone.

5. Describe air flow around an upper air trough and ridge.

6. How have upper air data been acquired?

7. Explain the relationship between a jet stream and a frontal system.

8. Give an example of a subjective synoptic classification.

9. What are the two fundamental orbits of satellites?

10. What type of satellite provides your home TV signal? Why is this fortunate?

CHAPTER

9

The Nature and Hazard of Atmospheric Extreme Events

CHAPTER OUTLINE

156

C limatology traditionally has dealt with the more stable aspects of weather, such as mean temperatures or mean amounts of precipitation over a particular month or year. Some events occur in our environment that are often sporadic in space and time. They represent major deviations from normal conditions. They are frequently violent manifestations of natural processes that often result in disaster for the human species.

These events may be atmospheric, hydrologic, geologic, or biologic, or they may come from space. They are also extremely difficult to predict. The duration of short-lived events varies over a wide range. Lightning strokes last milliseconds and are among the most instantaneous of phenomena. Floods may last for a few hours or for weeks. Many of these extreme events are atmospheric and just as much a part of the climate of a place. Table 9.1 lists many of the atmospheric hazards and the mean time span within which each phenomenon occurs. In some cases, there is considerable variation in duration, and so the mean life span of the event represents the most frequent duration.

There are a series of disturbances in the atmosphere called *severe storms*. These include thunderstorms, hailstorms, tropical cyclones, and tornadoes. These storms represent atmospheric response to unequal distribution of energy in the atmosphere. They are thus an integral part of dynamic energy exchange rather than simple isolated events. The energy involved in such storms is prodigious (Fig. 9.1). The concentration of this energy in hurricanes and tornadoes causes the loss of life and devastation associated with severe storms. Storms are mechanisms for global energy exchange, which makes them an important climatological element as well as a meteorological event. Analysis of the storms provides equally important knowledge about their role in distributing energy.

The human species has been exposed to these extreme events from the time the first of the species appeared on the planet. In prehistoric times, the number of people affected by any single event were small because there were few people and they were scattered over one or more continents. Because of the sparse nature of the population, these events probably affected only individuals or families. Nearly all groups of people in historic times had legends of natural disasters. Most groups have legends about the creation and the great flood. The story of the Biblical flood

Table 9.1   Atmospheric Extreme Events

Event	Mean Duration
Lightning	Milliseconds
Tornadoes	Minutes
Hailstorms	Minutes
Thunderstorms	Minutes
Blizzards	Hours
Flash floods	Hours
Heat waves	Days
Tropical cyclones	Days
Arctic cyclones	Days
Cold waves	Days
Nor'easters	Days
Regional floods	Weeks
Drought	Months

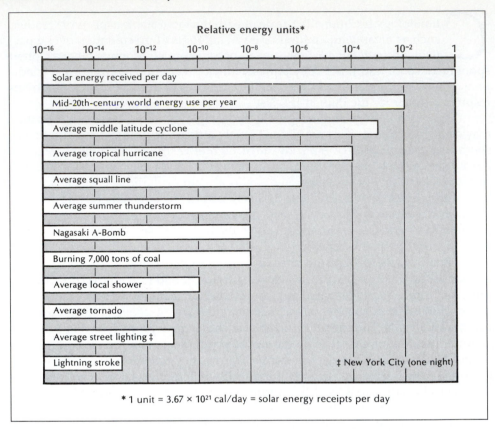

Figure 9.1
Energy equivalents of various phenomena relative to the amount of solar energy received by the earth in one day (data from Sellers W. D., *Physical Climatology*, Chicago: The University of Chicago Press, © 1965).

is replicated in folklore around the world. Written stories of natural disasters began to appear in different parts of the world as soon as writing was developed. One such written story known as the *Gilgamish* existed long before the *Holy Bible*. This story from the middle east contains the story of the deluge. The story may have existed long before the *Gilgamish* was written and handed down by word of mouth.

To explain these natural disasters in the past, they were referred to as *acts of God* or attributed to the spirits. Now we know that these events result from natural processes that operate on the planet. In today's world, these events consistently make the headlines of newspapers and news magazines and occupy prime-time TV. When they affect humans adversely, we refer to them as *hazards*.

The impact of these extreme events can take many forms and can be measured in terms of human stress. Both physical and mental health can be impaired by stress associated with extreme events. Some atmospheric events take lives instantly. Lives can also be lost in the aftermath of these events due to injury, disease, or starvation. There may be long-lasting and far-reaching effects of these events that cannot be measured by fatalities alone. The ultimate extent of the impact depends on a variety of factors, including the magnitude of the event, the number of people affected, and the cultural stage of the people affected.

Although short-lived phenomena are similar in some respects, they differ widely in others. The size, frequency, and spatial and temporal characteristics are different. Some short-lived phenomena have a similar size with each occurrence. Tropical cyclones, although varying in size and intensity, have wind velocities and central pressures that are within a range of a factor of one. Other events vary widely in size. Earthquakes are an example of this class. The range of size that humans can feel varies by a factor of one million. There is the same kind of variation in flood magnitude. The longer the period, the greater the chance of having a more extreme case of a short-lived phenomenon.

Table 9.2   Major Natural Hazards Globally and in the United States. The Events Are Not Listed in Any Order of Magnitude

Global	United States
Tropical cyclones	Tropical cyclones
Earthquakes	Earthquakes
Regional floods	Regional floods
Drought	Tornadoes

The frequency is a measure of how often an event of a given size may occur. It may also be how often the event occurs if there is little difference in size from one event to the next. The concept of frequency is far more important in the analysis of some events than others. Data show that the frequency of short-lived phenomena such as floods and earthquakes is increasing. It is virtually impossible to find out if this is in fact the case since records of these events exist for only a limited time. Records of floods on the Nile River go back many centuries, but this is an exception. Few nations began to compile data on extreme events until recent years. It is only the giants of extreme events for which there are records that have existed for more than a few decades.

Geographical distribution refers to the pattern of a phenomenon over the area in which it occurs. The areal extent ranges widely from a lightning bolt on the one hand to a drought on the other. Some phenomena have very well-defined limits, such as floods and avalanches. Others, such as hurricanes, are less well defined on the margins.

The temporal pattern of short-lived phenomena also varies widely. Although most have a random element, there may be a daily, seasonal, or annual fluctuation in the probability of the event. There may also be a pattern or repetition that is irregular in occurrence, which makes forecasting even more difficult.

Table 9.2 lists the top four hazards on a global scale and for the United States. On a global basis, droughts, tropical cyclones, regional floods, and earthquakes are the four greatest hazards measured in terms of casualties and economic cost. In the United States, the four top hazards are the same as on a global basis except tornados replace droughts. These hazards rise above many others when considered over a period of many years. The major hazards in any given year or even for a few years might look different.

TROPICAL CYCLONES

The tropical cyclone is the typhoon of the Pacific Ocean and hurricane of the Atlantic. It is a vortex or circular storm that rotates counterclockwise in the northern hemisphere and clockwise in the southern hemisphere. The tropical cyclone gets its energy from the latent heat of condensation. The energy in an average hurricane may be equivalent to more than 10,000 atomic bombs the size of the Nagasaki bomb. These storms range in size from a few kilometers to several hundred kilometers in diameter. In the middle is an eye that can be as large as 65 kilometers across. The total area involved may be as much as 52,000 square kilometers.

The atmospheric pressure is nearly symmetrical about the center (Fig. 9.2). The pressure may go as low as 650 millimeters, but this is rare. Such a drop in pressure represents a change of about 13 percent from normal. It is not an explosive drop since it takes place over a distance of many kilometers. The wind system associated with

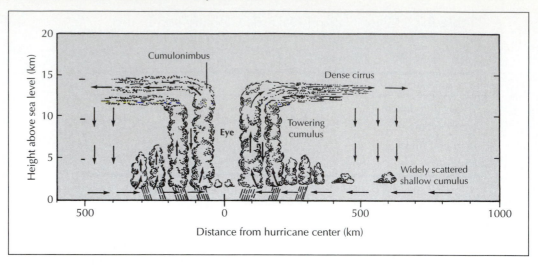

Figure 9.2
Vertical structure of a hurricane showing the cloud pattern and wind flow.

hurricanes is one of contrasts. In the eye of the storm, the winds are light and variable and of velocities not usually exceeding 25 kilometers. The wind velocities increase rapidly away from the eye, reaching their highest velocities just outside the eye and at a height of about 0.8 kilometers. To be classified as a hurricane, winds must exceed 125 kilometers per hour. The maximum winds in most well-developed hurricanes reach 200 kilometers per hour, but in extreme cases may reach 300 kilometers per hour. Hurricane-velocity winds may extend over an area of 300 to 500 kilometers in diameter, and gale force winds (65 km/hr) over an area of 600 to 800 kilometers in diameter. Anemometers reinforced against hurricanes have measured such winds at velocities greater than 250 kilometers per hour. Although the winds within the storm area are of very high velocity, the storm moves at a relatively slow speed, averaging only 15 to 30 kilometers per hour. In Hurricane Gilbert of 1988, the central pressure dropped to 885 millibars (26.13 in). This is the lowest pressure ever recorded in a western hemisphere hurricane. It produced 152 knot (175 mph) winds and pushed a hurricane surge of 4.5 meters (15 ft) in front of it.

Some of the heaviest rains on record in low latitudes came from these storms. Records of 500 millimeters of rain over a 48-hour period are relatively common. One typhoon in the Philippines produced over 1600 millimeters of rain. Radar observations show that 500 to 800 kilometers ahead of the storm there are often fairly well-defined lines of thunderstorms. These storms generate waves as high as 13 meters. Long sea swells as far as 1600 kilometers from the center are evidence of the storms' strength. Near the wall of the hurricane eye, wind often blows the tops off the waves, but in the eye the waves are often very high.

Formation of Tropical Cyclones

There is no agreement yet as to why these storms form. The weather conditions required to produce them are known; once formed, the storms can be tracked. The precise set of factors that trigger tropical storms requires further research. The storms only form over large bodies of warm water when both the air and water temperatures are higher than normal. Thus, they only form in summer over tropical oceans. Temperatures of only a few degrees above normal are enough. Hurricanes form in an atmosphere of essentially uniform pressure and are not associated with atmospheric fronts. They appear to build on a wave of low pressure or on a minor disturbance in wind circulation. These atmospheric ripples may result from local differences in water heating or instability in masses of tropical air.

The trade wind belt is typically a relatively shallow layer of warm, moist air above which is a deep layer of warmer, dry subsiding air. This forms the Trade Wind Inversion—a characteristic that limits the vertical development of clouds. The inversion is sometimes interrupted by a low-pressure trough, which allows thunderstorm development behind the wave. Increased convection and the normal pattern of high-altitude winds cause the trough to deepen so an isolated low-pressure system is formed. If the pressure continues to fall, winds accelerate and a tropical storm is born. The change in status from tropical storm to hurricane requires a mechanism to stimulate vertical air motion and convergence of air. Several possible trigger mechanisms exist, with an intruding high-altitude, low-pressure system being the most often cited cause. Derived from the remains of an upper tropospheric cyclone wave, these abandoned waves act in two ways to promote instability. First, there is divergence on the eastern side of the abandoned system; second, the low has a cold core so the lapse rate below is changed. Both the divergence and altered stability enhance surface pressure differences enough to generate a tropical cyclone. Once a hurricane begins to form, moist air from all sides converges toward the storm center. Condensation supplies the energy needed to develop the storm, and therefore a constant supply of water vapor is essential for the storm's continued existence.

The storm moves slowly at first, usually moving from east to west in low latitudes. As it gains strength, the speed increases and its path curves gradually toward the pole. As long as the storm remains over warm water, it can grow in intensity. A storm can travel far north along the East Coast of the United States because it follows the warm Gulf Stream. When a storm moves over the cold Labrador Current, it dissipates rapidly. If it moves over land, increased friction with the land surface and loss of the energy supply causes the storm to quickly dissolve.

The Time and Place of Occurrence

Tropical cyclones only occur in certain regions (Fig. 9.3). They start over the tropical oceans between latitudes of 5° and 20°, and most form over the western sides of the oceans. They are absent along the equator due to the weakness of the Coriolis effect. There are six general regions of occurrence: the Caribbean Sea and the Gulf of Mexico, the Northwest Pacific from the Philippines to the China Sea, the Pacific Ocean west of Mexico, the South Indian Ocean east of Madagascar, the North Indian Ocean in the Bay of Bengal, and the Arabian Sea. Over the central and western portions of the Pacific Ocean, an average of 20 tropical cyclones occur each year, mostly from June to October. In the eastern Pacific, southwest of Mexico, an average of three hurricanes form each season, but most rarely reach land. Some hurricanes occur outside these regions, but these are the areas of highest frequency. It is of some interest that they do not occur south of the equator in the Atlantic Ocean. Apparently the equatorial convergence zone does not migrate far enough south to provide the necessary convergence.

Hurricanes also occur at particular times. The peak frequency corresponds to the period of highest sea temperature during the year and the time of the maximum displacement of the convergence zone. Thus, late summer and early fall are the seasons of maximum occurrence.

Hurricane Surge and Coastal Flooding

Losses from hurricanes are mainly due to the high water and accompanying waves. Winds produce heavy seas, with waves higher than 13 meters reported. Long sea swells as far as 1600 kilometers from the storm center are evidence of both the

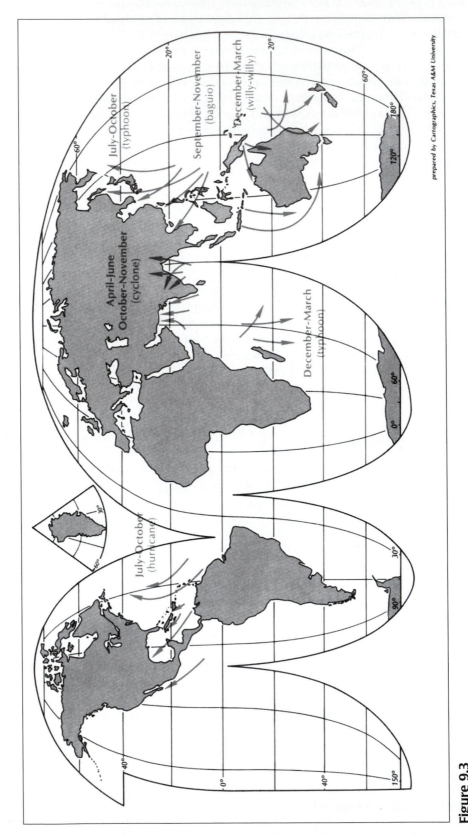

Figure 9.3
Areas of the world that experience the most frequent hurricanes.

strength of the storms and their far-reaching effects. Surges of water two to three meters in height may form ahead of the storm, and the pounding of the waves on top of the surge do most of the damage. It compounds the problem if the storm arrives at the time of high tide. Additional flooding results from the high amounts of rainfall that usually accompany the storm.

When a hurricane forms, its size, intensity, and path determine the potential for damage. The size and intensity of tropical cyclones provide a basis for classifying them. A widely used scale is the **Saffir/Simpson scale** (Table 9.3). This scale ranges from a value of 1 for the weakest storm to a value of 5 for the most severe. The scale is based on atmospheric pressure, wind speed, size of the waves generated, and relative damage the storm may cause.

Table 9.3 Saffir–Simpson Hurricane Damage Scale

Scale Number	Central Pressure		Winds		Storm Surge		Damage
	Mm	*In.*	*Mi/Hr*	*Knots*	*Ft*	*M*	
1	980	28.94	74–95	64–82	4–5	1.5	*Minimal*: damage mainly to trees, shrubbery, and unanchored mobile homes
2	965–979	28.50–28.91	96–110	83–95	6–8	2.0–2.5	*Moderate*: some trees blown down; major damage to exposed mobile homes; some damage to roofs of buildings
3	945–964	27.91–28.47	111–130	96–113	9–12	2.5–4.0	*Extensive*: foliage removed from trees; large trees blown down; mobile homes destroyed; some structural damage to small buildings
4	920–944	27.17–27.88	131–155	114–135	13–18	4.0–5.5	*Extreme*: all signs blown down; extensive damage to roofs, windows, and doors; complete destruction of mobile homes; flooding inland as far as 10 km (6 mi); major damage to lower floors of structures near shore
5	920	27.17	155	135	18	5.5	*Catastrophic*: severe damage to windows and doors; extensive damage to roofs of homes and industrial buildings; small buildings overturned and blown away; major damage to lower floors of all structures less than 4.5 m (15 ft) above sea level within 500 m offshore

North American Hurricanes

Major hurricanes classify as 3, 4, and 5 on the Saffir/Simpson scale. Of the hurricanes influencing the United States for the period 1900–1996, fewer than half classified as major. The majority were Class 1 storms. For the same period, more than 50 percent of all hurricanes affecting the United States occurred in September (Table 9.4).

To consider losses, it is necessary to distinguish between deadliest and costliest. Fatalities in the United States have been declining in recent years as a result of fewer severe hurricanes and better warnings. In contrast, dollar damages have been increasing due to the continued construction on the coastal lands.

Hurricane Hazard in the United States

In the United States, there has been a major movement of population into the coastal areas along the Gulf of Mexico and the Atlantic Ocean. This places many people in jeopardy from hurricanes. The population along the coast has almost doubled in the years since 1930. In Florida alone, over five million people live close enough to the sea to feel the effects of a hurricane surge. Nearly 90 percent of the Florida population is in urban areas, and a large proportion of these people are living in mobile homes. Coastal cities in Florida are among those that have a high chance of being struck by hurricanes (Table 9.5).

THUNDERSTORMS

Observers for the National Weather Service consider a thunderstorm to begin when thunder is heard or overhead lightning or hail is observed. Severe thunderstorms require three-quarter-inch hail and/or wind gusts of 50 knots. The storm is considered ended 15 minutes after the last thunderclap is heard. Note that this definition makes no mention of rainfall; in fact, in dry climates, thunderstorms often occur without measurable precipitation.

Regardless of whether precipitation occurs, thunderstorms form from an initial uplift of moist, unstable air and the release of sufficient latent heat to cause continued uplift. Thus, the keys to storm formation are a source of moist air and a mechanism to produce the required uplift. Given moisture and convective heat sources, the greatest number of thunderstorms occur in moist, tropical realms, especially in Africa. The second requirement, which is a mechanism to initiate cumulonimbus cloud development and, hence, thunderstorms, is obviously more varied; it can result from intense convective activity to forced uplift at fronts or squall lines.

Isolated thunderstorms that occur in summer in the mid-latitudes typify the air-mass thunderstorm. They occur in a disorganized manner, often consisting of a single cell or several distinct cells less than 10 kilometers wide. While air-mass thunderstorms sometimes occur because of differences in surface heating, other trigger effects often cause their development. Alternate mechanisms responsible for growth include converging winds and topography. An example of the former occurs in Florida, where convergence of the moist ocean air along both coasts produces frequent thunderstorms over the peninsula. The role of topography is apparent on the slopes of the Rockies. There air near the slope is heated more intensely than air at similar levels over flat land, resulting in a distinct upslope movement and the potential for the formation of cumulonimbus clouds.

Single-cell thunderstorms are generally much less violent than those associated with the forced upward motion of air that occurs along cold fronts and squall lines. Some of the most severe thunderstorms are associated with squall lines. These occur mostly ahead of cold fronts and, unlike the isolated air-mass type, are an integral part of large-scale circulation patterns. (The schematic

Table 9.4 Number of Hurricanes by Category, Distribution by Month, and Casualties

Hurricanes Affecting United States
(1900–1996) by Category
(Saffir/Simpson Scale Number)

	1	2	3	4	5	All
United States	58	36	47	15	2	158
Florida	17	16	17	6	1	57
Texas	12	9	9	6	0	36
Louisiana	8	5	8	3	1	25
North Carolina	10	4	10	1	0	25

Hurricanes by Month
(More Than 3 on Saffir/Simpson Scale)

	June	July	August	September	October	All
United States	2	3	15	36	8	64
Florida	0	1	2	15	6	24
Texas	1	1	7	6	0	15
Louisiana	2	0	4	5	1	12
North Carolina	0	0	2	8	1	11

Deadliest Hurricanes

Location (Name)	Year	Category	Deaths
TX (Galveston)	1900	4	8000[†]
FL (Lake Okeechobee)	1928	4	1836
FL (Keys) S. TX	1919	4	600[‡]
New England	1938	3*	600
FL (Keys)	1935	5	408

*Moving more than 30 miles an hour.
[†]May actually been as high as 10,000 to 12,000.
[‡]Over 500 of these lost on ships at sea; 600–900 estimated deaths.

Costliest Hurricanes

Location	Year	Category	Cost ($ billions)
Andrew (SE FL/SE LA)	1992	4	26.5
Hugo (SC)	1989	4	7.0
Fran (NC)	1996	3	3.2
Opal (NW FL/AL)	1995	3	3.0
Frederic (AL/MS)	1979	3	2.3

Hurricane data obtained from the National Hurricane Center *http://www.nhc.noaa.gov.*

Table 9.5 States with the Most Hurricanes and the Most Costly Hurricanes (1899–1999)

States with the Most Hurricane Landfalls		
1. Florida		60
2. Texas		37
3. North Carolina		29
4. Louisiana		26
5. South Carolina		15

Most Expensive Hurricanes (Property Damage in Billions of 1996 Dollars)		
1. Andrew	1992	$30.5
2. Hugo	1989	8.5
3. Agnes	1972	7.5
4. Betsy	1965	7.4
5. Camille	1969	6.1
6. Floyd	1999	6.0

sequence given in Figure 9.4 provides the essential details of the storm.) Warm, moist air at surface levels lies ahead of an advancing cold front. At the 850-millibars level (about 1500 m), this air flows from a warm, southerly source. In contrast, at 500 millibars (about 5500 m), a westerly stream of cool, dry air—with divergence—flows across the surface systems. The combination of the unstable surface air and the divergence aloft leads to extensive vertical development of clouds and a line of thunderstoms. The differential flow of air at varying altitudes adds to the severity of the storm. The westerly high-level flow tilts the top of the storm clouds so that falling precipitation does not slow the updrafts, as it does at the mature stage of air-mass thunderstorms, thus extending the storm's life and increasing the potential for hail to form. At times the size and severity of the storms allow them to be classed as supercells. These are enormous storms whose updrafts and downdrafts are sufficiently balanced to enable cells to last many hours. Note that the squall line may be located hundreds of kilometers ahead of the cold front. In fact, squall lines in tropical climates do not require the presence of a front for their formation.

When a number of individual thunderstorms grow in size and organize into a large convective system it is termed a *mesoscale convective complex*. These are often large enough to cover an entire state in the Mid-west and can persist for more than 12 hours.

Associated Hazards

Because thunder is the resulting sound from the violent expansion of air close to the lightning, the definition of a thunderstorm includes the presence of lightning. Thus, lightning is an integral part of severe storms and is itself a distinctive hazard. Lightning is an electrical discharge resulting from separation of positive and negative charges within clouds and between clouds and the ground. When the difference in charge is great enough to overcome the insulating effect of air, a lightning flash results. While the reason for the separation of charges is still imperfectly understood, a number of plausible theories try to explain it. One of these suggests that soft hail

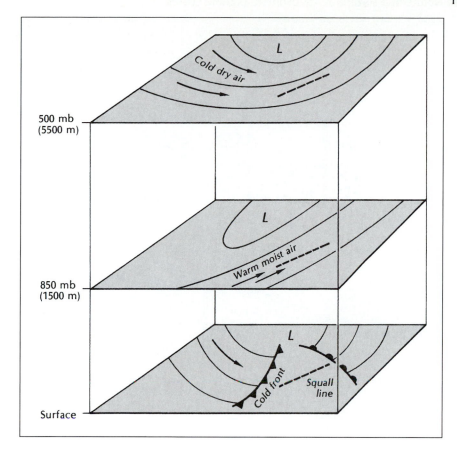

Figure 9.4
Schematic depiction of airflows, shown at three levels of the atmosphere, that give rise to severe weather.

(graupel) in the cloud becomes polarized, with negative charges at the top and positive charges at the bottom of the particles. At this stage, the graupel has no net charge, but collisions result in larger falling pellets acquiring a negative charge while smaller, positively charged graupel is carried aloft. The separation of charges leads to lightning inside the cloud and between the cloud and the positively charged earth below.

The distribution of lightning is obviously associated with the distribution of thunderstorms. However, deaths resulting from lightning in the United States do not correlate perfectly with this distribution, since lightning-caused deaths are not concentrated in one area. This is probably due to the time of thunderstorm occurrence. A study of injuries and fatalities from lightning shows that 70% occur in the afternoon, and only 1% occurs between midnight and 6:00 A.M. Thus, where the proportion of night storms is high, injuries are fewer because people aren't generally exposed to them.

Hail also results from thunderstorm activity. (Its formation is illustrated in Figure 9.5.) Essentially, hailstones form as a result of adding supercooled water to an initial nucleus; the eventual size of the stone depends upon the length of time spent on its passage through the cloud.

Hail exists in three forms. Graupel, or soft hail, is usually less than 5 millimeters (0.2 in) in diameter and has a crisp texture that causes it to be crushed easily when it strikes the ground. Graupel may serve as the nucleus for small hail, which is often mixed with rain. Because of the thin layer of exterior ice, a small hailstone remains intact on reaching the ground. Although graupel and small hail are about the same size, neither is large enough to cause much destruction. Destructive hail, called true or severe hail, can attain large sizes: The remarkable Coffeyville, Kansas, hailstone weighed 3.7 kilograms (8.2 lb). Neither the size nor the

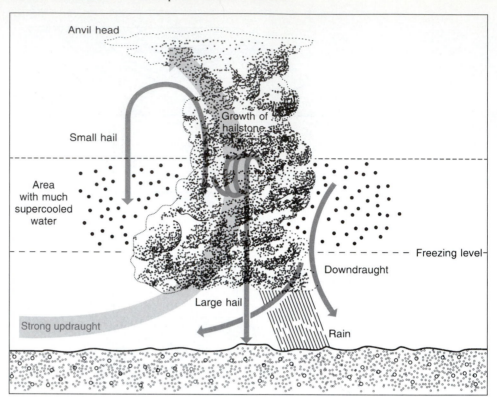

Figure 9.5
Simplifed representation of the formation of hail in cumulonimbus clouds.

number of stones associated with severe hail have been systematically measured for long periods throughout the United States. However, based on all occurrences of damaging hail, whether large or small, Texas, Oklahoma, Kansas, Nebraska, and Missouri (ranked in order of the most to least severe) experienced over one-half of the total severe-hail occurrences in the 48 continental states over a period of 12 years. Studies have shown that the diurnal distribution of hailstorms peaks between 3:00 P.M. and 6:00 P.M. local standard time. Of the severe hail that falls in the continental United States, 40% occurs within these three hours.

Downbursts are another hazard associated with thunderstorms. These are localized downdrafts at the base of a cumulonimbus cloud that strike the ground and spread out horizontally. When the extent of the downburst is less than 4 kilometers, it is called a microburst. An intense microburst can have winds as high as 160 mph. This short-lived wind can cause much damage at the surface and, because of the wind shear associated with it, is a major hazard to aircraft. In fact, some of the worst commercial air craft disasters have been related to landing in downburst conditions.

TORNADOES

The tornado is the most intense vortex that occurs in the atmosphere. It is a converging spiral of air with wind speeds estimated at several hundred kilometers per hour. It is the most violent of atmospheric storms, but it is seldom larger than a kilometer in diameter. The direction of rotation is counterclockwise except in rare instances. The tornado depends on moisture as an energy supply and occurs mostly in moist tropical air. Funnel clouds form out of other clouds. They work downward toward the ground and classify as a tornado only if they reach the ground. The funnel cloud of condensed moisture hanging from the storm cloud makes the tornado readily visible. The cloud may vary widely in thickness and sometimes may be larger at the bottom than at the top. Most often it is gray in color due to condensed water vapor. Yet as the tip contacts the ground, the appearance changes due to dirt and debris picked up from the surface. In winter, a tornado may touch down on a field of snow and become a brilliant white.

The storm develops extremely low atmospheric pressure in the center. The record known drop in pressure occurred in Minnesota in 1904, when the pressure dropped to 813 millibars (against the mean sea level pressure of 1013 millibars). Wind velocities reach as high as 400 kilometers per hour (250 mph). In most tornadoes, wind velocities are less than 145 kilometers per hour (90 mph). Direct measurement is difficult as the sudden pressure changes and high wind velocities destroy the instruments.

Movement of the funnels over the ground surface is erratic. Although they normally move parallel to cold fronts, they occasionally move in circles and figure eights and may even stay in one spot. Their speed over the ground ranges from nearly stationary to as much as 110 kilometers per hour (65 mph). The average is between 40 and 65 kilometers per hour (25–40 mph). The surface path of most tornadoes is short and narrow, varying from a few meters to two kilometers (1.2 mi) in width. The path averages less than 40 kilometers (25 mi) in length. They stay on the ground an average of 15 to 20 minutes. In May 1977, a tornado traveled 570 kilometers (340 mi) across Illinois and Indiana, existing for 7 hours and 20 minutes. They stay on the ground longest and travel the straightest over flat, open land. Hills, large man–made structures, and other wind barriers alter their course.

Formation of Tornadoes

Tornadoes occur most often with thunderstorm activity. Ideal conditions for tornado formation are those found ahead of a cold front. Tornadoes form from the collision of a mass of warm, very moist air with cooler, drier air from polar regions. Extreme turbulence develops along the air–mass boundary, and eddies occasionally develop into strong whirls through which warm air escapes upward. If the air is extremely unstable, the convergence intensifies and the storm forms.

For tornadoes to form, several prerequisites are essential. (1) There must be a mass of very warm, moist air present at the surface. (2) There must be an unstable vertical temperature structure. (3) There must be a mechanism present to start rotation.

The Great Plains is the foremost tornado region in the world, and in this region a set of weather conditions often provides these three elements. Low-pressure centers develop east of the Rocky Mountains. They typically have a cold front extending to the south and the warm front extending east from the low. In the warm sector, there is a south–north flow of warm moist air from the Gulf of Mexico. Above this air is a stream of cold dry air from the west. This air comes from the Pacific Ocean as rather cool moist air. As it crosses the western mountain ranges, it loses its moisture.

The boundary between the warm moist air from the Gulf of Mexico and the dry air from the west is a zone of great turbulence and is termed the *dry line*. If the air pouring over the Rocky Mountains is warm enough, it will move out over the warm moist air from the Gulf. The result is that the dry line extends more horizontally rather than vertically. This provides a stable, but potentially explosive condition. If there is some disturbance, either from the jet stream above or the surface below, there may be a violent exchange of air.

One such disturbance is heat from the ground surface. As the ground heats during the daylight hours, it steadily radiates more and more heat to the air above. The warm moist air moving northward heats during the daylight hours; by late afternoon, the surface air gets quite hot. The air near the surface becomes hot enough to break through the dry line above. This results in the explosive development of thunderstorms. The upward rush of air reaches velocities as high as 165 kilometers per hour (100 mph). Under favorable conditions, the storms will grow through the tropopause to heights of 18,000 meters (60,000 ft). These **supercells** produce heavy rainfall and large hail. Violent updrafts develop in large thunderstorms and are one reason that commercial aircraft try to avoid flying through them.

For a tornado to develop, something must start the column rotating. The mechanism exists in the form of wind shear. **Wind shear** is a change in wind speed and direction with height. The upper level winds are blowing across the path of the lower level winds and at higher speeds. It is this wind shear aloft that starts the rising column of air to rotating counterclockwise. As more air flows in and the storm stretches in height, the rate of rotation increases. Once the center of the system begins to spin, it becomes a *mesocyclone* and may be as much as 10 kilometers (6 mi) across (Fig. 9.6). The rotation starts in the middle level of the storm and works downward. If the process continues, the mesocyclone grows vertically through the thunderstorm and intensifies. It is unfortunate in a sense that the core of the storm begins to rotate first. The rotation is not visible on the outside of the thunderstorm. In most thunderstorms, the supply of warm moist air gradually shuts off, and the storm dies as the energy dissipates. If the supply of moisture continues, the rotating core stretches both upward and downward toward the ground. The strengthening results in higher wind velocities, a smaller **vortex**, and lower pressure.

When turbulence and vertical motion become excessive in a severe thunderstorm, mamantus clouds appear at the cloud base. These clouds occur only when there is extreme turbulence and are a good sign of tornado potential. Rotating clouds at the base of a thunderstorm are an indicator that a mesocyclone exists. As vorticity increases, the rotating clouds drop below the rest of the thunderstorm to form a wall cloud. Within this spinning mass of air, a funnel cloud may form and drop

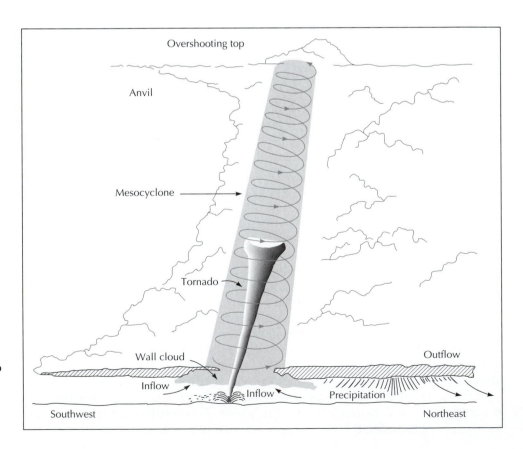

Figure 9.6
Diagram of the development of a mesocyclone and tornado in a "supercell" (from Donald Ahrens C., *Meteorology Today*, West Publishing Co., 1988).

below the cloud. If a tornado forms, it comes out of this wall cloud. Hail is an indication of the severity of the thunderstorm and the high potential for a tornado forming. Heavy falls of hail accompany most, but not all, tornadoes.

Distribution of Tornadoes

Every state in the United States has experienced tornadoes, but their occurrence in regions north of 45° and west of the Rocky Mountains is infrequent. Most of them occur in the Great Plains, a region often called Tornado Alley. Nine states in the United States—Kansas, Iowa, Texas, Arkansas, Oklahoma, Missouri, Alabama, Mississippi, and Nebraska—report an average of over five tornadoes per year. Figure 9.7 shows tornado hazard potential. The maps provide the average of all and strong/violent tornadoes for each 10,000 square miles by state. By expressing the numbers per unit area, it becomes possible to compare the relative occurrence in states of different sizes. Table 9.6 provides data on distribution of tornadoes and tornado fatalities.

　　The time that tornadoes most often occur in the United States is similar to that of thunderstorms. The daily pattern consists of a maximum concentration in a

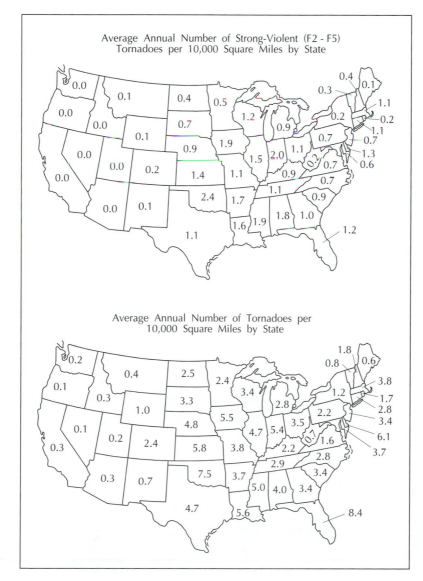

Figure 9.7
Upper: Average annual number of strong/violent (F2 to F5) tornadoes per 10,000 square miles by state.
Lower: Annual average number of tornadoes per 10,000 square miles by state, 1950–1995 (NOAA).

Table 9.6   Number of Tornadoes, Fatalities (1950–1994)

Tornado Totals Ranked by State (1950–1994)

Rank	State	Number of Tornadoes
1	TX	5490
2	OK	2300
3	KS	2110
4	FL	2009
5	NE	1673

Tornado Deaths Ranked by State

Rank	State	Number of Deaths
1	TX	475
2	MS	386
3	AR	279
4	AL	275
5	MI	237

2-hour period between 4 p.m. and 6 p.m. About one fourth of tornadoes occur from 4 p.m. to 6 p.m., and two thirds of all tornadoes occur in the 6-hour period from 2 p.m. to 8 p.m. This is the time when Earth radiation to the lower atmosphere is greatest. They are least frequent around dawn, when the air near the surface has cooled and is relatively stable.

Although they occur in all months, the annual pattern contains a maximum from May to September (Table 9.7). The number of tornadoes that occurs varies from year to year. There is also a seasonal pattern as to where tornadoes occur (Fig. 9.8).

Table 9.7   Number of Tornadoes and Fatalities in the United States by Month (1950–1997)

Month	Tornadoes Total	Tornadoes Mean	Deaths Total	Deaths Mean
January	716	15	106	2
February	975	20	274	6
March	2514	52	602	13
April	5002	104	1245	26
May	8185	171	883	18
June	7715	161	521	11
July	4509	94	63	1
August	2806	58	112	2
September	1829	38	75	2
October	1298	27	82	2
November	1398	29	149	3
December	810	17	124	3

Tornado data obtained from National Storm Prediction Center.

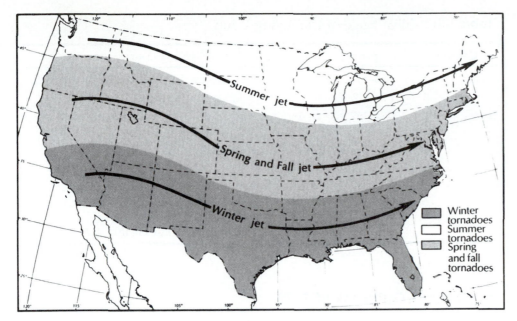

Figure 9.8
Migration of jet stream over the year causes areas of major tornado activity to follow a seasonal pattern (from Eagleman J. R., *Meteorology*, © 1980, Wadsworth Publishing Co.).

Hazards of Tornadoes

Among atmospheric events, tornadoes are the biggest killers in the United States. From 1916 to 1953, an average of 230 people died each year from tornadoes. The low was 36 in 1931 and the high was 842 in 1925.

Property damage from tornadoes on an annual basis is particularly large. From 1916 to 1953, property damage in the United States averaged $14 million per year. When a tornado is large and well developed, it destroys most ground objects touched by the lower end of the funnel. If the funnel lifts off the ground, surface destruction decreases rapidly. Considerable damage to roofs takes place because during construction most roofs are not fastened securely to the frame.

Other factors that increase the hazard potential are the development of second homes and the increased use of mobile homes in North America. Second homes are often built in high-risk areas. These sites include beach homes, mountain cabins, and homes along rivers. There has also been a major increase in the number of mobile homes sold in the United States. More than half of all mobile homes are outside standard metropolitan areas. Many mobile home parks are in locations where they do not have to meet building codes as stringent as they would have to meet within city limits. Mobile homes are more vulnerable to damage from wind than standard dwellings built on a foundation.

T. Theodore Fujita of the University of Chicago introduced a method for assessing the relative severity of tornadoes in the late 1960s. This scale is now known as the **Fujita Scale** (Table 9.8). The severity of the storm is determined only after the tornado is over and so is not a forecasting device as is the hurricane scale. The rating is based on the worst damage. This qualitative measure provides approximate wind speeds. There is an inverse relationship between the number of tornadoes in each class and the severity of the storm. Most of the storms are in the weakest class. Only 2 percent are in the most violent classes. This 2 percent, however, is responsible for more than 60 percent of the deaths from tornadoes. Deaths and damage from tornadoes will continue to increase. This is due to the increased density of population, increased property values, and lack of adequate warning systems.

Table 9.8   Fujita Scale of Damaging Wind

Category		mi per hr	Knots	Expected Damage
Weak	0	40–72	35–62	*Light*: tree branches broken, signs damaged
	1	73–112	63–67	*Moderate*: trees snapped, windows broken
Strong	2	113–157	98–136	*Considerable*: large trees uprooted, weak structures destroyed
	3	158–206	137–179	*Severe*: trees leveled, cars overturned, walls removed from buildings
Violent	4	207–260	180–226	*Devastating*: frame houses destroyed
	5	261–318	227–276	*Incredible*: structures the size of autos moved more than 100 m, steel-reinforced structures highly damaged

The Tornado Outbreak of 1974

The most lethal tornado outbreak in North America occurred in the spring of 1925. Once thought to be a single tornado, it actually consisted of a series of perhaps seven tornadoes. They developed over Missouri and traveled northeast across Illinois and Indiana for a distance of 703 kilometers (437 mi). In the path of the storms, over 600 people lost their lives.

On the third of April 1974, a strong cold front moved across the eastern half of the United States. On the fourth, an intense squall line formed ahead of the cold front (Fig. 9.9a). An extremely violent outbreak of severe thunderstorms and tornadoes accompanied the squall line. Over a period of 16 hours, 148 tornadoes and an unknown number of severe thunderstorms spread death and destruction. Tornadoes formed from Michigan south through Alabama and Georgia. Figure 9.9b shows the counties in which tornadoes occurred. There were 307 confirmed deaths, 6000 injuries, and property damage of more than $600 million. The storms moved along a combined path of 4180 kilometers (2598 mi).

Tornado Forecasting

The main factor in the high death rate is the problem of prediction and detection. Tornadoes are so localized in extent and random in distribution that it is very difficult to forecast them. Once located, radar can determine the general direction and rate of movement. Because of the random nature of tornadoes, the National Severe Storms Forecast Center in Kansas City, Missouri, issues tornado watches. When conditions are such that tornadoes can occur, the Center issues a tornado watch for the area concerned. When a tornado is actually sighted or detected by radar, a tornado warning is broadcast by the nearest National Weather Service Office. This warning gives the location and direction of travel of the storm.

Radar—A Tool for Analysis and Forecasting

Radar provides a means of examining what is happening inside storm clouds such as mesocyclones. A radar transmitter sends out radiation that ranges from 1 to 20 centimeters in length and are much longer than visible light. When the waves strike an object, part of the beam scatters back to the radar antenna. Cloud particles are very small and are detected with very short radar waves. Longer radar waves penetrate cloud particles and reflect from the larger raindrop-size particles.

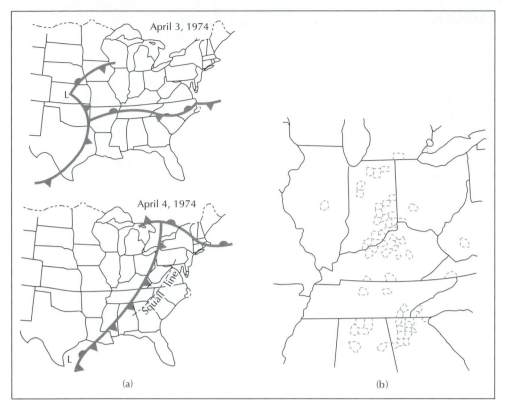

Figure 9.9
Severe weather systems of April 3–4, 1974.
(a) Weather maps showing the location of the fronts on April 3 and 4. Most of the severe thunderstorms and tornadoes developed along the squall line ahead of the cold front.
(b) Counties in which tornadoes touched down (from Hidore J. J., *A Workbook of Weather Maps*, © 1975, Wm. C. Brown Co.).

The harder it rains, the more of the beam reflects. The radar provides a map of where precipitation is occurring and how intense the precipitation is. When a tornado takes place within a mesocyclone, a hook-shaped pattern of rainfall often appears. Although the presence of this hook in a mesocyclone is a good indicator of the possible presence of a tornado, it is not completely reliable. The hook-shaped pattern of rainfall may appear and no tornado exist. Or there may be a tornado and no hook-shaped pattern of precipitation. It is not a very useful forecasting device because of these limitations. In addition, a tornado may be on the ground by the time the hook develops.

Doppler Radar

A more recent development in tornado research and forecasting makes use of **Doppler radar**. Doppler radar makes use of a phenomenon known as the Doppler effect. The frequency of waves from an object coming toward a person is higher than that of an object going away from a person. A train whistle is a good example. As a train approaches, the pitch of the whistle is higher than when it passes and is going away. The frequency of radar waves reflected from rain varies with the direction the rain is moving. By converting the frequency of the radiation to a color on a computer screen, it is possible to determine the direction of wind within a mesocyclone. If a storm develops into a mesocyclone with winds forming a vortex, the pattern of wind velocities shows this on the radar image. A mesocyclone develops before it spawns a tornado, sometimes as much as 20 minutes before. Therefore, when the mesocyclone forms with its rotating winds, Doppler radar can provide up to 20 minutes of warning for the affected area. Since radar can establish the speed and direction of the storm, it is possible to make a fairly accurate forecast of the likely path of the tornado. Once a tornado forms, a distinct pattern of very rapidly changing wind direction appears on the radar screen.

FLOODS

The uneven distribution of precipitation over both time and space causes large inequalities over the earth's surface. Although some places are perpetually dry and others wet, sometimes the precipitation reaches extremes of too much or too little. Floods and droughts can create major hazards for the human population.

Technically, a *flood* may be defined as the condition in any stream or lake when it rises above bank full. Although arbitrary, this is a useful definition of a flood. All natural stream floods are due primarily to surface runoff, which may result from heavy rainfall, the melting of snow, or a combination of both. Floods caused by rain can result from either short periods of high-intensity rainfall (such as rates of 2.5 cm/h or 25–45 cm/day) or from prolonged periods of steady rains lasting for several days or weeks. Flood runoff from small watersheds usually results from different causes than floods on large drainage basins. Small watersheds are defined somewhat arbitrarily; the average would be 25 km^2 or less. Watersheds of this size can be completely covered by a single convective storm, and most floods on small drainage basins are caused from cloudbursts. The rainfall is so intense that the stream channels cannot carry off the water as fast as it falls. The floods that isolated New Orleans and created havoc in other parts of Louisiana and Mississippi in April 1983 were caused by a series of thunderstorms over several days. The flood that destroyed parts of Rapid City, South Dakota, in 1972 was caused by an extremely intensive storm that covered a very restricted area. The rainstorm lasted less than 8 hours and produced flooding on only a few small watersheds.

On large drainage basins, extended precipitation from cyclonic storms or massive snowmelt is required to produce flooding. The Mississippi River floods of the spring of 1973 are a good example. The winter in the upper valley was very wet. The ground was saturated. When the spring melt occurred, almost all of the water ran off the land surface. To add to the problem, a series of cyclonic storms moved across the **drainage basin**. The result was one of the highest floods in recent history on the lower Mississippi River.

The time of occurrence of floods varies over the earth's surface. All around the Red Sea is a complete network of stream beds that are dry most of the time. During short rainy periods, they often flood. The Amazon River Basin receives large amounts of rain in all parts some time during the year, and somewhere in the large basin there is usually an area in flood at any given time. The Ganges River of India is frequently flooded during the monsoon season, from April onward, because melting snow in the Himalaya Mountains and excessive rain combine to produce overflow.

DROUGHT

Droughts that create a major problem for humans at this stage in our history are not just dry conditions, but abnormally dry conditions; the absence of precipitation when it normally can be expected, and a demand exists for it; or much less precipitation than can be expected at a given time. In some locations where daily rainfall is the usual condition, a week without rain would be considered a drought. In parts of Libya, only a period of 2 or more years without rain would be considered a drought. Along the **flood plain** of the Nile River, rainfall is unimportant in determining drought. Prior to the construction of the High Aswan Dam, a drought was any year when the Nile failed to flood. The annual flood provided the soil moisture for agriculture along the river all the way from Khartoum, Sudan, to Alexandria, Egypt. In the monsoon lands of the world, a rainy season producing half the normal precipitation may bring drought; in areas that normally have two rainy seasons, the failure of one would be considered a drought. Since precipitation and water supply vary greatly over the earth, the term *drought* has different meanings in different places. Thus, drought is a relative term, and the total rainfall in a place is not a suitable indication of its presence. It is for this same reason that it is so difficult to devise a widely applicable quantitative index for drought. Nearly every location has its own criteria for drought conditions.

Drought also has different meanings depending on user demands. There are often distinctions made among meteorological drought, agricultural drought, and hydrologic drought. Meteorological droughts are irregular intervals of time, most often of months or years in duration, when the water supply falls unusually far below that expected on the basis of the prevailing climate. Agricultural droughts exist when soil moisture is so depleted as to affect plant growth. Since agricultural systems vary widely, drought must be related to the water needs of the particular animals or crops in that particular system. There are degrees of agricultural drought that depend on whether only shallow-rooted plants are affected or whether deep-rooted plants are affected as well. Both growing season and dry season precipitation can affect crop yields. The Thornthwaite water budget method is often used to assess relative drought. An alternate quantitative index is the Palmer Meteorological Drought Index. A slightly modified Palmer index is used to develop a generalized map of abnormally dry conditions for crops that is published as part of the Weekly Weather and Crop Bulletin.

The temporal and spatial scales of drought vary widely. The duration of droughts varies as much as the timing of their occurrence. How long a drought will last still cannot be predicted for the same reasons that other irregular oscillations cannot be predicted with accuracy. Drought simply ends when the rains come and the streams rise. At present we are unable to predict when they will occur or how long they will last or to prevent them from occurring. All that is certain is that they are a part of the natural system and they will occur again, perhaps with even greater duration and intensity than in the past. They may be very local in extent, covering only a few square miles, or they may be widespread, covering major sections of continents. Even in large-scale droughts, the intensity is likely to vary considerably.

SUMMARY

There are many atmospheric events that in most cases are not typical of usual day-to-day weather. They range over a wide variety of weather conditions. Many, if not most, of these events may cause casualties or property damage. The extent of these losses varies greatly from one event to another. Some of these events occur in very limited areas and others occur over much of Earth.

Drought is an event that may last for months or years and may cause massive starvation, death, and extensive property damage. Drought in developing countries may be extremely hard to combat as both food and water may become scarce. Relief efforts are often hampered by the shear scale of the event. Drought may cover a large part of a continent and affect millions of people. At this stage in history, it is not possible to forecast drought in many areas. The increasing knowledge of changes in atmospheric circulation allows some forecasting of probable drought conditions. An example is drought associated with El Niño.

At the other extreme are floods. Flash floods that occur on small watersheds are very hard to forecast and there is little warning time. As a result, fatalities often result from these storm-produced floods. Floods on large watersheds take a longer time to develop, and forecasting is more reliable and evacuation and other preparations are possible.

Tornadoes and hurricanes are two storms that affect the United States on a relatively frequent basis. Both are dangerous storms and often cause fatalities and extensive damage. Hurricanes affect coastal areas primarily, and the Gulf Coast and Atlantic Coast are the most frequently hit. Tornadoes have occurred in all 50 states. The frequency is much greater in the Great Plains than elsewhere, however. Each has a season of maximum occurrence. Forecasting of hurricanes is more reliable and earlier due to their size and speed of travel.

KEY TERMS

Doppler radar
Drainage basin
Flood plain

Fujita Scale
Mesocyclone
Saffir-Simpson Scale

Supercell
Vortex
Wind shear

REVIEW QUESTIONS

1. Which natural hazards are most destructive in the United States?

2. On a global basis, the major hazards are not the same as in the United States. How and why do the two groups differ?

3. What are the requisite conditions for the formation of hurricanes?

4. What are the requisite conditions for the formation of tornadoes?

5. Why is Doppler radar better than standard radar for detecting tornadoes?

6. How does a hurricane surge differ from waves generated by hurricanes?

7. Drought causes large numbers of casualties on a global basis, but not in the United States. Why is this the case?

8. Regional floods result in casualties and extensive damage even in arid regions. How can this be?

9. The amount of damage caused by both tornadoes and hurricanes is increasing in the United States. Why is this the case when there is no indication that the frequency of either is increasing?

10. Why is evacuation from tornadoes and floods on small watersheds not a practical means of reducing casualties?

PART II
Regional Climatology

Hurricane, Bahamas by Winslow Homer
The Metropolitan Museum of Art, Amelia B. Lazarus Fund, 1910. (10.228.7)
Photograph © 1980 The Metropolitan Museum of Art

Regional Climates: Scales of Study

CHAPTER OUTLINE

C limates can be identified and analyzed over a broad range of areal units from the small vegetable garden up to the continental size regions. The climate over a plowed field is different from that over a field of clover or pasture. The climate of a city differs from that of a rural area, and the climate of a desert differs from that of a rain forest. Significant differences occur in the climate at all of these different scales. Climatologists must deal with these differences at all levels. Often the analytic approach used varies with the relative size of the study area. Before considering major world climates, it is desirable to briefly examine some other scales of study.

DEFINITIONS

There has been much confusion in identifying and naming the range of scale in climatic studies largely as a result of the difficulty of separating the atmospheric continuum into discrete units. The identification problem has been compounded by the historic evolution of climatic studies. Researchers in different countries have used names for areas that, in other countries, are described by different terms. Thus, we find such terms as *gelendeklima*, *ecoclimate*, and *topoclimate* referring to areas of somewhat similar size. At the same time, the dimensions suggested as suitable boundaries for identified scales frequently differ from one source to another.

In a discussion of scale in climatology, the Japanese climatologist M. M. Yoshino derived a general consensus of definitions; the description given here follows his grouping. Figure 10.1 provides examples of the scales that he suggests. The major subdivisions include the following:

Microclimate: Characterized by the climate that might occur in an individual field or around a single building, this describes the climates of an area that may extend horizontally from less than 1 meter to 100 meters. Vertically the area may extend from the surface up to 100 meters.

Local Climate: This category comprises a number of microclimatic areas that make up a distinctive group. The climate in and above a forest or that of a city may be classed in this division. Horizontal dimensions may extend from 100 meters to 10,000 meters, and the vertical extent is up to 1000 meters.

Mesoclimate: Such climates may range horizontally from 100 meters to 20,000 meters and vertically from the surface to 6000 meters. As Figure 10.1 illustrates, a great variety of individual landscapes are considered in this category.

Macroclimate: The largest of the areas studied in climatology, extending horizontally for distances more than 20,000 meters and vertically to heights in excess of 6000 meters. Such areas can be continental in extent.

The dimensions provided in the listing are basic guides, with many area studies overlapping the identified groups. Perhaps the best way to illustrate the role of scale in regional studies is to examine specific examples.

MICROCLIMATES

Studies of these small areal units inevitably begin with fieldwork. The preliminary step is required for accumulation of data and generally includes the measurement of variables. Such a procedure is needed because published climatological data are almost entirely comprised of readings taken at the standard height of the instrument shelter. The microclimatologist is often concerned with the state of the atmosphere below that level. Furthermore, the stations that are used to report climatological data are widely spread, with perhaps one station being the representative of many square

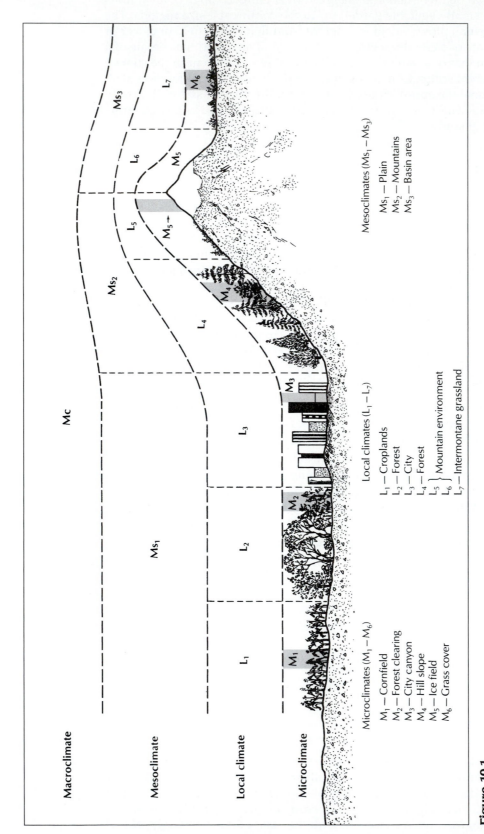

Figure 10.1

Area scales of climatic investigation.

Macroclimate

Mc

Mesoclimate

Ms$_1$

Local climate

L$_1$ L$_2$ L$_3$ L$_4$ L$_5$ L$_6$ L$_7$

Microclimate

M$_1$ M$_2$ M$_3$ M$_4$ M$_5$ M$_6$

Ms$_2$ Ms$_3$

Mesoclimates (Ms$_1$ – Ms$_3$)

Ms$_1$ — Plain
Ms$_2$ — Mountains
Ms$_3$ — Basin area

Local climates (L$_1$ – L$_7$)

L$_1$ — Croplands
L$_2$ — Forest
L$_3$ — City
L$_4$ — Forest
L$_5$ — Mountain environment
L$_6$ —
L$_7$ — Intermontane grassland

Microclimates (M$_1$ – M$_6$)

M$_1$ — Cornfield
M$_2$ — Forest clearing
M$_3$ — City canyon
M$_4$ — Hill slope
M$_5$ — Ice field
M$_6$ — Grass cover

miles. To obtain data such that differences over small distances can be derived, it is necessary to place a fairly dense network of instruments within a small area.

Many studies of microclimatic environments have provided a generalized picture of climatic conditions near the ground. These findings indicate the types of conditions that will occur over a bare surface. Through analysis of surfaces covered by vegetation or synthetic materials, it is possible to generate a more complete understanding of the variation of climatological processes that occurs at the microscale.

General Characteristics of Microclimates

Figure 10.2 shows the temperature characteristics at the interface of the earth and atmosphere. A large diurnal temperature range occurs at the near-Earth levels; during the day, interface temperatures have been found to be as much as 10°C warmer than the air only 1 meter above the surface. At night, the situation is reversed; an inverted lapse rate occurs, with temperatures at the surface cooler than those immediately above. Such a response is to be anticipated. During daylight hours, incoming solar radiation warms the surface and heat diffuses into the atmosphere, raising the temperature of the air by smaller amounts at increasing altitude above the ground. After sunset, under clear-sky conditions, the surface cools rapidly by radiation and the atmosphere loses heat by diffusion to the cold surface.

Temperature changes in soils decrease with depth. As Figure 10.3 illustrates, the amplitude of diurnal temperature change is greater near the surface and decreases with depth until equilibrium is attained. A similar pattern is obtained for the annual cycle, although, of course, it follows the seasonal rather than the diurnal cycle.

Figure 10.4 is an idealized representation of wind near the surface. At the interface between the air and ground, a thin layer of air adheres to the surface; in this layer, the flow is **laminar**; streamlines are parallel to the surface and lack the cross-stream component of convective currents. The depth of this layer depends on the surface roughness and wind speed, but it is seldom more than 1 millimeter in

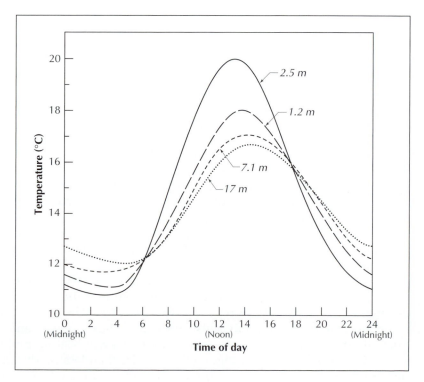

Figure 10.2
The daily course of temperature on a summer day at four elevations. Data from Geiger (1950).

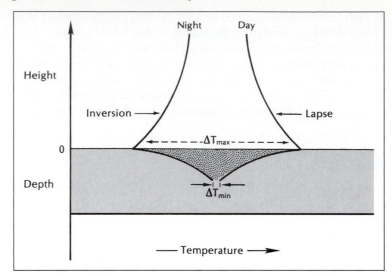

Figure 10.3
Idealized profiles of temperatures near the boundary layer in clear weather (after Oke, 1978).

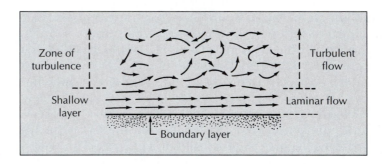

Figure 10.4
Flow of air at the boundary layer from laminar to turbulent flow.

thickness. Its significance lies in its role as an insulating barrier in which all nonradiative transfer is by molecular diffusion rather than the turbulent transfer typical of most of the lower atmosphere. The turbulent surface layer comprises a complex flow of swirling eddies extending up some 50 meters. In this zone, a general increase of wind speed occurs with height.

The exchange of moisture between the surface and air above is reflected in humidity measurements at various levels. Figure 10.5 provides a generalized profile of water vapor concentration day and night. During the day, the concentration decreases away from the surface in a similar way to the temperature profile. At night, if the dewpoint of the air is not attained, the humidity profile is somewhat similar to that in daytime. This occurs because evaporation will continue during the night hours but at a lower rate than during the day. If the dewpoint is reached, an inverted moisture profile will be found. The deposition of dew causes a lowering of near-surface moisture content; if the moisture is replaced by downward movement of air, dew formation will continue. Downward movement will occur through slight turbulence. If this turbulence does not occur, the near-surface air is not replenished with moisture and dewfall ceases.

Although it is possible to generalize about the nature of the mircoclimate above a bare soil surface, it is important to note that the actual characteristics depend, at least in part, on the type of soil surface exposed and the amount of water it contains. For example, sandy soils have a lower heat capacity and thermal conductivity than clays. This means that a sandy soil will heat up rapidly in its top layers during the day, but at night will be cooler than less sandy soils because it undergoes rapid radiational cooling. Organic matter in the soil reduces heat capacity

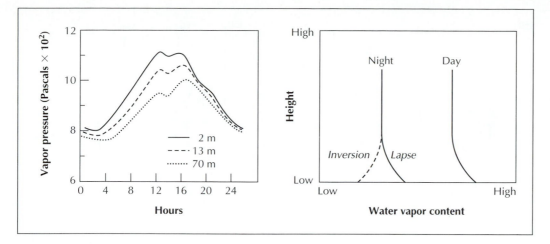

Figure 10.5
Generalized profile of water vapor concentration near the boundary layer (after Oke, 1978).

and thermal conductivity, and the dark color increases the absorption of solar radiation. At night, such soils are relatively warm.

The presence of water in the soil or at the surface greatly modifies the exchange of energy that occurs. When water is present, incoming energy is used for evaporation, making it less available for sensible heat. A comparison of the energy budgets for an irrigated field and one that is dry shows that the temperatures over a moist soil are usually lower than those over dry ground.

These general principles apply to all microclimates. However, modification of the bare surface through vegetative growth or human interference alters the intensity and rates at which ongoing processes occur. Such changes are described in the following section.

The Role of Surface Cover

The role of surface covering in creating microclimates is a response to the way in which incoming energy is disposed at the surface and the way in which the surface modifies airflow. The differences that exist can best be shown through illustrative data. Figure 10.6a shows, in schematic form, some modifications that occur when plants cover the surface. Of particular significance is the creation of a **canopy layer**, in which the microenvironment reflects the nature and extent of the canopy. Clearly, the relative continuity of the canopy plays a major role so that plants with large leaves horizontal to the surface form a more effective canopy than those whose leaves are small and aligned vertically. The canopy becomes most effective when a plant stand has grown to the extent that the ground is shaded. This growth causes the highest temperature zone to move away from the surface to the canopy area. Hence, the top of the canopy, rather than the soil surface, becomes the energy exchange layer.

For example, a building or fence has a marked effect on the microclimate. A building creates a **climatic sheath** (Fig. 10.6b), in which temperature variations occur as a result of shading, humidity anomalies are found, and even a rain shadow effect may occur. When the changes in the microenvironment of a single building are assessed, it is not surprising that cities consisting of buildings can create their own climate.

The modification of wind in the microenvironment is well demonstrated by effects of a vertical structure on air flow. As the diagram in Figure 10.6d shows, the patterns change with the direction from which the wind is blowing. Such modification has been put to good use in the construction of snow fences to keep roads and other used areas clear of drifting snow.

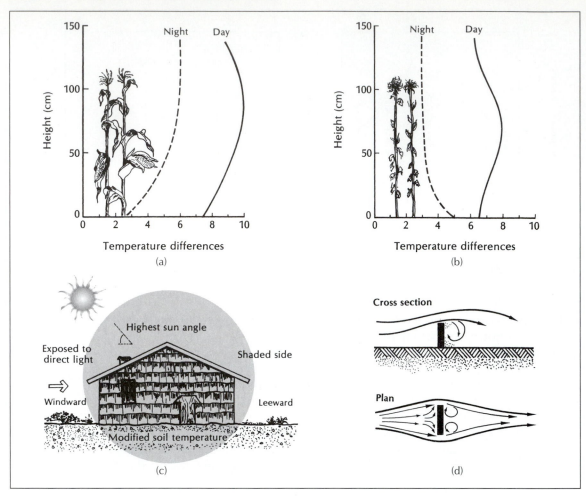

Figure 10.6
Schematic representation of microclimatic modifications caused by surface cover.
(a) Temperature profiles in a crop whose growth shades the ground. (b) Temperature profiles for crops with essentially vertical growth. (c) Climatic sheath around a building. In the sheath, many microclimatic variations occur. Examples show creation of windward-leeward sides, sunlit and shaded sides, and soil modification. (d) Airflow across a barrier at right angles to wind direction. Cross-section shows how such an obstruction is used as a snow fence.

The Role of Topography

It was noted earlier that topographic aspect is of importance in determining the distribution of temperature at a given location. The height of the snow line on the equatorward facing slopes compared to those facing poleward was stressed. The differences can be translated to the microlevel. Most people have seen the effect during winter when snow on one side of a small hill remains in place long after that on the opposite slope has melted.

Topographic variations also modify thermal patterns in small areas. One of the best-known effects of this is the creation of an inversion that, at times, can create a distinctive air flow. On cool, still nights, air close to the surface becomes cooler than that above. If this cooling occurs in areas of uneven topography, the cold dense surface air tends to flow downslope and accumulate in the bottom of valleys and depressions. Such air flow is called a **katabatic wind**.

If the temperature of the ground is at or just below freezing, the collection of cold air in the valley bottom causes a localized frost. Such an effect in citrus growing areas can cause appreciable damage to crops. Another negative effect of topographically induced air drainage is the creation of an air pollution potential. If the pool of cold air becomes deep enough, any effluent carried into the inversion layer is trapped in it. Should the inversion remain in place for more than a few days, the stagnant air can become highly polluted.

LOCAL AND MESOSCALE CLIMATES

Local climates consist of a number of microclimate environments because they comprise a larger area. There is, of course, overlap in the organization framework, but a further differentiation can be made because local climatic studies stress horizontal rather than vertical differences in climate. There are many examples of local climate studies, and here the climates of forest and urban areas are used to illustrate this level of analysis. One example is natural and the other created through human activities.

Forest Climates

Forest climates differ from those of surrounding nonforested areas. The **boundary layer** of the forest is its canopy; it is at this level that energy exchanges occur. Some insolation is returned directly to space; the amount depends on the albedo, or reflectivity, of the canopy layer. In some forests, the quantity varies enormously from season to season. Some energy is trapped within the canopy layer and some penetrates to the forest floor. As illustrated by the data in Table 10.1, the amount of penetration is characteristically low. The evaluation of the data is further complicated because the amounts involved in the disposition of radiant energy vary with both the state of the sky and amount of foliage that exists (Table 10.1).

Table 10.1   Variations in Radiation Receipts Within a Forest

Daily Totals of Net Radiation In and Above a Young Pine Forest (ly/day)						
Height (m)	*10.0*	*5.0*	*4.1*	*3.3*	*2.1*	*0.2*
July 7, 1952						
total	566	555	223	36	–	35
percentage	100	98	39	6	–	6
Nov. 9, 1951						
total	291	–	104	–	14	–
percentage	100	–	35	–	5	–

Solar Radiation Received on a Horizontal Surface at the 1-m Level in an Oak Forest as a Percentage of Incident Radiation Above the Canopy		
	Clear Sky	*Overcast Sky*
Foliaged	9%	11%
Defoliaged	27%	56%

Source: After Munn R. E., *Descriptive Meteorology* (New York: Academic Press, 1966).

Compared with the flow of air over open areas, wind inside the forest is reduced. The amount of decrease depends on the type and structure of the forest. For example, in a deciduous forest, the wind velocity is reduced by as much as 60 percent to 80 percent at a distance of 30 meters (100 ft) inside the forest. Similarly, in a Brazilian forest, the wind speed has been found to decrease from about 8 kilometers per hour (5 mph) to 1.6 kilometers per hour (1 mph) with the same distance. The flow of air is, of course, highly complex and varies in the vertical and horizontal dimensions. Studies have shown that winds from 8 to 24 kilometers per hour (5–15 mph) above the canopy are often less than 3.2 kilometers per hour (2 mph) at the surface.

The forest environment also modifies local moisture conditions. Evaporation from the forest floor is relatively low because of reduced insolation and wind velocity. This effect is counterbalanced by the fact that, with the profuse vegetation, high transpiration occurs. As such, the humidity within a forest depends on the density of the forest and the rates of transpiration that occur. It is generally found that the relative humidity in the forest may be from 2 percent to l0 percent higher than that of nonforested areas, with the highest humidities occurring during the high-sun season. Note that comparisons of humidity using relative humidity values are not always meaningful because the modified thermal environment directly influences the water-holding capability expressed by using relative humidity.

The thermal differences that occur result from a combination of the factors already outlined: shelter from direct rays of the sun, heat modification through water transfer, and the blanket effect of the canopy. The essential result of the interaction of these factors is that temperatures inside a forest are moderated, the maximum is lower, and the minimum higher than those of nonforested areas experiencing a similar climate regime. The amount of variation is seasonal: the main difference in summer being as much as 2.8°C (5°F) within a low-altitude, mid-latitude forest. Exceptions to such moderating influences do occur; the Forteto oak forests of the Mediterranean, for example, experience higher temperatures than neighboring, nonforested areas. Such trees transpire slowly; therefore, the usual hygric and thermal conditions of the forest are modified.

Forest temperatures vary vertically as well as horizontally. For the most part, temperature increases with height during the day and lapse conditions prevail at night, reflecting the role of the canopy in the energy exchange.

Mesoscale Climates

These climates are frequently identified with a distinctive geographic region. In such a region, the physical controls of climate are similar and not modified by major differences within the region. Thus, a mesoscale climatic study might concern the climates of areas such as the Central Valley of California, the lands in the vicinity of the Great Lakes, or the Mississippi Delta lowlands.

Climatologists have completed extensive research on mesoscale climates. The type of research has evolved over time as newer methodologies became available. Early studies often comprised an inventory approach of a region's climate. Classic works on such locations as the Paris and Thames basins provided the basis for the understanding of much regional climatology.

More recently, mesoscale climatic studies have often concerned scales of motion rather than scales of area. This development corresponds to the meteorological concept of mesoscale phenomena, which includes analysis of such features as severe storms, mountain-valley winds, and the like. The climatological equivalents are seen in studies ranging from the cause of the Sahel drought to precipitation variations resulting from circulation patterns in the Midwest of the United States.

Although the distinction among various scales of climate is not always clear, the identification of **macroclimates** is aided by the concept of **filtering**—the averaging of circulation features over successively longer time periods. As the time scale becomes longer, the smaller scale patterns are filtered out and only the general characteristics remain. Thus, a mesoscale feature such as a hurricane or land-sea breeze would not appear on a macroscale representation.

The climatic analysis of macroclimates can follow two methods of approach. First, the analysis can consider the major surface climatic characteristics, ranging from temperature and precipitation patterns to statistical analysis of other elements. Such an approach provides the descriptive climatology of large climatic regions or even continents. Second, the analysis can deal with the dominant circulation patterns that influence the climate and provide the key to understanding its cause. This process is accomplished through analysis of the state of the atmosphere as inferred by pressure maps showing patterns and winds at various levels of the atmosphere. Alternatively, the patterns can be depicted as cross-sections through the troposphere and stratosphere. This approach obtains a synopsis, or condensed view, of the atmosphere at a given time and is referred to as the *synoptic climatology*.

As noted, one approach to describing the climates of the world involves the descriptive climatology of large regions. Historically, the boundaries of such regions have been defined in terms of the surface conditions, and a number of classification schemes have been implemented. Such classification procedures are considered in the next chapter. The remaining chapters of Part II describe and analyze the climates of the region. The treatment is in terms of both the traditional descriptive regional climatology, in which the major surface climatic characteristics are utilized, and the dynamic analysis that uses the principles of synoptic climatology. Of necessity, the scale used to provide an overview of the entire climate of the world is at the macrolevel.

CLASSIFICATION OF WORLD CLIMATES

The classification of climates on a formal basis began with macroscale climatic regions, which cover large geographical areas, some being subcontinental in size. To produce a useful classification of any data, it is necessary to first group together those items that present the greatest number of common characteristics and then subdivide those groups on a uniform basis until a satisfactory degree of subdivision is reached. In most areas of science, attempts have been made to produce classifications based on the most fundamental characteristics possible, rather than on elements that might be more easily observed but of less intrinsic importance. Sometimes items do not lend themselves easily to grouping or classifying. Attempting to group similar regional environments on the earth is much like trying to group students on the basis of height or weight. There are no clear-cut divisions.

The climatic elements of a region do not distinguish a region by their presence or absence, but by the differences in their character. Change from place to place is a basic assumption in regional study. This is not to imply that the differences from place to place are not without order. There is a systematic variation in climatic elements from place to place, sometimes rather abruptly and in other cases over considerable distance. Since change with time and place is an integral part of the earth's environment, it is appropriate that a classification of climates should be based on the patterns of variation that occur.

Throughout this book, there have been repeated references to the annual and diurnal **periodicities** that exist. The motions of the earth in space give rise to periodic climatic, hydrologic, biological, and geological events. If the basic pattern of energy fluctuation is considered, we find that it is responsible for the largest share of the periodicity and spatial variation in the earth environment. The changing intensity and duration of solar radiation bring about changes in the atmosphere and seas, influence migratory and hibernation habits in animals, and set the life cycle of much of the biota including humans. The changing intensity of solar radiation between the northern and southern hemispheres produces a shifting of the general circulation of the atmosphere, which in turn is responsible for periodic changes in the hydrologic cycle over the earth. Thus, an energy flow that varies systemically through time and space results in an environment that also varies systemically through time and space. It is the purpose of this chapter to consider the major regions that result from periodic patterns of energy and moisture and to show how other environmental variables are related to these seasonal patterns.

Attempts to classify regional differences date back thousands of years. During the period of Greek civilization, the earth was divided into three broad temperature zones: torrid, temperate, and frigid. The use of the word *temperate* to describe mid-latitude weather may not have been a particularly good choice, but nevertheless the classification persisted through the centuries. Since the recording of that very early division of the earth, classification schemes for organizing regional differences on the earth have appeared from necessity with ever greater frequency.

The division of the earth into three temperature zones as formulated by the Greeks centuries ago is appropriate. The Tropics are essentially winterless areas as they are not directly affected by the outreach of cold polar air; they are dominated by air currents that originate in the warm areas between 30° N and 30° S. The polar regions are areas that lack a warm summer with the incursion of tropical air currents. The *mid-latitudes* are characterized primarily by the very marked summer and winter seasons since these areas are dominated seasonally by tropical and polar air currents.

A second element that varies markedly over the earth is the seasonal pattern of moisture. Some areas have a nearly equal probability of rain every day of the year. Most of the earth is subjected to a seasonal probability of rainfall as the primary circulation shifts back and forth with the solar energy supply. Some desert areas have nearly equal probability of having no rain on any day of the year. This breakdown gives perennial precipitation regimes, seasonal precipitation regimes, and dry regions. Such basic divisions of global climate using temperature and precipitation form the basis of a number of well-known classification systems.

Approaches to Classification

Of basic importance in the classification procedure is the selection of the variables used in the delineation process. The use of temperature and precipitation as the major variables has been historically determined. Reliable data have only been available over the last 100 years (much less for most climatic stations), and many early records consist of only temperature and precipitation. Much of the early work in classification was limited to the use of these two variables. Furthermore, much of the early work was carried out by plant physiologists and plant geographers who found a correlation between vegetation and the temperature and moisture. Mid-19th-century researchers were influential in the development of climatic classification. It is not surprising, in view of the botanical training of these men, that the distribution of climate and natural vegetation should be treated simultaneously. As an example of the influence of plant geographers on climatic classification, many climatic regions are identified by plant association. This procedure is still followed by some writers. It is not unusual to find a climate type described as savanna, taiga, or tundra. Note that both plant and animal names are given to climatic regions. However, it would be quite unusual to find a climate described as yak or penguin climate in modern literature. The correlation between climate and vegetation is still prevalent, and climates of the world are still described in terms of natural vegetation distribution.

It is evident that, in relating the distribution of natural vegetation to the distribution of climate, the effect of climate is being measured, instead of the climate. It is assumed that a given climate gives rise to a distinctive vegetation association. To identify the climatic type, it is first necessary to determine the vegetation and then infer the climate. If climate can be so identified—that is, by expressing it as the result of the distribution of one selected component of the environment— then equally useful climatic regions may be identified through other measures of the climate's effects. It becomes possible to devise climatic schemes using factors ranging from the human response to climate, to the effect of climate on rock weathering. Such systems would be based on the observed effects of climate, and the criteria used to delimit their boundaries established by best-fit properties. Systems derived through such methodology may be collectively termed *empiric classifications*. The use of the qualifying term *empiric* connotes identification through the observed effects of climate.

The empiric systems concern themselves with identifying similar climate types. An equally valid approach to climatic distribution employs the study of why climatic types occur in distinctive locations. Systems that attempt to deal with the question "why" must concern the cause of climate variation. As such, they may be collectively termed *genetic classifications*. As in the empiric systems, there are a number of methods of examining the causes of climates so that the basis of genetic systems can vary appreciably.

Much controversy concerns the relative merits of the genetic and empiric approaches to classification. The view presented here is that two great groups of climatic divisions exist: those proceeding from observation and those from explanation. If it is accepted that there are two approaches and both are valid, then the genetic versus empiric argument is not of concern. What is important is which classification should be selected for a given distributional problem. A discussion of the climatological implications of the migration of the polar front may not use an empiric system; a genetic approach may not prove of great value in the discussion of specific temperature requirements for plant types. In effect, the approach used and the classification selected depend on the purpose for which they are designed.

Because it is possible to observe the effects of climate on a whole range of environmental phenomena, there are many bases that can be utilized in the formulation of an empiric system. The following list illustrates some of the innumerable interrelationships that can be examined.

1. The human response to climate
2. Climatic requirements for crop growth
3. Water needs and precipitation effectiveness related to vegetation
4. Study and identification of climatic analogs (e.g., agricultural analogs)
5. Vegetation distribution related to climatic controls
6. Geomorphic processes acting under different climatic conditions
7. Climate and soil-forming processes
8. Synoptic conditions and satellite imagery

A little thought could provide many other relationships, and for each there is probably a climatic classification available.

In recent years, the availability of extensive databases and electronic computing devices has seen the development of what may be termed **numerical classification**. This method uses many variables and numeric procedures, such as correlation and cluster analysis, to classify climates. Numerical classification is entirely objective in that the data and selected analytic method influence the way in which the classification is organized. It is becoming an increasingly important method, especially in special purpose classifications, and it is anticipated that most future systems will be of this type.

THE EMPIRIC SYSTEMS

Of the many classifications that have been devised, it is inevitable that one or two develop into what might be termed *standard systems*. Such classifications become standard as a function of their wide usage. It follows that the most widely used systems are those that facilitate an orderly description of world climates. To achieve any prominence, such a system must, of course, be acceptable conceptually.

One system that has developed along these lines is that formulated by Köppen. In its various forms, it is probably the most widely used of all climatic systems. As a result of its innovative methodology, a second scheme that is also widely used is that devised by C. W. Thornthwaite.

The Köppen System

Wladimir Köppen made one of the most lasting and important contributions in the field of climatic classification. He was trained as a botanist. In the early stages of his work, he was strongly influenced by the writings of botanists. The systems he formulated range from a highly descriptive vegetation zonal scheme to a classification in which boundaries are defined in relatively precise mathematical terms.

Beginning with his doctoral dissertation (Leipzig) in 1870 and continuing up to his death in 1940, Köppen proposed, modified, and remodified his system. By 1951, it had become so established that F. Kenneth Hare reported to the Royal Meteorological Society of Canada that some regarded the system "as an international standard, to depart from which is scientific heresy." Such rigorous interpretation was not probably intended by Köppen, to whom the scheme was never completely satisfactory. The evolution of the system shows that Köppen was not so concerned with the precise boundaries as he was with attempting to use simple observations of selected climatic elements to provide a first-order world pattern of climates.

Köppen's early work was completed at a time when plant geographers were first compiling vegetation maps of the world. His early publications (1870, 1884) were concerned with temperature distribution in relation to plant growth, and it was not until 1900 that any of his publications were really concerned with world climatic classification. The 1900 system, which did not get much notice, is a highly descriptive scheme making use of plant and animal names to characterize climate. In 1918, Köppen produced a system that is substantially the one in use at the present time. Boundary values have changed and new symbols have been introduced, but the framework of the present system was clearly evident. The scheme demonstrates Köppen's major contribution to the systematic treatment of the climates of the world. He recognized a pattern underlying world climatic regions and introduced a quantitative method that allows any set of data to be categorized within the system. The classification is considerably enhanced by the introduction of a unique set of letter symbols that obviates the necessity of long descriptive terms.

The classification is based on the distribution of vegetation. Köppen's assumption was that the type of vegetation found in an area is very closely related to the temperature and moisture characteristics of the region. These general relationships were already known at the time Köppen's classification was produced, but he attempted to translate the boundaries of selected plant types into climatic equivalents. The Köppen system is based on monthly mean temperatures, monthly mean precipitation, and mean annual temperature.

Table 10.2   Köppen's Major Climates

A	Tropical rainy climates
B	Dry climates
C	Mid-latitude rainy climates, mild winter
D	Mid-latitude rainy climates, cold winter
E	Polar climates

Principal Climatic Types According to Köppen's Classification

Af	Tropical rainy	Cw	Mid-latitude wet-and-dry, mild winter
Aw	Tropical wet-and-dry	Cf	Mid-latitude rainy, mild winter
Am	Tropical monsoon	Dw	Mid-latitude wet-and-dry, cold winter
BS	Steppe	Df	Mid-latitude rainy, cold winter
BW	Desert	ET	Tundra
Cs	Mediterranean	EF	Ice cap

Köppen recognized four major temperature regimes: one tropical, two mid-latitude, and one polar (Table 10.2). After identifying the four regimes, he assigned numerical values to the boundaries (Table 10.3). The tropical climate is delimited by a cool month temperature average of at least 18°C (64.4°F). This temperature was selected because it approximates the poleward limit of certain tropical plants. The two mid-latitude climates are distinguished on the basis of the mean temperature of the coolest month. If the mean temperature of the coolest month is below −3°C (26.6°F), it is **microthermal**. If the temperature is above −3°C (26.6°F), the climate is **mesothermal**. The fourth major temperature category is the polar climate. The boundary between the microthermal and polar climates is set at 10°C (50°F) for the average of the warmest month, which roughly corresponds to the northern limit of tree growth. A fifth major regime, the dry climates, was based not on temperature criteria, but lack of moisture. Dry climate boundaries are obtained using derived formulas. Figure 10.7 shows the distribution of the identified climatic types.

The system has been subjected to criticism from two aspects: (1) There is no complete agreement between the distribution of natural vegetation and climate. This is to be expected since factors other than average climatic conditions (e.g., soils) affect the distribution of vegetation. (2) The system is also criticized on the basis of the rigidity with which the boundaries are fixed. Temperatures at any site differ from year to year as does rainfall, and the boundary based on a given value of temperature changes location from year to year. Despite the criticisms and empiric basis of the classification, it has proved quite usable as a general system.

The Thornthwaite Classification

In a paper published in 1931, C. W. Thornthwaite proposed a climatic classification that is a marked departure from previous systems. Unlike most classifications available at the time, Thornthwaite based his system on the concepts of moisture and thermal efficiency. Many authors prior to Thornthwaite suggested that the relationship between precipitation and evaporation provides a useful measure of precipitation effectiveness, but few utilized the concept because of lack of evaporation data. Faced by the same problem, Thornthwaite produced a precipitation-evaporation index that could be determined empirically from available data. Using this index, Thornthwaite devised humidity provinces. These formed the first-order division of his classification

Table 10.3 The Köppen Classification

A	*Temperature of the coolest month above 18°C (64.4°F)*	

Subcategories:

f:	rainfall in driest month at least 6 cm (2.4 in)	
m:	rainfall in driest month greater than $10 - r/25$, but less than 6 cm when r = annual rainfall in cm OR	
	rainfall in driest month greater than $3.94 - r/25$, but less than 2.4 in., when r = annual rainfall in inches	
w:	rainfall in driest month less than 6 cm (2.4 in), but insufficient for *m* and dry season in low sun period	
s:	rainfall in driest month less than 6 cm (2.4 in), but insufficient for *m* and dry season in high sun period	
w':	maximum rainfall in autumn	
w":	two rainfall maxima, with intervening dry periods	
i:	annual temperature range less than 5°C (9°F)	
g:	warmest month precedes summer solstice	

B	*Evaporation exceeds precipitation for the year*

Subcategories:

BS (Steppe): Derived by the following, when	r = annual rainfall in cm and
	t = annual average temperature in °C
70% of rainfall in summer six months:	$r = 2(t + 14)$
70% of rainfall in winter six months:	$r = 2t$
Even rainfall distribution or neither of above:	$r = 2(t + 7)$
OR, when	r = annual rainfall in inches
and	t = annual average temp. in °F
70% of rainfall in summer six months:	$r = .44t - 3.5$
70% of rainfall in winter six months:	$r = .44t - 14$
Even rainfall distribution or neither of above:	$r = .44t - 8.5$

The value r is the BS/humid boundary. When the derived r is greater than the value on the right of the equation, the climate is humid; when less it is *B*. If *B*, then determine if BS by dividing the answer by 2. If, after dividing, r is greater than value on right, climate is BS; if less, climate is BW (desert).

scheme. It differs from the Köppen system in that boundaries between provinces are not related to any practical vegetation or soil criteria. Instead, they are based on regular arithmetic intervals of values. The 1931 system has been adequately described and analyzed by a number of authors. Despite some adverse criticism, it is considered a major contribution to the process of climatic classification.

Of more significance at present (because it has superseded the earlier systems) is Thornthwaite's 1948 classification, which is a radical departure from the 1931 system because it makes use of the important concept of evapotranspiration. The earlier system had been concerned with the loss of moisture through

Table 10.3   *(continued)*

BW (Desert): Derived as indicated above

Subcategories of BW:

h:	average annual temperature above 18°C (64.4°F)
k:	average annual temperature below 18°C (64.4°F)
k':	average of warmest month below 18°C (64.4°F)
n:	high frequency of fog
s:	70% of rainfall in winter six months (summer dry season)
w:	70% of rainfall in summer six months (winter dry season)

C   *Coolest month temperature averages below 18°C (64.4°F) and above −3°C (26.6°F); warmest month is above 10°C (50°F)*

Subcategories:

f:	at least 3 cm (1.2 in) of precipitation in each month; or, neither *w* nor *s*
w:	minimum of 10 times as much precipitation in a summer month as in driest winter month
s:	minimum of 3 times as much precipitation in a winter month as in driest summer month, and one month with less than 3 cm (1.2 in) of precipitation
x:	rainfall maximum in late spring or early summer; dry in late summer
n:	high frequency of fog
a:	warmest month over 22°C (71.6°F)
b:	warmest month under 22°C, but at least four months over 10°C (50°F)
c:	only one to three months above 10°C
i:	mean annual temperature range less than 5°C (9°F)
g:	warmest month precedes summer solstice
t':	hottest month delayed until autumn
s':	maximum rainfall in autumn

D   *Coolest month temperature averages below −3°C (26.6°F) and warmest month over 10°C*

Subcategories:

d:	coldest month below −38°C (−36.4°F)

Other subcategories same as for C

E   *Warmest month temperature averages less than 10°C (50°F)*

Subcategories:

ET:	average temperature of warmest month between 0°C (32°F) and 10°C (50°F)
EF:	average temperature of warmest month below 0°C (32°F)

evaporation, whereas the new approach considers loss through the combined process of evaporation and transpiration. Plants are considered physical mechanisms by which moisture is returned to the air. The combined loss is termed *evapotranspiration*. When the amount of moisture available is nonlimiting, the term *potential evapotranspiration* is used.

As with any widely used system, the 1948 classification has been subject to criticism. Many of the criticisms were unrelated to the empiric formula used to express evapotranspiration and to the way in which the water budget of a station is manipulated. Such criticisms are dealt with in a later section of the chapter.

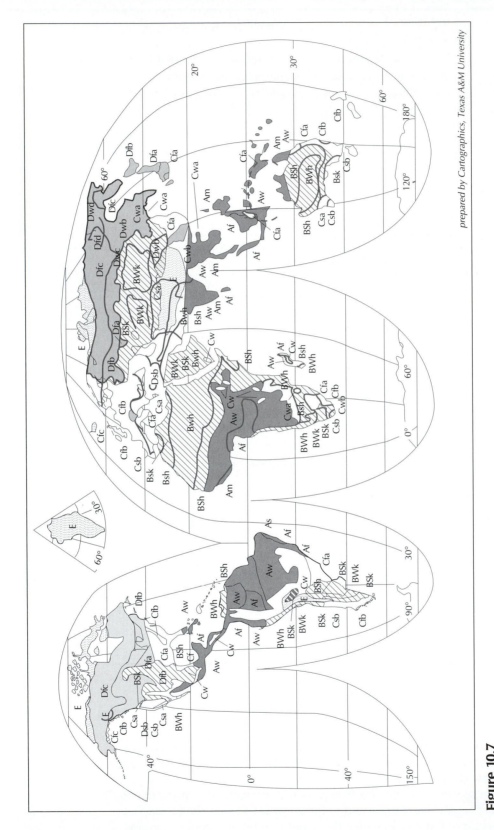

Figure 10.7
The Köppen classification of climate (from Griffiths, J. F. and Driscoll D. M., *Survey of Climatology*, © 1982, Prentice Hall, Upper Saddle River, N.J.).

prepared by Cartographics, Texas A&M University

The two major aspects of the system are the uses of precipitation effectiveness and temperature efficiency. Precipitation effectiveness was designed as an indicator of net moisture supply, taking into account both the actual amount of precipitation and the estimated consumption of moisture by evaporation. The precipitation effectiveness is determined by calculating the ratio of the precipitation to evaporation (P/E) ratio for each month of the year and summing them to form the precipitation effectiveness (P-E) index. Temperature efficiency (T-E) in this classification is used as an indicator of the energy or heat supply relative to evaporation rates. The T-E index is calculated in the same fashion as the P-E index using temperature and evaporation data.

On the basis of the precipitation effectiveness index, nine moisture provinces were established (Table 10.4). The boundary of each progressively more humid

Table 10.4   Thornthwaite's 1948 Classification of Climate

Nine Divisions Based on Moisture Efficiency					Nine Divisions Based on Temperature Efficiency	
			T/E Index			
	Climatic type	*Moisture index*	*(cm)*	*(in)*		*Climatic type*
A	Perhumid	100 and above	14.2	5.61	E'	Frost
B_4	Humid	80 to 100	28.5	11.22	D'	Tundra
B_3	Humid	60 to 80	42.7	16.83	C_1'	Microthermal
B_2	Humid	40 to 60	57.0	22.44	C_2'	
B_1	Humid	20 to 40	71.2	28.05	B_1'	
C_2	Moist subhumid	0 to 20	85.5	33.66	B_2'	Mesothermal
C_1	Dry subhumid	−20 to 0	99.7	39.27	B_3'	
D	Semiarid	−40 to −20	114.0	44.88	B_4'	
E	Arid	−60 to −40			A'	Megathermal

Subdivisions Based on Seasonality of Precipitation		
Moist Climates (A, B, C_2)		*Aridity index*
r	Little or no water deficiency	0–16.7
s	Moderate summer water deficiency	16.7–33.3
w	Moderate winter water deficiency	16.7–33.3
s_2	Large summer water deficiency	33.3+
w_2	Large winter water deficiency	33.3+
Dry Climates (C_1, D, E)		*Humidity index*
d	Little or no water surplus	0–10
s	Moderate winter water surplus	10–20
w	Moderate summer water surplus	10–20
s_2	Large winter water surplus	20+
w_2	Large summer water surplus	20+

Aridity index = water deficit/water need.
Humidity index = water surplus/water need.

region was established at a doubling of the P-E index. On a comparable scale of the temperature efficiency index, nine major temperature provinces are recognized. As in the case of the P-E index, each progressively warmer province is bounded by an index double that of the preceding province. Thornthwaite used many of the same alphabetic symbols that Köppen used in his classification. To the two indexes of moisture and temperature are added a letter designation for rainfall distribution through the year. The initial classification yields 32 different climatic types.

Many other classifications of climate have been proposed, and numerous modifications of these schemes have been suggested. In classifying continuous variables, it must be realized that any classification is going to be arbitrary and there are an infinite variety of climatic regions found on the earth. No two square miles of Earth's surface are likely to have the same atmospheric conditions.

GENETIC SYSTEMS

In comparison to the empiric classifications, genetic systems are generally less formally defined and often less well developed. As a result, they are less widely used in general descriptive climatology. The types that have been proposed are based on either identified physical determinants or air mass dominance. The scheme described here and used as a basis for classification of regional climates in the following chapters is an air mass approach first formulated by Hidore in 1969.

On the basis of seasonal patterns of radiant energy and precipitation, the earth's environments can be identified as belonging to nine basic types, as shown schematically in Figure 10.8. It can be noted in the figure that each row and column of three has the same seasonal characteristics of either temperature or moisture. Thus, for each of the three types of tropical systems, there is a corresponding mid-latitude and polar system with a similar seasonal moisture pattern.

Figure 10.9 illustrates several characteristics of four of the regional systems. Each of these four areas tends to lack any strong seasonal variation in energy or moisture. Some differences between seasons do exist, of course. In the polar ice caps, radiation increases in the summer season; but even so, it remains cold throughout. The increase in radiant energy simply is not sufficient to bring about

Figure 10.8
Schematic representation of the nine basic types of climate found on Earth. Shown are seasonal patterns of temperature and moisture, the types of air masses most common to the area, and the location in the general circulation.

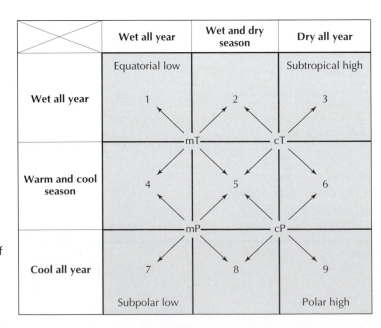

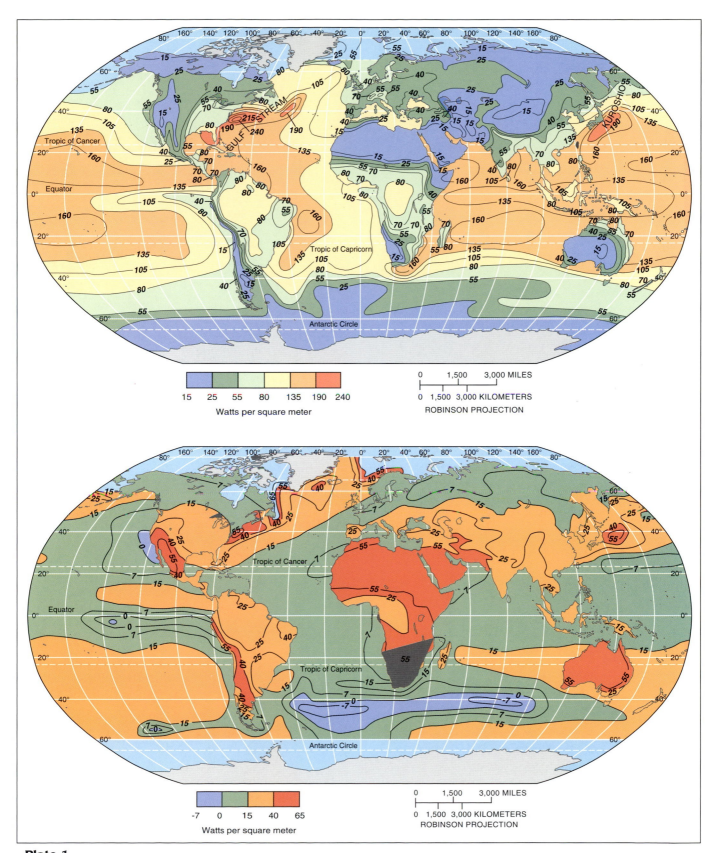

Plate 1

Upper: Global heat of evaporation

Lower: Global sensible heat (from Christopherson R. W., *Geosystems*, 4e., © 1999, Prentice Hall, Upper Saddle River, N.J.).

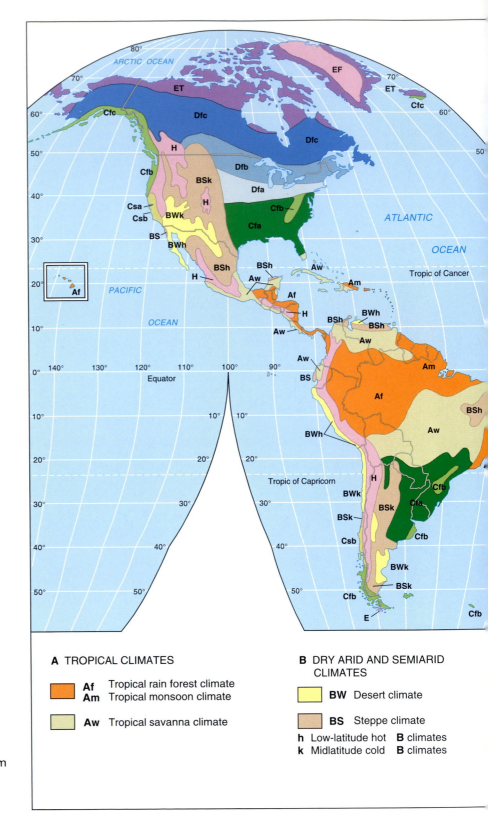

Plate 2

World climates according to the Köppen-Geiger classification system (from Christopherson R. W., *Geosystems*, 4e., © 1999, Prentice Hall, Upper Saddle River, N.J.).

A TROPICAL CLIMATES

Af	Tropical rain forest climate
Am	Tropical monsoon climate
Aw	Tropical savanna climate

B DRY ARID AND SEMIARID CLIMATES

BW	Desert climate
BS	Steppe climate

h Low-latitude hot **B** climates
k Midlatitude cold **B** climates

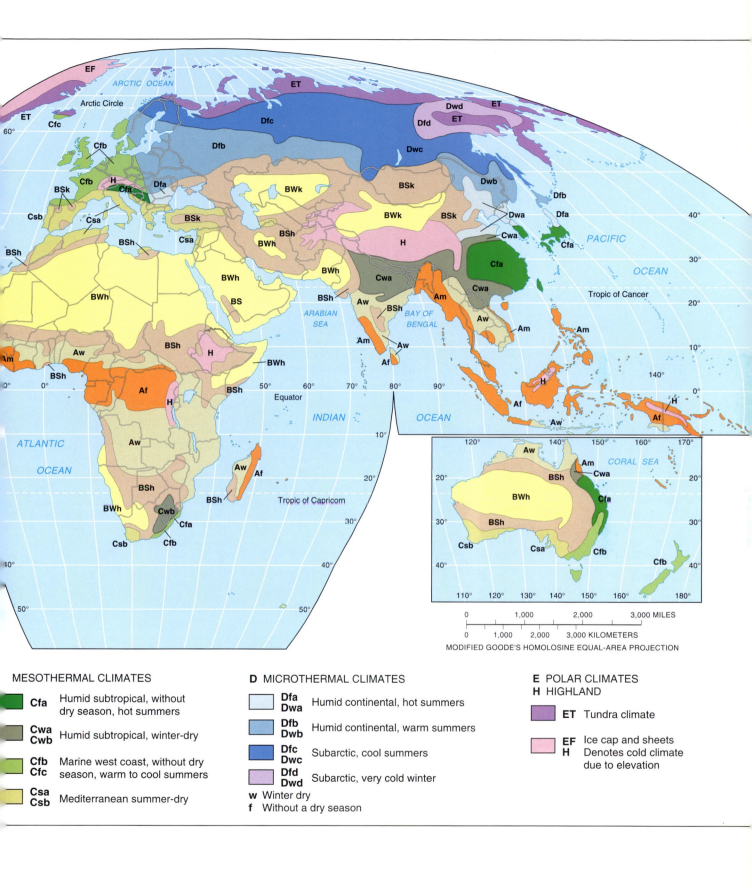

MODIFIED GOODE'S HOMOLOSINE EQUAL-AREA PROJECTION

MESOTHERMAL CLIMATES

Cfa	Humid subtropical, without dry season, hot summers
Cwa Cwb	Humid subtropical, winter-dry
Cfb Cfc	Marine west coast, without dry season, warm to cool summers
Csa Csb	Mediterranean summer-dry

D MICROTHERMAL CLIMATES

Dfa Dwa	Humid continental, hot summers
Dfb Dwb	Humid continental, warm summers
Dfc Dwc	Subarctic, cool summers
Dfd Dwd	Subarctic, very cold winter

w Winter dry
f Without a dry season

E POLAR CLIMATES
H HIGHLAND

ET	Tundra climate
EF H	Ice cap and sheets Denotes cold climate due to elevation

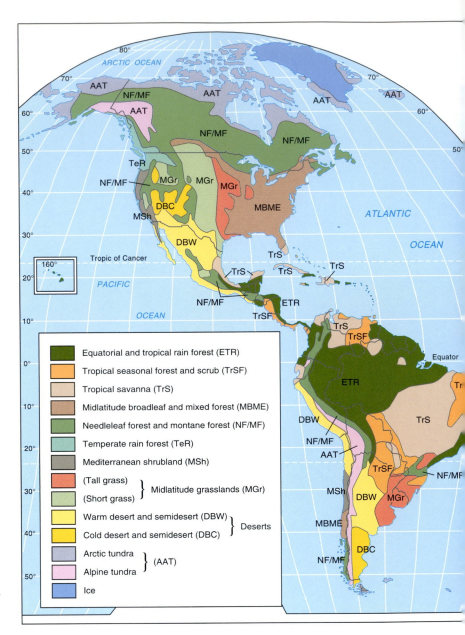

Plate 3
The 10 major global terrestrial biomes (from Christopherson R. W., *Geosystems*, 4e., © 1999, Prentice Hall, Upper Saddle River, N.J.).

Equatorial and tropical rain forest (ETR)

Tropical seasonal forest and scrub (TrSF)

Tropical savanna (TrS)

Midlatitude broadleaf and mixed forest (MBME)

Needleleaf forest and montane forest (NF/MF)

Temperate rain forest (TeR)

Mediterranean shrubland (MSh)

(Tall grass) } Midlatitude grasslands (MGr)
(Short grass)

Warm desert and semidesert (DBW) } Deserts
Cold desert and semidesert (DBC)

Arctic tundra } (AAT)
Alpine tundra

Ice

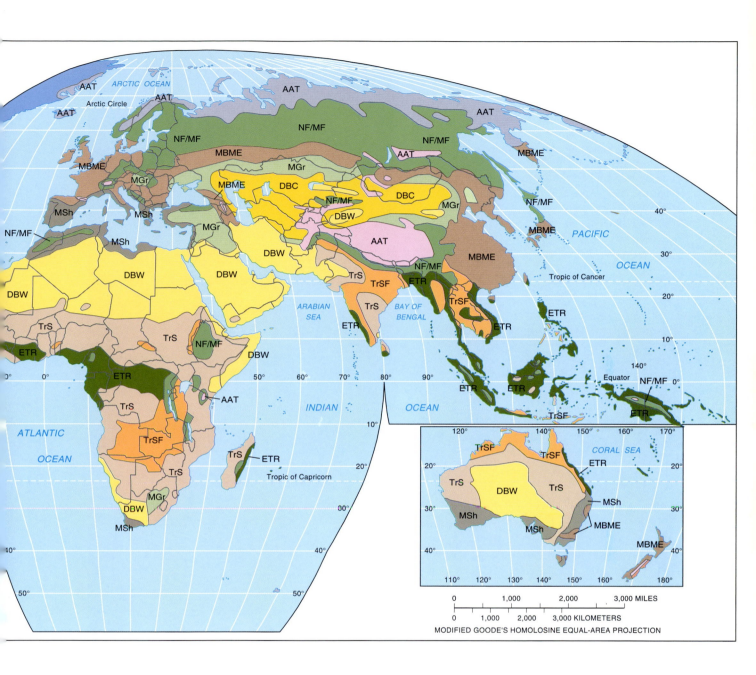

MODIFIED GOODE'S HOMOLOSINE EQUAL-AREA PROJECTION

Plate 4

Right: A sequence of images of the sun in ultraviolet light was taken by the Solar and Heliospheric Observatory (SOHO) spacecraft in 1996. An "eruptive prominence" of 60,000°C gas over 80,000 miles long is seen on the left of each image. Eruptions of this sort affect communications, navigation system and power grids on Earth.

Left: A close up of the region around a sunspot. The granulation is the result the turbolent eruptions of energy at the surface (from JPL/NASA).

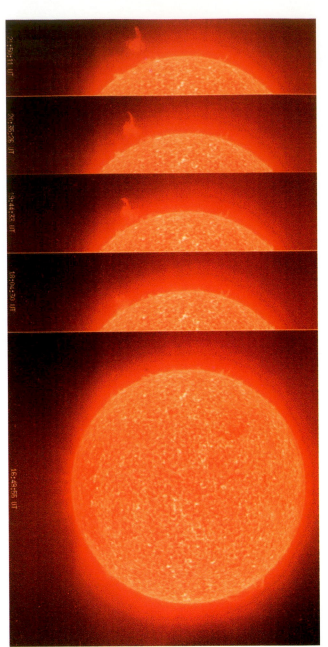

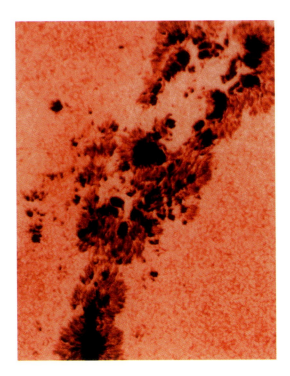

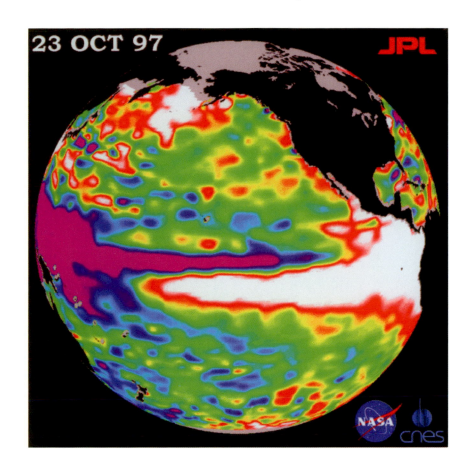

23 OCT 97   JPL

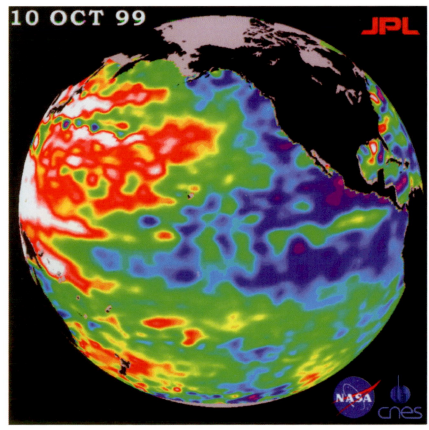

10 OCT 99   JPL

Plate 5
Upper: A strong El Niño is shown in this October, 1997 image taken by the US/French TOPEX/POSEIDON satellite. The image shows sea surface height relative to normal ocean conditions as the warm water associated with El Niño (in white) spreads both north and west. Lower: La Niña conditions are seen in this TOPEX/POSEIDON image for October 1999. The height of the sea surface, an indicator of ocean temperature, shows the normal (green) temperatures and areas of cooler water (blu/purple) (from JPL/NASA).

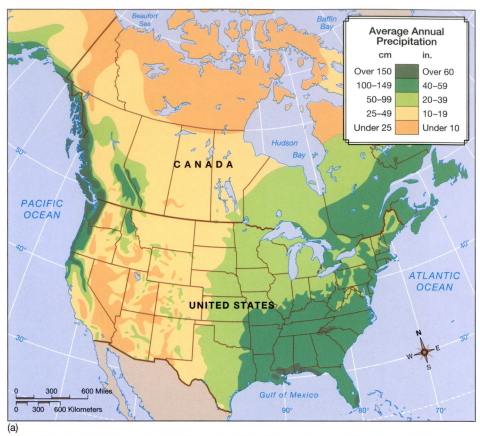

(a)

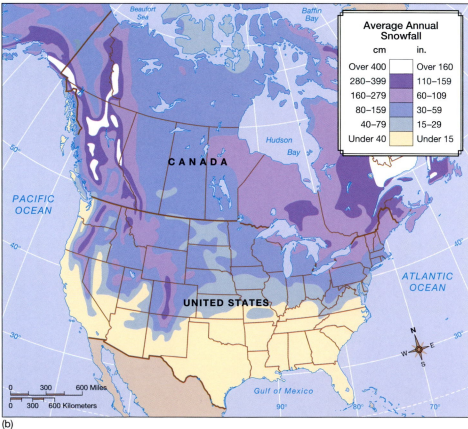

(b)

Plate 6
(a) Average annual
snowfall (cm) and
(b) average annual
precipitation in Canada
and the United States
(from Aguado and Burt,
*Understanding Weather and
Climate,* 1e., © 1999,
Prentice Hall, Upper
Saddle River, N.J.).

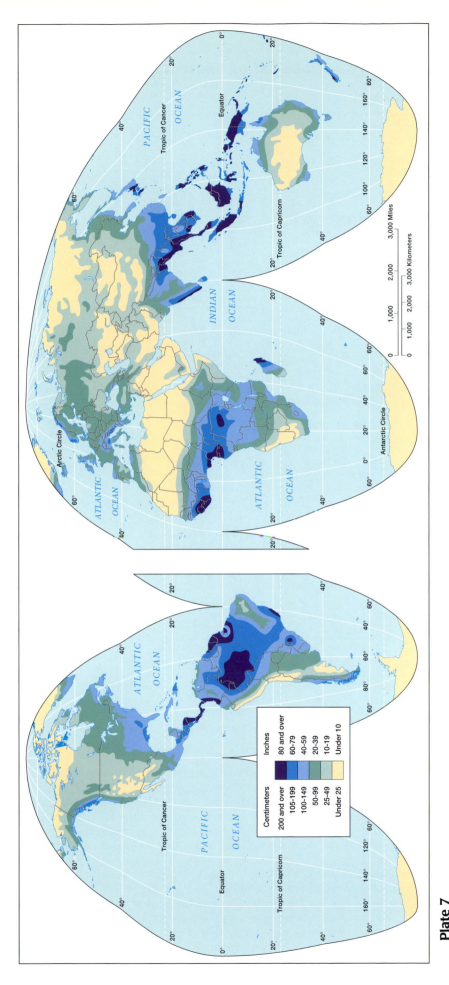

Plate 7
Average annual precipitation over the land areas of the world (from McKnight, *Physical Geography*, 6e., © 2000, Prentice Hall, Upper Saddle River, N.J.).

Centimeters	Inches
200 and over	80 and over
105-199	60-79
100-149	40-59
50-99	20-39
25-49	10-19
Under 25	Under 10

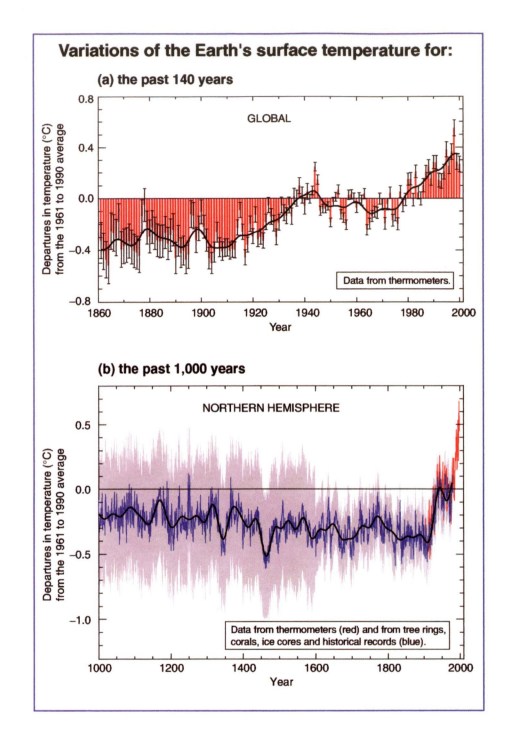

Variations of the Earth's surface temperature for:

(a) the past 140 years

GLOBAL

Data from thermometers.

(b) the past 1,000 years

NORTHERN HEMISPHERE

Data from thermometers (red) and from tree rings, corals, ice cores and historical records (blue).

Plate 8
Variations of the Earth's surface temperature over the last 140 years and the last millennium.
(a) The Earth's surface temperature is shown year by year (red bars) and approximately decade by decade (black line, a filtered annual curve suppressing fluctuations below near decadal time-scales). There are uncertainties in the annual data (thin black whisker bars represent the 95% confidence range) due to data gaps, random instrumental errors and uncertainties, uncertainties in bias corrections in the ocean surface temperature data and also in adjustments for urbanisation over the land. Over both the last 140 years and 100 years, the best estimate is that the global average surface temperature has increased by 0.6 ± 0.2°C.
(b) Additionally, the year by year (blue curve) and 50 year average (black curve) variations of the average surface temperature of the Northern Hemisphere for the past 1000 years have been reconstructed from "proxy" data calibrated against thermometer data (see list of the main proxy data in the diagram). The 95% confidence range in the annual data is represented by the grey region. These uncertainties increase in more distant times and are always much larger than in the instrumental record due to the use of relatively sparse proxy data. Nevertheless the rate and duration of warming of the 20th century has been much greater than in any of the previous nine centuries. Similarly, it is likely that the 1990s have been the warmest decade and 1998 the warmest year of the millennium (from www.ipcc/ch Scientific Assessment).

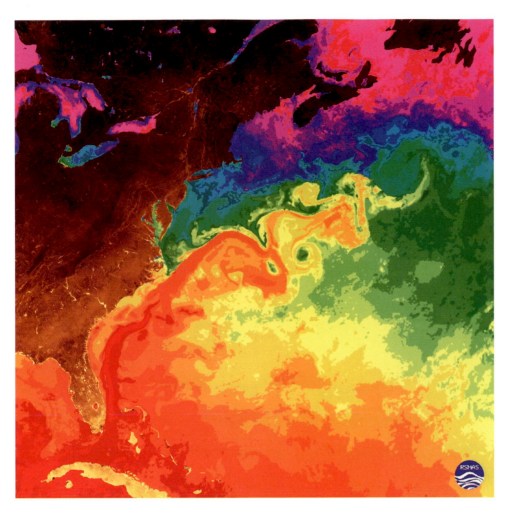

Plate 9
An infrared satellite image of the Gulf Stream shows that it flows in a complex pattern with eddies of different size superimposed (from NASA/NOAA).

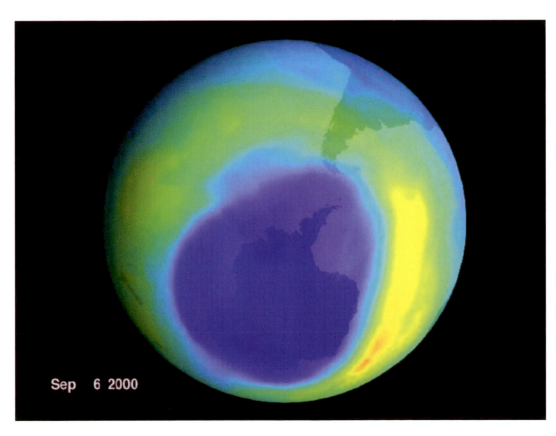

Sep 6 2000

Plate 10
The largest-ever ozone hole, roughly three times the size of the U.S., was detected in September 2000 by NASA's Total Ozone Mapping Spectometer (from NASA: Goddard Space Flight Center).

Plate 11
Political/Physical map of the
world (from Clawson, *World
Regional Geography,* 7e., ©
2001, Prentice Hall, Upper
Saddle River, N.J.).

Plate 12
A tropical rainforest biome typical of wet climates. This example is in La Amistad Biosphere Preserve in Costa Rica (from Tom Stack & Associates © Chip and Jill Isenhart).

Plate 13
A tropical grassland biome associated with wet-and-dry climates. This is the Samburu Reserve, Kenya, East Africa (from DRK Photo © Stephen J. Krasemann).

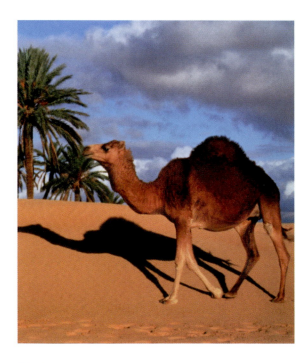

Plate 14
A tropical desert. In the photo are date palms and camels, both of which require a lot of water. The palm trees, which are in some cases immersed in sand, have their roots in water in an oasis. Camels can store water in their humps and can go for a number of days without drinking (from the Stock Market © Jose Fuste Raga).

Plate 15
A coastal desert in the tropics. Here in the Peruvian desert it very rarely rains even though adjacent to the ocean. Subsidence of dry air in the subtropical high produces a very stable lower atmosphere (© Michael Fogden).

Plate 16
A midlatitude forest in Olympic National Park, Washington. Frequent midlatitude lows from over the Pacific Ocean provide abundant rains to the area (© D. Cavagnaro).

Plate 17
A midlatitude grassland in National Bison Refuge, Montana. There is enough rainfall here to support a tall-grass prairie (from Photo Researchers, Inc. © Barry Griffiths).

Plate 18
A midlatitude grassland in a low rainfall area. This short-grass prairie is in Wind Cave National Park, South Dakota (from Bruce Coleman, Inc. © John Shaw).

Plate 19
A midlatitude desert in Oregan Pipe National Monument, Arizona. Most desert vegetation blooms in the spring (from Bruce Coleman, Inc. © Martin W. Grosnick).

Plate 20
The polar desert of Antarctica. Adelie penguins nest along the coast (from Bruce Coleman, Inc. © Des & Jen Bartlett).

Plate 21
The foreground shows the tundra-heath vegetation with boulders in the Lac de Gras area of the Northwest Territories, Canada. (Reproduced with the permission of the Minister of Public Works and Government Services Canada, 2001 and Courtesy of the Geologic Survey of Canada).

Plate 22
The boreal forest covers the glacied Canadian Shield of northern Manitoba, Canada. (Reproduced with the permission of the Minister of Public Works and Government Services Canada, 2001 and Courtesy of the Geologic Survey of Canada).

Plate 23
The barren, windswept Nightingale Island just off the coast of Tristan de Cuhna in the South Atlantic Ocean illustrates the wet polar-type climate (© John Ekwall).

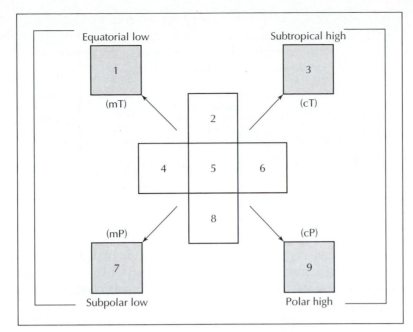

Figure 10.9
Four of the climates are dominated primarily by single types of air masses. Each of these is found in the core area of one of the semipermanent high-pressure zones.

real warming, and tropical air currents do not penetrate that far poleward. Each of the four systems is influenced primarily by one kind of air mass, and each is associated with one of the four semipermanent pressure zones in the atmosphere.

Four of the remaining five types of regions have a marked seasonal variation in either temperature or moisture, but not both. Two have pronounced seasonal moisture regimes and two pronounced seasonal temperature regimes. These four environmental groups are situated between the major pressure zones and are affected by the seasonal migration of the primary circulation. They are subjected to air currents considerably different from one season to the next (Fig. 10.10). One

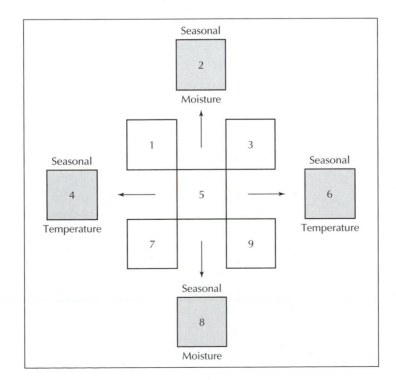

Figure 10.10
Four of the climates have a marked seasonal change in temperature or moisture associated with a seasonal change in air mass control. The other climate, shown in the center of the diagram, is subject to conditions associated with all four basic kinds of air masses. It has both wet and dry seasons and summer and winter seasons.

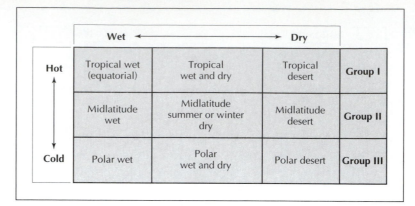

Figure 10.11
Climates differentiated in the air mass model.

of the nine systems, the mid-latitude seasonal rainfall regime, represents the maximum in seasonal variation of weather systems. Periodic invasion of air currents developing in each of the major types of source regions brings a variety of weather.

Using the model, it is now possible to identify the climatic types that occur within the model (Fig. 10.11) and to list their characteristics (Table 10.5). When the identified types are located on a world map, a highly generalized distribution results, with large areas grouped as a single climatic type. Given that the distribution is based on only nine major climatic types, such might be expected. The resulting map (Fig. 10.12) provides a useful guide to the classes identified within the model. For each of the regions, it is possible to subdivide the climates to whatever scale is needed by using a selected criterion. Thus, in Figure 10.12, the mid-latitude wet climate has been divided with summer conditions as a criterion.

Table 10.5 Classification of Climates by Air Mass Control

	Type	Air Masses
Group I	Tropical air masses dominate	
	1. Tropical wet	mT or mE air masses dominate
	2. Tropical wet and dry	mT/cT seasonally
	3. Tropical dry	cT dominates
Group II	Tropical/polar air masses seasonally	
	4. Mid-latitude wet	mT or mP dominates
	5. Mid-latitude wet and dry	
	5S summer dry	cT/mP
	5W winter dry	mT/cP
	6. Mid-latitude dry	cT/cP seasonally
Group III	Polar air masses dominate	
	7. Polar wet	mP dominate
	8. Polar wet and dry	mP/cP seasonally
	9. Polar dry	cP dominate

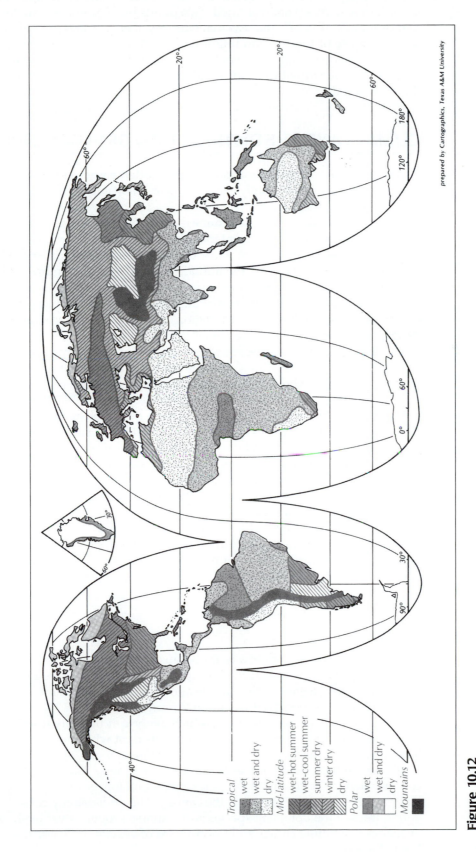

Figure 10.12
World climatic regions differentiated using the air-mass model.

prepared by Cartographics, Texas A&M University

Tropical
wet
wet and dry
dry
Mid-latitude
wet-hot summer
wet-cool summer
summer dry
winter dry
dry
Polar
wet
wet and dry
dry
Mountains

Each type of climate is detailed in the chapters that follow. It can be seen that mountain regions are not identified. Two points must be remembered with regard to mountain regions: (1) the seasonal patterns of energy and precipitation are the same as those of the surrounding lowlands, and (2) mountain regions may be and probably are found in each of the regions differentiated in the previous paragraphs. The altitudinal variations that occur in mountains are similar whenever they are found.

CLIMATE AND THE DISTRIBUTION OF VEGETATION

Given that early climate classifications were often based on vegetation distribution, it is appropriate to briefly examine the relationship between climate and the distribution of natural vegetation.

Among the factors that determine the distribution of flora are moisture availability, radiant energy, soils, parent material, slope, and other biota. Clearly climatic factors are a major determinant in the distribution of individual species and communities. At the **biome** level, the implications of climate are most readily visible. Where the climatic conditions are fairly similar from year to year in terms of temperature and moisture, a fairly distinct climax plant formation has evolved. In the classification of climates, four such regions were identified: tropical rainy, tropical desert, polar rainy, and polar desert or ice caps.

A recognized biome associated with the tropical rainy climate is the tropical rain forest. The relationship between the rain forest and climate is close enough so that, until recent years, the distribution of rain forest has been used to map the areas with a tropical rainy climate. It is now clear that the association between the two is not that perfect. In the tropical deserts is found the desert biome, which covers large areas on the earth's surface. Associated with the polar rainy is to be found the coniferous forest (**taiga**). In the vicinity of the ice caps, little or no vegetation is found as a result of very harsh conditions. In the intervening areas between these four very different biomes, classification is more difficult and often confused by the intermix of communities and species. In these regions, the biomes present transitions from tropical to polar and humid to dry. For instance, some of the different biomes found in the tropical seasonal rainfall regime include seasonal forest, woodland, savanna, and steppe. Since the communities form a continuum over the landscape, placing the boundaries is generally difficult. A brief description of some of the more widely recognized biomes follows.

Forests

There are many definitions of a forest because of the many varieties of forest communities. Here forests include those formations in which trees are a prevalent plant form. As already suggested, this classification covers a multitude of different ecosystems, and here it is possible to mention but a few.

Rain forests are formations in which evergreen trees are dominant and the canopy is more or less continuous. Rain forests exist where moisture is abundant, if not on a year-round basis, at least during the greater part of the year. Most of the vegetation in these communities is found in the canopy. The foliage of the trees is concentrated in the crowns; hence, the canopy and foliage of the lianas, epiphytes, and parasites are concentrated in the canopy. The under storey consists largely of young trees of the dominant species and a sparse ground cover of shade-tolerant or shade-demanding shrubs and lower forms of vegetation. These forests are very limited in extent.

The majority of the world's forests occur where there is a dry season long enough to affect a seasonal change in the forest community. The seasonal forest may include evergreen, semideciduous, deciduous trees, or some combination of these. Where a mixture exists, it is not usually a random mixture of individuals of each species, but mixed stands of one type or the other. Local differences in soil or other site characteristics often determine which community persists. Since the seasonal forests exist where there is seasonal precipitation, the character of the forest is closely associated with the length of the rainy season. As the length of the rainy season decreases, the density of the canopy decreases. The most dense forests, or true jungles, are found in the seasonal forests. Here the dry season is long enough to spread the canopy and allow sunlight to reach the ground, but there is not such a long dry season that edaphic drought can occur with any regularity.

Deserts

A desert is characterized by a discontinuous plant cover. The total vegetal coverage is usually small. Some desert plants, such as the creosote bush, produce their own population control devices. They produce hormones through their roots or leaves that inhibit the sprouting of other individuals within the proximity of the parent plant. This controls spacing of the plants and nutrient resources in a fashion that increases the probability of survival of some individuals of the species.

Many of the animals found in the deserts are the same as those found in the grasslands, but in greatly reduced numbers. There are some animals, including the kangaroo rats, which have adapted specifically to the deserts, but such species are relatively few in number.

Grasslands

The world's grasslands consist of formations made up of communities in which the herbs and shrubs are dominant. The communities are dominated by grasses and legumes, some of which reach considerable size. Grasses ranging upward of 3 meters (10 ft) in height are not uncommon in the more humid grasslands. Grazing animals live in their greatest numbers in the grasslands; among their attributes are their tendencies to live in groups and depend on speed for defense against predators. Some creatures have developed the abilities to leap high above the grasses for easier traveling and to see over the grass to watch for predators. The jack rabbit of the American and Australian plains is a good example. Many small animals burrow for shelter and concealment. Among this group is the prairie dog of the Great Plains of North America, which exhibits both the group social structure and the construction of extensive underground towns.

Savannas are tropical grasslands with scattered trees or clumps of trees. Isolated trees are often found right at the desert edge. It is this scattering of drought-resistant trees that gives the tropical grasslands a distinctly different appearance from the mid-latitude grasslands. Fires are a recurrent phenomenon in the savannas, and both the grasses and trees are fire-tolerant. The varieties of species of trees and grasses are few compared with the tropical forests. Although the species are fewer because of the necessary adaptation to drought and fire, they are also very hardy and respond rapidly after a disturbance. Acacia and baobab trees are among the species that spread through the savannas.

Very few areas of natural grassland remain on the face of the earth. The grasslands have proved to be the most useful of the biomes when measured in terms of agricultural purposes. The natural communities of the grasslands have been either burned off or plowed up and replaced by the simpler communities of the domestic cereals such as corn, rice, wheat, and barley.

The grasslands are found where there is a seasonal moisture regime. They occur on all of the continents except Antarctica, and convergent adaptation has led to similar species in each and strange forms in some. Grasslands have been subdivided by secondary structural characteristics; grasses have been divided between steppe and prairie, tall and short, and sod and bunch grass. The grasslands may border forests, seasonal forests, woodlands, or deserts.

Colder Realms: Taiga and Tundra

Taiga is a word used to describe the great northern forests. The dominant trees of this forest are the needle-leaved evergreen trees. The members of the spruce, pine, and fir families are most common. Under a mature coniferous forest, there is very little under storey as a result of the dense shade. The ground is often covered with a fairly thick layer of undecomposed and partially decomposed needles. This forest is associated with cool, moist conditions poleward to the limits of tree growth. The forest extends equatorward considerable distances along mountain ranges, where favorable temperature and moisture conditions prevail.

Vegetation of the **tundra** consists largely of grasses, sedges, lichens, and dwarf woody plants. The tundra is associated with the seasonal rainfall areas around the Arctic Ocean and at high altitudes in mountains. It exists where there is a short summer season that is too cool for trees to thrive. The vegetation of the tundra has distinct characteristics that allow it to survive the cold temperatures, wind, and very long physiological drought. Almost all tundra plants are low growing and compact to escape the wind, reduce evaporation, and conserve heat. Perennials predominate, and many reproduce asexually by runners, bulbs, or rhizomes. Being perennial, the plants store food through the winter, and the buds are sheltered either underground or close to the surface.

Disturbed Formations

Some scientists maintain that the grassland, savanna, and brush formations are disturbed formations that have persisted because of repeated razing by fire. The sharp boundaries that exist, the variety of formation boundaries, and the lack of woody plants have led to this hypothesis. At this time, the debate seems to be a long way from resolution. The suggestion that grasslands are a product of disturbance does not in any way change the fact that disturbances are a very real factor in formation structure. Disturbance affects formations in several ways. It tends to sharpen the boundary between formations, at least in some areas, and often favors the intrusion of one formation into another. Local disturbances within a formation may cause an alternative formation to become established. The effects of a disturbance vary depending on the type of formation. Some communities in mid-latitude grasslands may completely reestablish a climax community in less than 50 years. The tundra, where ecological processes are very slow, may not recover from a disturbance for centuries.

SUMMARY

Climates can be studied at different scales—from the smallest microclimate through local climate and mesoclimate to the largest macroclimates. The study of microclimates usually requires instrumentation that is placed in the location under study. From the many unique microclimatic studies, a number of generalized observations have been made. It is shown that surface cover plays an important part in determining microclimate characteristics. The types of study associated with local and mesoscale climates are seen in forest climates where the canopy layer often serves as the boundary layer.

To study climates at the world scale, a classification system is required. Depending on the purpose of the classification, an empiric or genetic system may be used. The widely used Köppen classification originally based on vegetation distribution is an example of an empiric system; the described air mass classification is based on the cause of climate and is a genetic system. Of the many other available classifications, that devised by Thornthwaite is of special interest.

Natural vegetation of the earth is closely related to climate. A useful level of equating the two is at the biome level. The forest biomes consist of the rain forests and the seasonal forest with the latter occurring because of a seasonal change of climate. Deserts, tropical grasslands (savannas), grasslands, coniferous forests (taiga), and the tundra are the other identified biomes.

KEY TERMS

Biome
Boundary layer
Canopy layer
Climatic sheath
Filtering
Katabatic wind

Laminar flow
Local climates
Macroclimates
Mesoclimates
Mesothermal
Microclimates

Microthermal
Numerical classification
Periodicities
Savanna
Taiga
Tundra

REVIEW QUESTIONS

1. Describe the four scales at which climate may be studied.
2. Why is the study of microclimatology a field study?
3. How does topography influence wind? How is the wind modified by a forest?
4. What is filtering in relation to climatic studies?
5. What is the difference between an empiric and a genetic classification?
6. Outline the basis of the Köppen climate classification system.
7. How many climate regions are identified in the air mass system? Why is this number selected?
8. Both the rain forest and taiga are forest climates. Outline the main differences between them.
9. What is a biome? Give appropriate examples.
10. What are the main differences between tropical and temperate grasslands?

CHAPTER
11
Tropical Climates

CHAPTER OUTLINE

206

Tropical regions have long had a certain mystique for mid-latitude peoples. The rich flora and fauna of certain tropical areas have encouraged some to envision a utopian economic development. Diseases such as malaria and yellow fever, fears of the debilitating effects of the tropical climate, and fears of ferocious insects, animals, and people have retarded development. Many of the diseases are under control. The climate is tolerable and, with air conditioning, comfortable indoors.

RADIATION AND TEMPERATURE

There are some distinctive attributes that characterize tropical climates in general, although there is no single tropical climate as such. The most important of these include the energy balance. There are several aspects of the energy balance that distinguish these climates from those farther toward the poles. One is that the influx of solar energy is high throughout the year, although it does vary with the seasons. In this sense, there are indeed climates without winter. Intensity of solar radiation is high all year. There is very little variation in the length of the day from one part of the year to the next. The **photoperiod**—the relative lengths of day and night to which plants must adjust—varies between 11 and 13 hours from winter to summer. Contrary to popular notion, although solar radiation is relatively high all year, sometimes it is not as high as it is in mid-latitudes, particularly in summer. In the equatorial lowlands, for instances, clouds can and usually do reflect more than half of the total solar radiation. In addition to the large amount of sunlight reflected by clouds, the high humidity and smoke from widespread burning further reduce solar radiation near the ground. At no time during the year do the tropical regions receive as many hours of sunlight as mid-latitude locations receive in the summer.

Annual temperatures average about the same throughout the tropical regions. There are slightly higher averages in the drier areas due to more intense radiation at the surface. The annual range in temperature depends on the length of the dry season. Where there is no dry season, the annual range in mean monthly temperatures may be as little as one or two degrees. Where there is dry weather in the winter months, the mean temperatures drop and the annual range increases.

Another significant aspect of the energy balance of tropical climates is that the primary energy flux is diurnal (Fig. 11.1). The variation in temperature from day to night is greater than the variation from season to season (Fig. 11.2). Tropical regions are not places of continuous high temperatures. Nights can be rather cool. The diurnal radiation and temperature cycles are more important than the annual temperature cycle as a regulator of life cycles. So much solar radiation is reflected and scattered that a light-skinned person has difficulty getting a tan. The ultraviolet radiation is largely filtered out.

PRECIPITATION

In tropical areas, rainfall rather than temperature determines the seasons. It is the amount and timing of rainfall that form the chief criteria for distinguishing the various climates. Contrary to popular belief, only a very small portion of the tropical regions has a year-round rainy season. For the continent of Africa, for instance, the area with substantial rainfall in each month is less than 10 percent of the total land area. The largest portion of tropical environments has a marked seasonal regime of rainfall that governs the biological productivity of the system. The remaining areas

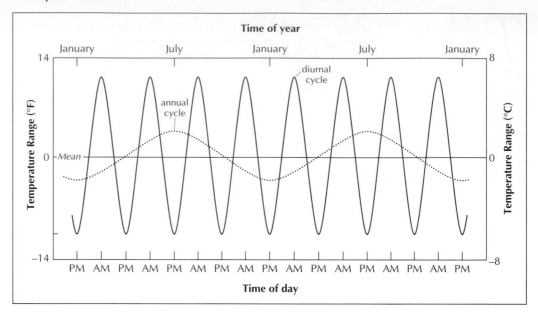

Figure 11.1
The relationship between the annual and diurnal energy periodicities in tropical regions.

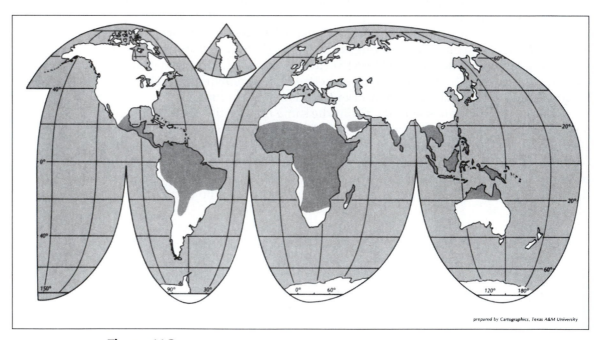

Figure 11.2
Shaded land areas experience a greater diurnal energy change than annual energy change.

are deserts, where rainfall is incidental throughout the year. In fact, it is the seasonal moisture pattern that distinguishes the major tropical environments—*rain forest*, *savannas*, and desert—from each other.

   The notion that these are monotonous, seasonless regions is far from the truth. Unlike the mid-latitudes, tropical areas lack snow storms, ice storms, and tornadoes. Change is an attribute of the tropical regions as well as anywhere else, and to the resident it as much a conversation item.

THE TROPICAL RAINFOREST CLIMATE

Tropical rainforests exist along the equator in Asia, Africa, and South America. These forests cover about 7 percent of Earth's land area. Extensive tracts of rainforest are in the Amazon River Basin, the Congo River Basin, and the East Indies. Half of the earth's tropical forests were located in four countries in 1990. These countries were Brazil, Indonesia, Peru, and Zaire. Smaller tracts exist at other sites. There are two different varieties of tropical rainforests: equatorial rainforest and tropical wet rainforest. Two thirds of the rainforests are classified as equatorial rainforests. The tropical wet rainforests are found on the margins of the equatorial rainforest and have a distinctly short dry season. The rainforest is also restricted to low elevations, usually below 1000 meters (3300 ft), because at higher altitudes temperatures are considerably lower. As altitude increases, the forest changes character. Trees become shorter in height, number of species decreases, and highland vegetation replaces the rainforest.

The rainforest climate refers to the luxurious evergreen forest typical of the wet tropical lowlands. The distribution of this climate is more limited than most people imagine because it is confined by a rather limited set of climatic conditions. The region has fairly even and high temperatures averaging between 20°C and 30°C (68°–86°F), a high frequency of precipitation year round, and total rainfall of over 200 centimeters (80 in). The climate is dominated by tropical maritime air masses (Table 11.1). So much solar radiation is reflected and scattered that a light-skinned person has difficulty getting a tan. The ultraviolet radiation is largely filtered out.

Part of the reason for the restricted geographical area is that the region must lie within the tropical convergence zone throughout the year. This does not happen over an extended amount of land because the convergence zone is continually shifting north and south. This migration is due to the passage of the overhead sun as schematically illustrated in Figure 11.3. Figure 11.4 shows the average extreme positions of the ITCZ. Where the migration of the ITCZ is maximum, the rainforest is the smallest in extent.

The Rainforest Biome

The abundant precipitation and high-intensity radiation year round in this region results in an evergreen forest of tremendous productivity and diversity. The tropical rainforests have been in existence for as long as 45 million years. In this time, millions of species of plants and animals have evolved to take advantage of every niche in the system.

The monthly average temperatures vary from 24°C to 30°C (75°–86°F), with an annual range of only 3°C (5°F) or so. Belem, Brazil, for instance, experiences ranges from the warmest to coolest month of only 1.6°C (3°F). Since there is greater variation in radiation within the day than from day to day, the diurnal range may be two to three times the annual range. The fairly even temperatures of the rainforest are due primarily to:

Table 11.1   Classification of Tropical Climates

	Köppen Type	Thornthwaite Type
Climates dominated by mT air masses	Af	AA′r, BA′r, CA′r
Climates with alternating seasons of mT and cT air masses	Aw, Am, BSh	BA′w, CA′w, DA′w
Climates dominated by cT air masses	BWh	EA′h, EB′d, EA′d, DA′d

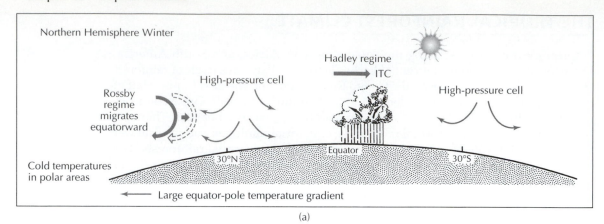

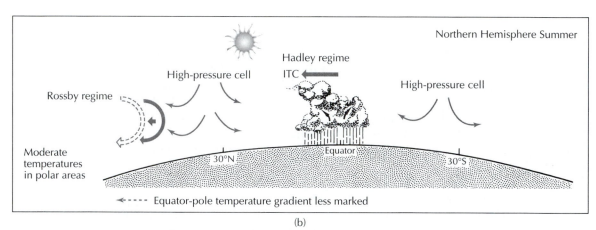

Figure 11.3
Schematic diagram showing the extent and migration of the Rossby and Hadley regimes in (a) winter and (b) summer.

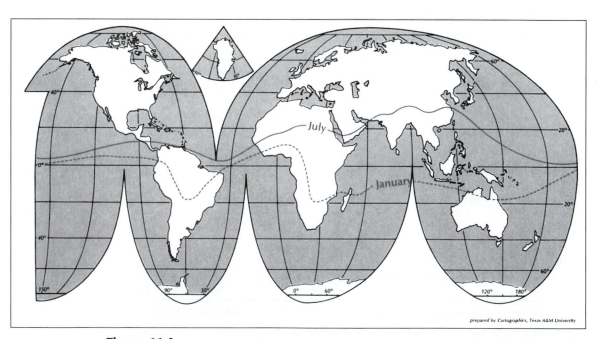

Figure 11.4
Mean extreme positions of the ITCZ (from Griffiths J. and Driscoll D., *Survey of Climatology*, Charles E. Merrill Publishing Co.).

1. the equal or nearly equal periods of daylight and darkness,
2. the uniformly high intensity of radiation throughout the year, and
3. the constantly high humidity.

Associated with the convergence of tropical maritime air are high humidity and cloudy skies. Since humidity is so high in the daytime, when nocturnal cooling occurs, early morning fogs and heavy dew are common. Dewpoint temperatures are in the range of 15°C to 20°C (59°–68°F).

The primary circulation determines the regional pattern of atmospheric circulation and the seasonal regime, and thunderstorms and easterly waves account for most of the day-to-day weather. Due to the degree of surface heating, local thermals, cumuloform clouds, and thunderstorms predominate. The cumulus clouds build during the daytime hours into towering cumulus and thunderstorms when the air becomes unstable. This process provides a distinctive diurnal rainfall pattern, such as that shown in Figure 11.5. Annual totals range up to two meters (6.6 ft). The highest amounts occur on mountain slopes such as Mount Waialeale, Hawaii, where the annual average is 11.8 meters (39 ft).

Average annual rainfall is not as important a factor as the frequency of precipitation. Precipitation occurs on more than 50 percent of the days. Duitenzorg, Java, averages 322 days a year with thunderstorms. Since the precipitation is largely from thunderstorms, it is of high intensity and short duration. Rainfall amounts of 50 to 65 millimeters (2.0–2.5 in) per hour occur at a return period of less than 2 years. Hourly rainfalls of 95 to 120 millimeters (3.7–4.7 in) are recorded in the East Indies. Although rainfall intensities for 1-hour periods are not significantly higher in Tropics than in mid-latitudes, sustained periods of high-intensity rainfall are significantly more common in tropical regions. The maximum 24-hour rainfall on record occurred at Cilaos, La Reunion, on March 16, 1952: 1870 millimeters (73.62 in) of rain fell (Table 11.2). Hail is fairly infrequent, but it does occur on occasion.

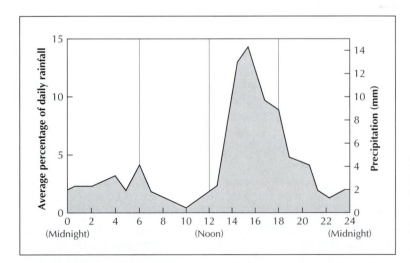

Figure 11.5
Diurnal distribution of rainfall at Kuala Lumpur, Malaysia. The midafternoon maximum is typical of many wet tropical locations.

Table 11.2 Precipitation Extremes Recorded in Tropical Wet Climates

Highest 12-hr total	Belouve, La Reunion Feb. 28–29, 1964	1340 mm (52.76 in)
Highest 24-hr total	Cilaos, La Reunion Mar. 16, 1952	1870 mm (73.62 in)
Highest 5-day total	Cilaos, La Reunion Mar. 13–18, 1952	3854 mm (151.73 in)
Highest number of rain days in a year	Cedral, Costa Rica 1968	355 days

In 1922, Mount Cameroon erupted in West Africa. The eruption produced a storm that left a snow cover on the summit, and at Debundscha at the foot of the mountain 14.53 meters (572 in) of rain were recorded during the year. Not all rainfall is convectional as there are periods of prolonged showers.

These forests produce a substantial part of the global oxygen supply and in the process remove carbon dioxide. These forests act to modify the climate in the regions where they are found and also influence the climate of the entire planet. An interesting aspect of this region is that the forest plays a significant part in the wetness of the climate. During the morning and early afternoon hours, this forest transpires huge quantities of water into the lower atmosphere. This added moisture aids in condensation of water in building cumulous clouds and afternoon and evening showers, and thunderstorms may drop copious amounts of precipitation back onto the canopy layer of the forest. Much remains there to start the process over again the next day. The regularity of these showers in some areas is such that one can almost set one's clock by them.

Species Diversity

The tropical rainforest is a multilayered forest of largely evergreen species. Most of the leaf matter in the rainforest is found in the canopy. The canopy is represented by the crowns of the dominant tree species. In an area of approximately one hectare ($2\frac{1}{2}$ acres) of equatorial rainforest, there may be as many as 200 tree species. Within the canopy is where most of the other vegetation and animal life are found. The canopy within a true rainforest is a mat of nearly continuous vegetation. Epiphytes such as orchids are most likely found in the canopy as are birds, monkeys, and tree snakes. This canopy may be over 30 meters (100 ft) above the ground. So thick is this layer that little light reaches the forest floor. The intensity of light on the forest floor may be as little as 10 percent that on top of the canopy. Extending above the canopy are the true giants of the rainforest whose crowns appear well above the surrounding canopy and are called emergents. It seems certain that over half of all species found on Earth are only found in the rainforests. It may be that the percentage is as high as 90.

Tropical forests are rich in species diversity. Tropical rainforests may include 50 to 80 percent of all plant and animal species on Earth. Half of Earth's bird species and 90 percent of primates inhabit the tropical rainforests. Half of the plant species that provide the basic food grains for humans originated in the tropical forests. These include rice and maize. Of an estimated 3 to 4 million species of organisms thought to exist in the tropical forests, only about 15 percent are yet classified.

Although there is little difference in the length of the photoperiod during the year, the variations that exist are important biologically. Even very small differences in length of day are enough to stimulate flowering or dormancy in plants. Evidence shows that many tropical plants are more sensitive to slight changes in day length than are plants found in higher latitudes. The common poinsettia is a tropical shrub for which reduced day length triggers flowering. In their natural habitat, these shrubs do not bloom at Christmas, but much later in the year. We force them to bloom before Christmas by controlling exposure to light and reducing the exposure artificially in early autumn.

Tropical Deforestation

The past, current, and projected destruction of this forest with all its species may be the most destructive act of humanity. These forests, which have been in existence for tens of millions of years, could effectively be destroyed to the place where they can no longer continue their present function by the year 2025. Destruction of the tropical rainforest is occurring rapidly. Some estimates suggest the annual loss to be an area equal in size to the state of West Virginia. In many tropical areas, 1 to 2 percent of the rainforest is cut or burned each year. This is equivalent to removal of 20 to 50 hectares each minute. The area of tropical rainforest in the Ivory Coast dropped 30 percent from 1956 to 1966. From 1980 to 1990, some 154 million hectares of forest land was cleared for other use. Studies now show that the area of degraded and fragmented tropical forest may be greater than the area deforested. This is extremely important in terms of loss of biodiversity.

The pressure to cut these forests comes from several sources. The need for land for farming and ranching is the major reason for removal. Second is the demand for tropical wood for lumber. Between 10 and 15 million hectares of land are cleared of forest in the Tropics each year for agriculture and pasture. Brazil and Indonesia account for about 45 percent of global rainforest depletion. In India between 1972 and 1982, nine million hectares of forest were cleared. This is about a fourth of the forest that existed in 1972.

Deforestation in the Amazon River Valley

The Amazon rainforest is the largest continuous stand of forest left on the face of the earth. In 1980, the area of the remaining forest was some 5 million km^2—an area nearly half the size of the United States. Brazil contains about 30 percent of Earth's tropical forest. The forest covers about 40 percent of the land in Brazil. It extends into Bolivia, Columbia, French Guyana, Guyana, Peru, Surinam, and Venezuela.

For several decades, the government of Brazil envisioned the economic development of this vast region. To begin development, the 2500-kilometer Trans-Amazon highway project was begun in 1970 and completed in 1974. The highway was funded by the World Bank. The highway, known as BR-324, runs from eastern Brazil to the western state of Rondonia. The highway is part of a scheme to develop the northwest section of the rainforest. Included in the plan was the clearing of 160,000 square miles of forest. The road was initially a 900-mile dirt road of poor construction. It often washed out during the rainy season. Even in its poorest form, it cut the travel time from the eastern part of the country to Rondonia from weeks to days. The road was paved in 1982 and a flood of people moved along the road, clearing land as they went. They built the highway partly to move people out of the overcrowded northeastern part of the country. The result of constructing this road and the thousands of kilometers of feeder road was disaster for the forest. A population explosion of humans and livestock followed the construction of the road. Between 1966 and 1978, settlers cleared 80,000 km^2 of forest. Most of the land was planted in grass for cattle ranching. Some estimates place the amount of original forest already cut or burned as high as 30 percent.

The government of Brazil encourages deforestation. To encourage development, the government provided incentives to foreign investors to explore for minerals and harvest the timber. The rate of deforestation has increased rapidly since 1980. Brazilian leadership recently announced a controversial new plan to harvest an additional 40 million hectares of timber. At this same rate of removal, the entire forest in South America will be gone by 2050.

The Galapagos—An Equatorial Dry Region

As with most generalizations, there are exceptions to the rules about where certain types of climate are actually found along the equator. Some extremely dry environments exist where we expect to find the humid tropics. Herman Melville wrote of "The Encantadas" (Galapagos Islands):

> But the special curse, as one may call it, of the Encantadas, that which exalts them in desolation above Idumea and the Pole, is that to them change never comes; neither the change of seasons nor of sorrows. Cut by the equator, they know not autumn and they

> know not spring; while already reduced to the lees of fire, ruin itself can work little
> more on them. The showers refresh the deserts but in these isles, rain never
> falls. ... Nowhere is the wind so light, baffling, and in every way unreliable, and so
> given to perplexing calms, as at the Encantadas. (p. 182)

A shack inhabited by a woman stranded on one island, used a system for collecting dew to provide fresh water:

> ... and here was a simple apparatus to collect the dews or rather doubly distilled
> and finest winnowed rains, which in mercy or in mockery, the night skies some-
> times drop upon these blighted Encantadas. All along beneath the eave, a spotted
> sheet, quite-weather-stained, was spread, pinned to short, upright stakes, set in
> shallow sand. A small clinker, thrown into the cloth, weights the middle down,
> thereby straining all moisture into a calabash placed below. This vessel supplied
> each drop of water ever drunk upon the site by the Choles. Hunilla told us the cal-
> abash would sometimes, but not often, be half-filled over-night. It held six quarts,
> perhaps. (p. 231)

Global Warming in the Tropics

Just as in most regions of the earth, the effects of a warming planet are apparent. Throughout the Tropics, there are mountains that support glaciers. Many of these mountains are famous as destinations for tourists and in literature. Melting is now taking place more rapidly than in the past. In Africa, Mt. Kenya supported a large glacier on its peak, and it is now all but gone. Mt. Kenya lies just a few miles north of the equator. Just to the south of the equator on the border between Kenya and Tanzania lies Mt. Kilimanjaro. Famous for its majestic shape with a snow- and ice-capped crown, it is now also undergoing change as the snow and ice are melting at an unprecedented rate. Around the base of the mountain and part way up the flanks is an intensive agricultural region based on the rich volcanic soil found there. Meltwater from the snow and ice provides a steady supply of irrigation water. As the snow and ice retreat to higher elevations, more of the precipitation occurs as rain. The rain runs off almost immediately and so is lost for irrigation purposes. The livelihood of thousands is in jeopardy.

In South America, the Quelccaya ice cap in the central Andes is melting at a more rapid rate than in the past. The same is true in Irian Jaya, where three glaciers have experienced a 50 percent increase in the rate of retreat of the glacier fronts since the 1920s.

THE TROPICAL WET AND DRY CLIMATE

The distinguishing feature of this climate is a very pronounced seasonal moisture pattern. Atmospheric humidity, precipitation, soil moisture, and stream flow change through the year in a rhythmic pattern. These environments are found next to the tropical rainforest environments and in most cases poleward of them. They occupy much of the area from the boundary with the tropical rainforest to the Tropics of Cancer and Capricorn. They occupy a much larger area than that of the rainforest. There are areas with this type of climate north and south of the Amazon Valley in South America, across Africa north and south of the Congo Basin, and in much of southeast Asia and parts of the Pacific islands. Their location in the general circulation is such that they experience the kinds of weather found in both the equatorial convergence zone and the subtropical divergence zone.

The tropical wet and dry climate has seasons dominated alternately by two maritime tropical and continental tropical air masses (Table 11.3). The seasonal pattern of moisture is due to the migration of the tropical convergence zone. The rainy season is concurrent with the high sun and the presence of the convergence zone. The dry season is a product of the more stable air from the subsidence in the subtropical highs. Seasonality increases away from the equator. In a traverse away from the equator, the low sun precipitation begins to decrease first, with the high sun precipitation remaining as high as in the tropical convergence zone. Between the equatorial convergence zone and the tropical desert, the winter precipitation drops to near zero. From that point poleward, the high sun precipitation declines until it is no longer significant and we find the tropical desert.

This climate has the most pronounced seasonality of precipitation of any climatic type. Rangoon provides an example of the extremes of precipitation that occur in these regions. Rangoon, Burma, has a 3-month winter average of 25 millimeters (1 in) and a 3-month summer average of 1880 millimeters (74 in). The extremes increase northward to Akyab, where the averages are 3.8 centimeters and 426 centimeters (1.5 in and 167 in). The greatest extremes in the Asian region are at Cherrapunji, India. In two winter months, they receive an average of only 2.5 centimeters (1 in) of precipitation, but in the two summer months they get 528 centimeters (207 in) of rain. Cherrapunji recorded a 5-day total of 405 centimeters (159 in) in August 1841 and a 1-month total of 915 centimeters (30 ft) in July 1861. In 1 year, from August 1860 to July 1861, they measured 26 meters (85 ft) of rain.

The average annual precipitation varies so much from place to place that it cannot be used as a criterion for distinguishing the region. On the wet margins where topography is favorable for orographic increase in rainfall, totals run over 1000 centimeters (400 in). On the dry margins, it drops to less than 25 centimeters (9.8 in). In fact, it is quite likely that the highest annual average precipitation totals on Earth occur where there is a strong seasonal pattern of precipitation. Variation in annual rainfall is higher here than in the tropical rainforest. The precipitation total in any year is subject to the extent of migration of the general circulation and the length of time the zone of convergence stays over an area. The farther a site is

Table 11.3 Monthly Data for Tropical Wet-and-Dry Climates

	Jan.	Feb.	Mar.	Apr.	May	June	July	Aug.	Sept.	Oct.	Nov.	Dec.	Yr.
Calcutta 22°32′ N 88°22′ E: 6 m													
Temp. (°C)	20.2	23.0	27.9	30.1	31.1	30.4	29.1	29.1	29.9	27.9	24.0	20.6	26.94
Precip. (mm)	13	24	27	43	121	259	301	306	290	160	35	3	1582
Cuiaba, Brazil 15°30′ S 56°03′ W: 165 m													
Temp. (°C)	27.2	27.2	27.2	26.2	25.5	23.8	24.4	25.5	27.7	27.7	27.7	27.2	26.5
Precip. (mm)	216	198	232	116	52	13	9	12	37	130	165	195	1375
Dakar, Senegal 14°39′ N 17°28′ W: 23 m													
Temp. (°C)	21.1	20.4	20.9	21.7	23.0	26.0	27.3	27.3	27.5	27.5	26.0	25.2	24.3
Precip. (mm)	0	2	0	0	1	15	88	249	163	49	5	6	578
Darwin, Australia 12°26′ S 131°00′ E: 27 mm													
Temp. (°C)	28.2	27.9	28.3	28.2	26.8	25.4	25.1	25.8	27.7	29.1	29.2	28.7	27.6
Precip. (mm)	341	338	274	121	9	1	2	5	17	66	156	233	1562

from the heart of the convergence zone, the lower the annual average precipitation, the shorter the rainy season, and the greater the annual variation. The weather is, of course, very different from season to season. During the rainy season, it is warm and humid, and there are frequent rainstorms. During the dry season, more or less desert conditions exist.

The lag of the rainy season behind the migration of the sun produces three seasons in many parts of the wet and dry Tropics. There is a cool season, hot season, and rainy season. The cool season is during the winter months when solar radiation is at a minimum. The hot season follows when temperatures rise as the sun moves higher in the sky and solar radiation increases. The onset of the wet season brings an increase in cloud cover, slightly reduced radiation, and cooler temperatures.

Where moist air of oceanic origin exists, precipitation occurs. Where the dry continental air is usually present, precipitation is largely absent. Often there is a sharp boundary between the two types of air. The **Hadley cells** north and south of the equator and the convergence zone between them shift north and south, following the migration of the vertical rays of solar energy. This migration produces the seasonal pattern of precipitation that characterizes so much of the tropical region. The system moves through 20° of latitudes through the year from about 5°S to 15°N.

THE TROPICAL DESERTS

World deserts cover more than one fourth of all the land area of the continents. Because they cover more area than any other single climatic type, there is concern that the current deserts may be expanding. Deserts exist at all latitudes between 50°N and 50°S, with the largest found between 30° north and south latitude (Table 11.4). The heart of the tropical deserts lies near the Tropics of Cancer and Capricorn primarily toward the west sides of the continents. They are less common on the east side of the land masses because the trade winds carry considerable amounts of moisture on shore. There are several characteristics that distinguish the tropical deserts. They are:

1. low relative humidity and cloud cover,
2. low frequency and amount of precipitation,
3. high mean annual temperature,
4. high mean monthly temperatures,
5. high diurnal temperature ranges, and
6. high wind velocities.

Table 11.4   Major Deserts of the World

Desert	Approximate Area	
	km² (thousands)	mi² (thousands)
Sahara	9100	3500
Central Asia	4510	2200
Australian	3400	1300
North American	1300	500
Patagonian	680	260
Indian	600	230
Kalahari-Namib	570	220
Atacama	360	140

The basic controls of climate in the tropical deserts are the upper air stability and subsidence. Divergence, general stability, and low relative humidity are basic characteristics of the tropical desert. Relative humidity averages 10 to 30 percent in the interior areas, but is slightly higher in coastal locations. It has been measured as low as 2 percent. With low relative humidity, cloud cover is also low. The Sahara averages 10 percent cloud cover in winter and 4 percent in the summer.

Precipitation is very low in amount and very sporadic in time and space. Low annual precipitation is the basic characteristic of these climates, but there are occasionally heavy downpours. Arica, Chile, receives an average of 5 millimeters (0.2 in) of precipitation per year. Iquique, Chile, experienced a 14-year period with no rainfall, and Wadi Halfa in the Sahara Desert experienced a 19-year period with no rainfall. Iquique, on another occasion, received 100 millimeters (4 in) of rain in 1 day. One rain may bring 125 to 250 millimeters (5–10 in) of precipitation; then for a period of years no precipitation will fall. One station in the Thar Desert with an annual average of 100 millimeters rainfall received 850 millimeters in 2 days. At Dakhla, in southern Egypt, where the average is 100 millimeters, one 11-year period elapsed with no rain. General rains over large areas of the desert are infrequent, but they do occur. In February 1980, a widespread rainstorm struck eastern Saudi Arabia with paralyzing results. Streets in the major cities flooded since there are no storm sewers to carry off the water, and water blocked the major highways from Riyadh to Dhahran. There is a greater frequency of precipitation on the equatorward margins of the deserts in the summertime as the ITCZ moves over the area. On the poleward sides, the precipitation is mainly in the winter because the major mechanism for precipitation is the mid-latitude cyclone. The precipitation that does fall is primarily convective on the equatorward side, with only occasional cyclonic precipitation. As latitude increases, the frequency of cyclonic precipitation also increases.

Seasonal weather in the deserts is more evident in temperature than with precipitation (Table 11.5). Winter has slightly cooler days and much cooler nights. The highest average annual temperatures on Earth are found in the tropical deserts. They vary between 29°C and 35°C (84°F–96°F). At Lugh Ferrandi in Somalia, the temperature averages 31°C (88°F), which is considerably higher than in any of the other tropical climates. The major factor controlling the average annual temperature is latitude, and the stations with the highest averages are those closest to the equator. The highest official temperature yet recorded is 58°C (136°F) at El Azizia near

Table 11.5 Monthly Data for Tropical Desert Climates

	Jan.	Feb.	Mar.	Apr.	May	June	July	Aug.	Sept.	Oct.	Nov.	Dec.	Yr.
Al-Hofuf, Saudi Arabia 25°22′ N 49°35′ E: 145 m													
Temp. (°C)	14.0	15.9	20.6	25.4	30.8	33.5	34.6	33.9	31.3	26.8	20.9	16.1	25.3
Precip. (mm)	23	8	16	16	1	0	0	0	0	1	1	6	49
Marrakesh, Morocco 31°37′ N, 08°00′ W: 458 m													
Temp. (°C)	11.5	13.4	16.1	18.6	21.3	24.8	28.7	28.7	25.4	21.2	16.5	12.5	19.9
Precip. (mm)	28	28	33	30	18	8	3	3	10	20	28	33	242
Alice Springs, Australia 23°38′ S, 133°35′ E: 570 m													
Temp. (°C)	28.6	27.8	24.7	19.7	15.3	12.2	11.7	14.4	18.3	22.8	25.8	27.8	20.8
Precip. (mm)	43	33	28	10	15	13	8	8	8	18	30	38	252

Tripoli in northern Africa on September 13, 1922. In the United States, the highest recorded temperature is 56.7°C (134°F) in Death Valley, California. Winter averages are below those of other parts of the tropical regions as Earth radiation at night rapidly cools these areas. Averages are as low as 15°C to 20°C (59°F–67°F). The lower temperatures of winter give the tropical deserts the highest annual range found in the tropical regions. Aswan, Egypt, located on the Tropic of Cancer, has an annual range of 19°C (34°F).

Diurnal ranges in the deserts are the highest of any of the climates; as in other tropical climates, they far exceed the annual range. The average diurnal range is 14°C to 25°C (25°F–45°F), but occasionally it is much greater. At Birmirha in the Sahara south of Tripoli, the temperature went from 37.2°C down to −0.6°C (99°F–31°F) in one 24-hour period—a range of 37.2°C (68°F). The largest known 24-hour drop in temperature occurred at the oasis of Salahin in the Sahara Desert on October 13, 1927. The temperature dropped from an afternoon high of 52.2°C to −3.3°C (126°F–26°F) the following morning—a range of 55.6°C (100°F).

COASTAL DESERTS

Along coastal sections of the tropical deserts there is a rather distinctive set of atmospheric conditions that are different from the rest of the tropical deserts. These desert areas have cool temperatures, shallow temperature inversions, cold water offshore, considerable fog and stratus cloud cover, and little rain (Table 11.6). In fact, the areas with the lowest annual rainfall of all the deserts are found here. Baja California, Ecuador, Peru, and the Sahara and *Namib deserts* of Africa have zones with this type of climate. The primary factor in producing these conditions is the flow of cool air onto the land. The air is chilled by crossing a cold current flowing along the coast. One effect of this onshore flow of air is to cool the area during the summer months, dropping the annual mean temperature by 5°C to 10°C. It also reduces the annual range in temperature. At Iquique, Chile, the relative humidity averages 81 percent in August, yet the rainfall averages only 28 millimeters (1.1 in) per year. During a 20-year period, there was no measurable precipitation. Caloma, Peru, averages 48 percent relative humidity in August, and no measurable precipitation has been recorded. These coastal deserts have extremely low precipitation, even for

Table 11.6 Comparison of Temperature and Relative Humidity in Tropical Deserts and Coastal Deserts

	Average of Warmest Month			Average of Coolest Month			Mean Annual Temp. (°F)	Annual Temp. Range (°F)
		R.H. (%)			R.H. (%)			
	T(°F)	0700	1400H	T(°F)	0700	1400H		
Tropical deserts								
In-Salah (Sahara)	98	36	25	56	63	37	77	42
Riyadh (Arabian)	93	47	31	58	70	44	75.5	35
Laverton (Gr. Australian)	87	36	24	52	60	43	74.5	35
West coast deserts								
Arica (Atacama)	72	74	61	60	83	74	66	12
Walvis Bay (Kalahari)	66	91	73	58	83	65	62	8
Villa Cisneros (Coastal Sahara)	72	88	63	63	75	51	67	9

desert areas, and a much higher percentage of cloud cover. Humidity is often in the 80 to 90 percent range, and extensive fog is common. Near Lima, Peru, in July and August, fog often covers the city for days at a time. When there are optimum conditions, the fog will pass over the city and penetrate for miles into the valleys of the Andes Mountains to the east of the city. These are unusual climatic areas and do not cover large areas of the earth's surface.

DESERT STORMS

Severe storms are not frequent in the deserts primarily because of the lack of moisture to supply the energy. However, deserts are very windy, particularly in the afternoon and during the hottest months. Common features of the air over deserts are a steep lapse rate and instability in the air near the surface. **Dust devils** are the most visible form of the turbulence. Developing in clear air with low humidity, they become visible due to the debris they carry. They can reach a height of nearly 2 kilometers. On occasion they will sustain velocities high enough to blow down shacks or blow screen doors off their hinges. These dust devils are very common in deserts, and it is not unusual to see a multitude of them at one time under the right atmospheric conditions. The dust devil results when there is an intense thermal at or near the ground surface, and surrounding air moves inward to replace the rising air. As the air moves in toward the thermal, the radius of curvature decreases and velocity increases.

Other storms of the tropical deserts are the dust storms and sand storms. Dust storms occur as a result of **deflation**. One of the readily visible aspects of much desert surface is the lack of fine sediments. The wind keeps the surface swept clean. Winds blowing out of the deserts are often dust-laden. On the south side of the Sahara, they call the dry wind the **harmattan**. Since it is dry wind, it is cooling and welcome, although it may limit visibility and leave household furnishings covered with dust. These same winds blow out of the north side of the Sahara Desert. Occasionally, they travel across the Mediterranean Sea bringing disaster to agricultural crops. They are important enough in the climate of Mediterranean countries to warrant local names such as **sirocco** in Italy and Yugoslavia and **leveche** in Spain.

In the sandy parts of the deserts, it is the wind's ability to move sand that is significant. Sand dunes cover only a small part of the tropical deserts. Sand dunes cover something less than 30 percent of the Sahara Desert and only about 2 percent of the Sonoran Desert of North America. Sand storms only occur when wind velocities reach a high enough value to move the sand. The velocity at which sand begins to move is the threshold velocity and depends on the size of the sand particles, wetness of the sand, and other less important variables. For medium-sized dry sand (0.25 mm), the threshold velocity is 5.4 meters per second and is exceeded some 30 percent of the time. **Sand drift** increases rapidly as wind velocities increase above threshold velocity. We measured an average rate of 12.0 liters per meter width for winds 5.5 to 6.4 meters per second and a rate of nearly 10 times that for wind velocities of 10 meters per second. The increase is rapid because above the threshold velocity the power of the wind increases as the cube of the wind velocity.

Two kinds of sand storms develop in the tropical deserts. One is the result of surface heating and the resultant turbulence. It is not unusual for sand temperatures at the surface to reach 85°C (180°F). This type of sand storm is chiefly a daytime phenomenon and is most pronounced in the hottest months. At Al-Hofuf, Saudi Arabia, the percentage of hours in which sand drift occurs between 0600 and 1800 hours ranges from 76 percent in February to 91 percent in June. The total number of hours per day when winds exceed the threshold velocity also has a seasonal pattern, increasing from 3 hours per day in February to 9.4 hours per day in June. As a result of the increasing number of hours of high-velocity winds in summer, sand movement increases in summer. In the Nafud, it increases from 116 liters per meter per day in February to 406 liters per meter per day in June. In summer, the high frequency of afternoon sand storms makes travel and other outdoor activity particularly difficult.

The second type of sand storm is brought about by low-pressure disturbances passing through the area. These storms often generate higher velocity winds than does diurnal heating, and they last for longer periods. One such sand storm in the Nafud lasted for 43 hours, during which sand blew constantly. The wind averaged 10.7 meters per second (24 mph) and exceeded 15.6 meters per second (35 mph) for hours at a time.

Both types of sand storms have the ability to move huge amounts of sand. In the Nafud, the wind moves an estimated 80 m^3 of sand across each meter width of the sand field each year. The dunes move at fairly high speed. In the Nafud, dunes with an average height of 10 meters (33 ft) move at a rate of 15 to 19 meters each year.

DESERTIFICATION

In many parts of the world, particularly on the margins of the deserts, a process of environmental degradation called **desertification** is taking place. The process consists of the breakdown in the vegetal cover and soil until the land is no longer productive, and erosion by water and wind produce relatively sterile land. Desertification does not involve the outward expansion of the climate conditions responsible for the core areas of the world deserts. Rather, it is the alteration of the land to desertlike character by human activity.

The nature of the ecosystems in much of the Sahelian zone is such that they will not support many people on a subsistence basis. Dry-land cultivation practices are mostly both primitive and damaging to the land. They clear areas and farm them for several years, during which the soil gradually deteriorates and loses fertility. When the soil is exhausted, they abandon it on the assumption that it will rejuvenate with time. If the amount of land cultivated in any given year is small enough and the number of livestock units on the grassland is below the carrying capacity, the ecosystem will suffer little long-range damage.

Desertification takes place when the numbers of people and livestock exceed the capacity of the precipitation to support them. It is not a new problem, but the rate at which it is taking place is unprecedented. Before colonization of Africa, there were intermittent periods of environmental degradation. Whenever a dry cycle began, there was too much grazing and cultivation, which was followed by loss of vegetation and soils. Population increased during wet periods and decreased during drought (Fig. 11.6).

Figure 11.6
Population and carrying capacity in the precolonial era. Population size followed the precipitation cycles. When favorable periods occurred, population expanded; when dry periods followed, the population died back. The shaded areas represent times of major environmental degradation.

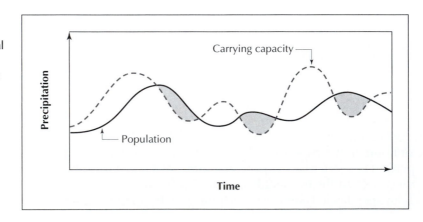

If an area became unproductive, the population either died back or moved away. The introduction of colonial administration brought some measures that altered the balance between the population and environment. They reduced intertribal warfare and introduced public health measures. These measures resulted in rapid population growth. In the year 2000, several of the Sahelian countries had the highest population growth rates in the world. Parallel with the increase in human numbers is an increase in animal populations.

The result of the rapid population growth in the face of varying rainfall is that the demands on the ecosystems become greater than the carrying capacity more frequently and for longer periods. As shown in Figure 11.7, the demand ultimately exceeds the carrying capacity on a continuing basis.

The result of the increasing pressure on the land is the expansion of the Sahara Desert southward into more humid areas. If human activity is not responsible for this phenomenon, it certainly accelerates it. Degeneration into desert usually occurs in scattered patches of bare ground from a few meters to several kilometers across. It is not desert expansion in the form of an even front of desert surface advancing across the landscape. Hence, it is not easy to measure. Desertification is taking place at a steady rate in many areas of the earth's grasslands and occurs on every continent except Antarctica. It becomes particularly rapid during periods of major drought.

When the plant cover falls below the minimum required for protection of the soil against erosion, the process becomes irreversible. The soil holds less water. It dries out. The temperature increases speeding the collapse. Bare soil appears and wind and water remove the fine soil particles. This loss of topsoil is large and very important. The movement of dust by wind has existed throughout historic times, and hence is by no means a recent phenomenon. Winds move the dust picked up from the Sahara and adjacent lands for great distances. The amount of dust carried by the atmosphere is large. Some of this dust moves across the Atlantic Ocean to the Americas. The total drift westward off the African coast may be as much as 60 million metric tons each year. It is extensive enough to be measurable in the Caribbean Sea.

Wind erosion is not restricted to dust-sized particles. When large patches of bare ground form, the soil is exposed to erosion of the larger sand-sized particles, humus, and mineral salts. Barren deflation pans then dot the surface. During the

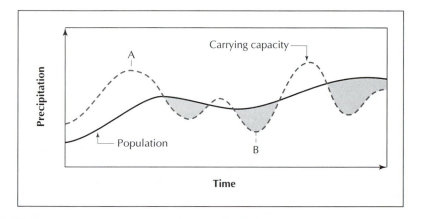

Figure 11.7
Population and carrying capacity following colonization when conditions changed to greatly increase population growth rates. At Point A, when the carrying capacity begins to decline, various forms of aid to the distressed areas keep down the death rate. Thus, when precipitation increases again at Point B, a much larger carry-over population remains from which to continue growth. The result is a population greater than the environment can support even under more favorable conditions. Desertification becomes even more accelerated.

height of the Sahelian drought, satellite images showed thousands of square kilometers of surface obscured by blowing sand. Blowing sand and silt further destroy the standing vegetation by stripping leaves or burying it under dunes.

SUMMARY

Tropical climates cover a large portion of the earth's surface between latitudes 25° north and south. In most of this region, temperatures change as much or more from day to night as from summer to winter. Radiation intensity is relatively high all year. Within the tropical realm, the precipitation regime defines the climatic regions. Some places have a high chance of precipitation every day of the year, and others have an equally low chance of precipitation on any given day of the year. Between these extremes are the vast areas with a rainy season and a dry season. In the wet and dry Tropics, there is wide range in both the length of the rainy season and in total precipitation received during the rainy season. The monsoon regions of Asia represent the extreme example of seasonal precipitation, and it is there that some of the highest amounts of annual precipitation occur. The driest areas of the world are in tropical regions. The coastal margins of the tropical deserts include stations with the lowest known annual total precipitation.

Deforestation of the tropical rainforests is a critical problem facing the human species today. The tropical forests not only affect the regional climate, but provide humanity with many crucial medicines and other products. If the forests are destroyed, a major planetary biome will cease to function.

KEY TERMS

Deflation	Harmattan	Sand drift
Desertification	Leveche	Savanna
Dust devils	Photoperiod	Sirocco
Hadley cells	Rainforest	

REVIEW QUESTIONS

1. Sand drift in deserts occurs under two different sets of conditions. What are these two different sets of conditions?
2. What is the primary reason that coastal deserts average less rainfall than interior deserts?
3. Why is the intensity of solar radiation high throughout the year in the Tropics?
4. What factors determine the difference between the different tropical climates?
5. Of the three main types of tropical climate, which is the smallest in geographical area?
6. Does most precipitation occur in the summer or winter season in the wet and dry Tropics?
7. In which tropical climates do the highest temperatures occur?
8. What is the difference between a dust devil and a sand storm?
9. In which of the tropical climates is desertification most common?
10. On which continents is desertification taking place?

CHAPTER 12

The Mid-Latitude Climates

CHAPTER OUTLINE

*T*he climates found in mid-latitudes differ from those of tropical regions in two major respects. First, in tropical regions, the variation in radiation and temperature is greater from day to night than it is from season to season. In mid-latitudes, the seasonal change is greater than diurnal variation (Fig. 12.1). In winter in mid-latitudes, there is much less solar radiation (see chap. 2). Second, in mid-latitudes, the primary circulation is quite different from that of the Tropics. The subtropical high over the oceans and polar high act as major source regions for air moving toward the convergence zone centered in the latitudinal range of 50° to 60°. These two source areas are very different in character. One is very warm and the other quite cold. Although this is enough to produce different air streams in mid-latitudes, differences in moisture further distinguish the air masses. Thus, in the tropical convergence zone, the converging air streams are quite similar in temperature and moisture, but such is not the case in mid-latitudes.

GENERAL CHARACTERISTICS

Mid-Latitude Circulation

Westerly upper air flows of the Rossby regime dominate the circulation in mid-latitudes. The equatorward limit of this regime is about at the axis of the zone of divergence associated with the semipermanent high-pressure cells. These are located beneath the **subtropical jet** stream. Equatorward of this axis, the Hadley circulation operates.

The temperature gradient between the equator and the poles determines the location of the subtropical jet stream. As polar areas become warmer in summer, the gradient lessens and the strength of the westerlies diminishes. The jet migrates poleward to attain the average position shown in Figure 12.2a. When

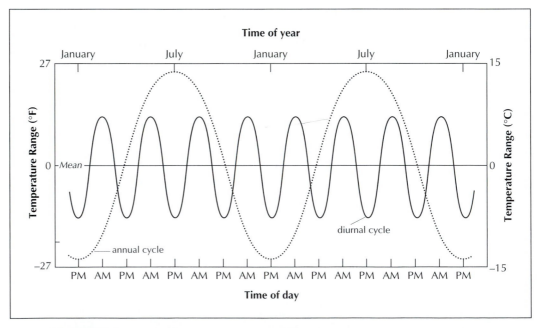

Figure 12.1
The relationship between the annual and diurnal energy cycles in mid-latitudes.

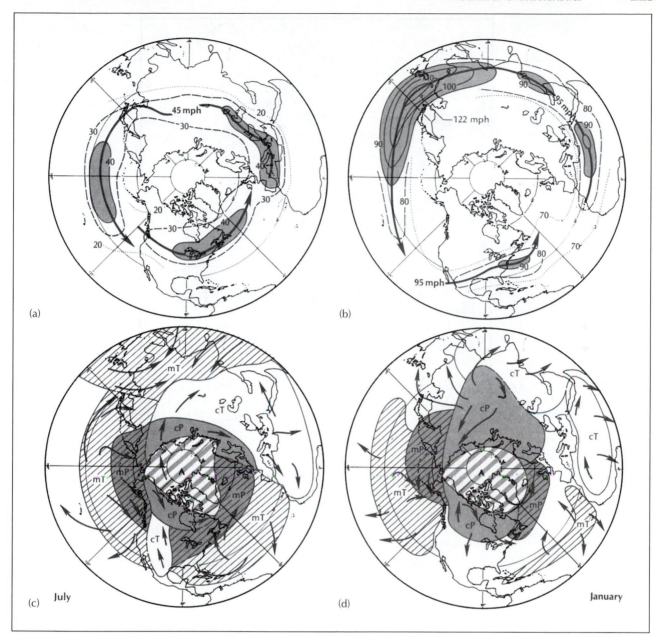

Figure 12.2
Upper: The mean jet stream in summer (a) and winter (b) (after Namias J. and Clapp
P. F., *Source: Introduction to Meterology*, 3e., © 1969, by S. Petterssen, N.Y., McGraw-Hill
Book Co.).
Lower: Principal northern hemisphere air mass source regions in (c) summer and
(d) winter (*Source: Introduction to Meteorology*, 3e., © 1980, Cole F. W., John Wiley &
Sons).

a very steep gradient exists, the westerly circulation is strong and the jet migrates closer to the equator. Figure 12.2b shows the mean jet stream for the northern hemisphere in winter.

Surface circulation patterns are closely related to the flow of the upper air
westerlies in mid-latitudes. The migration of the subtropical jet and the
weakening of the westerly circulation provide very different conditions from

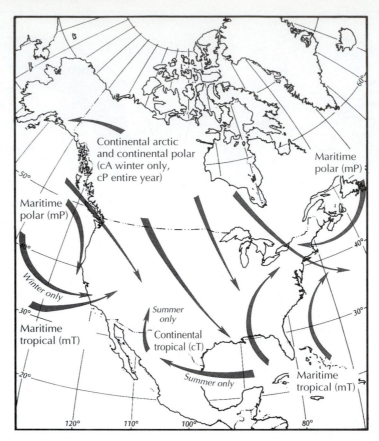

Figure 12.3
Source regions and paths
of air masses affecting
North America.

season to season. The movement and relative dominance of air masses varies, as does the frequency and path of traveling low-pressure systems. Figures 12.2c and 12.2d show some of the changes that occur in air mass dominance over the northern hemisphere. For example, note the difference in extent of cP air masses in winter and summer. Figure 12.3 illustrates the main source regions for North American air masses, and Table 12.1 provides some of the attributes of these air masses.

Summer

Summer in mid-latitudes has many of the characteristics of the tropical climates. At the time of the summer solstice, radiation intensity at 47°N and 47°S is as high as it is at the equator. Not only is the radiation intensity high, but the duration of radiation is longer than it is at the equator on this date (see Fig. 12.4). The inequalities in length of day and night increase with latitude. This is added to the variation in solar intensity in such a way that it compounds the seasonal differences. During the summer months, solar intensity and hours of daylight are high. The result is greater radiation reaching the surface in mid-latitudes than in the equatorial zone in summer. During summer, temperatures average in the 20° to 25°C (68°–77°F) range at most stations. Along the equatorward margin, the warm-month temperatures may average 26° to 30°C (78°–86°F)—comparable to rainy Tropics. The warm-month means drop slowly in a poleward direction. Only at latitudes greater than 45° will the summer averages drop below 20°C (68°F). July averages differ only slightly from the Gulf Coast to the Canadian border (Fig. 12.5).

Table 12.1   Weather Characteristics of North American Air Masses

Air Mass	Source Region	Temperature and Moisture Characteristics in Source Region	Stability in Source Region	Associated Weather
cA	Arctic basin and Greenland ice cap	Bitterly cold and very dry in winter	Stable	Cold waves in winter
cP	Interior Canada and Alaska	Very cold and dry in winter Cool and dry in summer	Stable entire year	a. Cold waves in winter b. Modified to cPk in winter over Great Lakes bringing "lake-effect" snow to leeward shores
mP	North Pacific	Mild (cool) and humid entire year	Unstable in winter Stable in summer	a. Low clouds and showers in winter b. Heavy orographic precipitation on windward side of western mountains in winter c. Low stratus and fog along coast in summer; modified to cP inland
mP	Northwestern Atlantic	Cold and humid in winter Cool and humid in summer	Unstable in winter Stable in summer	a. Occasional "nor'easter" in winter b. Occasional periods of clear, cool weather in summer
cT	Northern interior Mexico and southwestern U.S. (summer only)	Hot and dry	Unstable	a. Hot, dry, and clear, rarely influencing areas outside source region b. Occasional drought to southern Great Plains
mT	Gulf of Mexico, Caribbean Sea, western Atlantic	Warm and humid entire year	Unstable entire year	a. In winter it usually becomes mTw moving northward and brings occasional widespread precipitation or advection fog b. In summer, hot and humid conditions, frequent cumulus development and showers or thunderstorms
mT	Subtropical Pacific	Warm and humid entire year	Stable entire year	a. In winter it brings fog, drizzle, and occasional moderate precipitation to N.W. Mexico and S.W. United States b. In summer it occasionally reaches western U.S., providing moisture for infrequent conventional thunderstorms

Source: Lutgens F. K., Tarbuck E. J., *The Atmosphere,* 2e., © 1982, p. 203. Reprinted by permission of Prentice-Hall, Inc., Upper Saddle River, N.J.

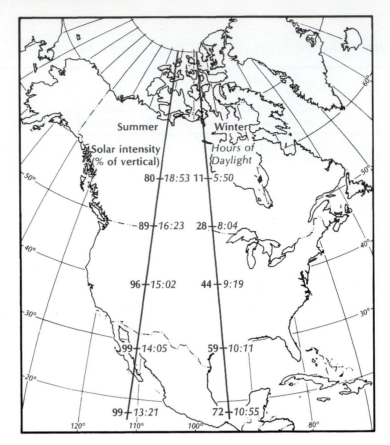

Figure 12.4
The length of daylight in summer and winter, and the relative intensity of solar radiation compared to a vertical solar beam.

Precipitation decreases toward the poles and toward the interior of the continents. The primary reason for this decrease is the increasing distance from the source of mT air. Areas on the west sides of the continents that have their precipitation from mP air normally have less total precipitation than areas receiving precipitation from mT air. This is because the mP air is cooler and contains less water vapor. The frequency of thunderstorms decreases rapidly from south to north. Cyclonic precipitation is most frequent at the polar margins and is more

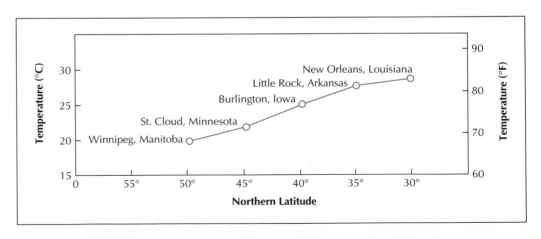

Figure 12.5
Mean July temperature along a north–south transect from New Orleans, Louisiana, to Winnipeg, Manitoba.

frequent in all sections of the mid-latitudes in the winter. Hurricanes provide another precipitation mechanism, and some of the heaviest rains along the coastal areas result from hurricanes.

Winter

Winter truly sets the mid-latitude zone apart from the Tropics. All areas in the mid-latitudes experience a range of 47° in the angle of the solar beam between the summer solstice and the winter solstice. This varying angle produces changes in radiation intensity at the surface. In winter, solar intensity is low and the hours of sunlight are short; temperatures reflect these conditions. Mean winter temperatures also decrease markedly as latitude increases. January temperatures average above 10°C (50°F) closer to the equator and drop to below −18°C (0°F) toward the pole. The range between mean July and mean January temperatures increases from about 8°C (14°F) in the warmer areas to about 40°C (72°F) in the Canadian Arctic. The coldest winter temperatures have a steep poleward gradient also, but not as steep as the average temperatures. Temperatures as low as −15°C (5°F) occur along the equatorward margin, and lows from −31°C to 46°C (−25°F−−50°F) characterize the more northerly locations. **Diurnal ranges** in winter are slightly greater than the summer ranges because the relative and absolute humidity are somewhat lower. The winter controls are, in essence, low solar radiation, short days, and moderate atmospheric humidity.

The equatorward migration of the polar front results in the dominance of polar air masses over much of the mid-latitudes (see Fig. 12.3). Periodically, the invasion of very cold arctic air gives rise to cold waves. These are any sudden drop in temperatures within a 24-hour period such that measures to prevent extreme distress to humans and animals may be necessary. In the United States, winter cold waves claim more lives than any other weather-related phenomenon. Statistics compiled by the National Center for Health show an average of 355 deaths per year from cold for the period 1949 to 1978. Deaths directly attributable to cold take the greatest toll. Evidence now indicates that cold substantially affects the death rate indirectly, particularly in areas where cold waves are infrequent.

Equally cold outbreaks of arctic air occur in Europe and Asia. In 763 A.D., a cold wave froze both the Black Sea and the Straits of Dardanelles in Europe. In 1236, the Danube River froze solid. In 1468, the winter in Flanders was so severe that wine rations issued to soldiers had to be cut with hatchets and distributed in frozen chunks. In 1691, crazed by the lack of food in the countryside, wolves invaded Vienna, Austria, killing humans for food.

CLIMATIC TYPES

Tropical climates were distinguished from each other on the basis of the seasonal pattern of precipitation. All mid-latitude locations have a seasonal temperature regime, stronger in some places than in others. Mid-latitude climates are also distinguished from each other on the basis of the seasonal pattern of precipitation (Fig. 12.6 and Table 12.2).

Mid-Latitude Wet Climates

The mid-latitude wet climates are associated with the subpolar lows over continental areas that spread toward the equator on the eastern sides of the continents. They tend to stretch nearly across the continents in the northern hemisphere near

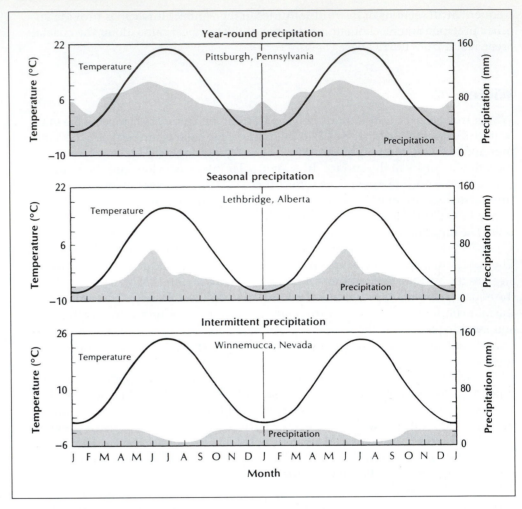

Figure 12.6
The three major types of mid-latitude climates based on the seasonal pattern of precipitation.

Table 12.2   Mid-Latitude Climatic Types

	Köppen Type	Thornthwaite Type
Climates dominated by mT and mP air	Cf, Df	BB', BC'
Winter-dry with seasonal mT and cP air	Cw, Dw	BB'w, CB'w, CC'w DB'w, CB'd, DC'd, CC'd
Summer-dry with seasonal mP and cT air	Cs, Ds	BB's, CB's, DB's
Climates with seasonal cT and cP air	BWk	Eb' Ec'

60° north and south. They reach equatorward as far as 25° or 30° along the eastern margins of land masses. In North America, this type of climate extends from the Pacific coast of Canada eastward in a crescent to the Atlantic coast. In the United States, it includes the area lying east of the Mississippi River and the first row of states west of the Mississippi. In Eurasia, this climate extends from the offshore islands of Great Britain and Ireland eastward into the Soviet Union. It also occurs along the Pacific coast of Asia in latitudes 30° to 60°. The climate also exists in South

America, Africa, and Australia. South America has two such areas: southern Chile, including the southern end of the central valley. The other includes the Pampas of Argentina and Uruguay and extends southward to include part of Patagonia. The area in Africa is very limited, consisting of a small area on the very southeast tip of the continent. Both Australia and New Zealand have sectors with this type of climate. Table 12.3 has sample climatic data.

Summer conditions are very similar to those found in the rainy Tropics. The primary differences are two: cyclonic storms that move across the region and temperatures that average from 21° (70°F) to 26°C (79°F). The tropical margins reach as high as 29°C (84°F), somewhat warmer than the humid Tropics.

The summer extremes exceed those of the tropical wet climate. Few stations in the humid Tropics have recorded temperatures in excess of 38°C (100°F). Most stations in the mid-latitude climate have recorded maximums this high or higher. The summer diurnal range is typically 8°C to 11°C (15°F–20°F), again quite comparable to the tropical humid climates. In the three summer months of June, July, and August, the temperatures of the southeastern United States average 1°C to 2°C (2°F–3°F) higher than in Belem, Brazil, and the rainfall is about the same. Nighttime temperatures are quite often within a few degrees of the daytime temperatures because the nights are short. Summer temperatures in mid-latitude humid climates are controlled mainly by the high solar intensity, long days, and high atmospheric humidity.

Precipitation is frequent throughout the year. The annual total is quite variable depending on the latitude and continental position. It varies from as little as 510 millimeters (20 in) upward to 1780 millimeters (70 in). Mobile, Alabama, is among the wettest of U.S. cities, with a mean annual precipitation of 1.73 meters (68 in). The variability of the annual precipitation is similar to that of the humid Tropics, normally less than 20 percent. The seasonal distribution is fairly even, but some areas have more in one season or the other. Frequency of precipitation varies. Bahia Felix, Chile, averages 325 days a year with rain. There is a 90 percent probability of rain on any and every day of the year. This not typical of these climates, of course, but it does show to what extent the precipitation mechanisms can persist.

Table 12.3   Monthly Data for Mid-Latitude Wet Climates

	Jan.	Feb.	Mar.	Apr.	May	June	July	Aug.	Sept.	Oct.	Nov.	Dec.	Yr.
New Orleans, LA 29°59′ N, 90°04′ W: 1 m													
Temp. (°C)	12	13	16	20	24	27	28	28	26	21	16	13	20
Precip. (mm)	98	101	136	116	111	113	171	136	128	72	85	104	1371
Montreal, Quebec 45°30′ N, 73°35′ W: 60 m													
Temp. (°C)	−11	−9	−4	5	13	18	21	19	15	8	1	−7	5
Precip. (mm)	83	81	78	72	72	85	89	77	82	78	85	89	971
Buenos Aires, Argentina 34°20′ S, 58°30′ W: 27 m													
Temp. (°C)	23	23	21	17	13	9	10	11	13	15	19	22	16
Precip. (mm)	103	82	122	90	79	68	61	68	80	100	90	83	1027
London, England 51°28′ N, 0°00′: 5 m													
Temp. (°C)	4	4	7	9	12	16	18	17	15	11	7	5	10
Precip. (mm)	54	40	37	38	46	46	56	59	50	57	64	48	595

The precipitation is more varied in form than in the Tropics. During summer and on the equatorial margins, convectional rainfall is the primary mechanism of precipitation. The southeast of the United States averages 40 to 60 days per year of thunderstorms. The frequency of thunderstorms decreases rapidly from south to north. Cyclonic precipitation is common on the polar margins and is more frequent in all sections of the region in the winter. Hurricanes provide another mechanism for inducing precipitation, and some of the heaviest rains along coastal areas result from hurricanes.

Although most of the annual precipitation is in the form of rain, snow is a factor to some extent in all places. In the United States, the mean snowfall varies from a trace along the Gulf to an average of about 1 meter (40 in) through the Great Lakes district. In January 1977, even Miami, Florida, had snow.

Humidity is generally high. On occasion it will drop quite low, but usually for short periods. High humidity and high temperatures make the sensible temperatures of summer quite high. As in the Tropics, radiation fogs are common on clear nights in summer and fall.

The summer is much like the rainy Tropics. Temperatures are high, humidity is high, and convectional showers are common. Summer storms such as tornadoes, thunderstorms, and hurricanes, and heat and cold waves bring variety to summer. Cold waves result from equatorward flow of summer continental polar air from the interior of high-latitude land masses. These cold waves may have temperatures above 10°C (50°F), but they represent much cooler air than is normal.

Within the wet climate area, there are differences in temperature that provide further divisions in the climate. Köppen distinguished between areas that have hot summers and mild winters (Cf) and those with cool summers and mild winters (Cbf). The latter region is mainly on the western side of the continents, where there are cooler summer temperatures due to the onshore winds from the oceans.

The mid-latitude wet climates are perhaps best developed on the west sides of the continents from about 45° to 65°. Here the westerlies from the oceans bring frequent mP air masses onshore, resulting in a high percentage of cloud cover, high humidity, and a high frequency of precipitation. The largest area of this type of climate is in Europe. Here it spreads inland a fair distance because there is no north–south mountain range to block the flow of the westerlies from the Atlantic Ocean. In North America, the zone extends from northern Oregon to Alaska. In South America, it extends from 40° to the tip of the land mass. Mountain ranges parallel to the coast restrict this type of climate in both North and South America. Small areas with this type of climate exist in South Africa, Australia, and New Zealand.

Winter is fairly long but mild, and summer is cool. The winter maritime air masses are usually at least as warm as the offshore ocean. The winter air is damp and chilling. Fogs are common because of the high humidity. The storm systems often become occluded, bringing extensive periods of rain and drizzle. For this reason, London, England, is regarded as a foggy, misty, and damp city. London averages 164 days each year with precipitation. This translates into a 45 percent probability of rain on any and every day. Yet the total annual rainfall in London is only 625 millimeters (25 in), so the amount of rain received on each rainy day averages less than 4 millimeters (0.15 in). Cloudy weather, fog, and drizzle are indeed typical.

Very cold weather is not common along these coasts. Cold cP air masses develop over land masses, and the westerlies carry them south and east most of the time. Only when an extremely large mass of cold air spreads out so it expands in a westerly direction do these cold waves affect coastal areas. Significantly, the warm ocean currents increase winter temperatures.

In summer, the frequency of precipitation is less because of the stabilizing effect of the subtropical highs as they move northward. Sites near sea level do not receive high total amounts of precipitation primarily because of the coolness of the air and cyclonic activity that produces the precipitation. Summers are subject to cool weather as the result of the influence of the adjacent oceans. The oceans in these latitudes never warm very much, and a cold current flows equatorward along most of these areas. Where highlands exist along these coasts, even summer precipitation is frequent, and rainforests evolved containing huge trees and dense surface vegetation interspersed with the trees. The famous redwoods of California and Oregon typify these forests. The Olympic Peninsula of Washington contains a magnificent forest of Douglas fir. These forests extend well into Canada. The green meadows of Ireland are also due to the frequency of the moist Atlantic air masses.

Mid-Latitude Winter-Dry Climates

The mid-latitude winter-dry climates (Table 12.4) follow a strong seasonal pattern of both temperature and moisture. The general location of this climate is the interior of the continents in mid-latitudes. This continental location provides the characteristic feature of a large average annual temperature range. Figure 12.7 shows the increase in summer temperatures and the decrease in those of winter for selected Eurasian stations. Box 12.1 shows how this continental effect varies in Eurasia. As the map shows, Asia is a region of maximum continental effect on climate.

Specific areas with this type of climate are the Great Plains of the United States and Canada, the grasslands of Eurasia, and small areas of Australia, Africa, and South America. These are continental locations lying where there is seasonal reversal of wind systems. There is frequent convergence during the summer and thus frequent precipitation. In winter, there is more divergence and a drier atmosphere. This is the most variable climate. These areas receive maritime tropical air masses and occasional tropical continental air masses from the adjacent deserts in summer. In winter, they have frequent outbreaks of polar continental air masses and an occasional maritime polar air mass. Each type of air stream brings its particular variety of weather.

A major difference in the response to solar input in this region is the result of lower atmospheric humidity. There are more clear days, and so there are higher rates of incoming and outgoing radiation. Both summer and winter extremes are greater as the result of more frequent dry air masses. Extreme high temperatures range from 38° to 44°C (100°F–110°F) in the wet and dry regions. They range up to 49°C (120°F) in the warmer sections as they did in the Great Plains in 1980.

Table 12.4 Monthly Data for Winter-Dry Climates

	Jan.	Feb.	Mar.	Apr.	May	June	July	Aug.	Sept.	Oct.	Nov.	Dec.	Yr.
Denver, CO 39°22′ N, 104°59′ W: 1588 m													
Temp. (°C)	0	1	4	9	14	20	24	23	18	12	5	2	11
Precip. (mm)	12	16	27	47	61	32	31	28	23	24	16	10	327
Bismarck, ND 46°46′ N, 100°45′ W: 507 m													
Temp. (°C)	−13	−10	−5	6	13	18	22	21	14	8	−2	−8	5
Precip. (mm)	11	11	20	31	50	86	56	44	30	22	15	9	385
Ulan Bator, Mongolia 47°55′ N, 106°50′ E: 1311 m													
Temp. (°C)	−26	−21	−12	−1	6	14	16	14	9	−1	−13	−22	−3
Precip. (mm)	1	2	3	5	10	28	76	51	23	7	4	3	213

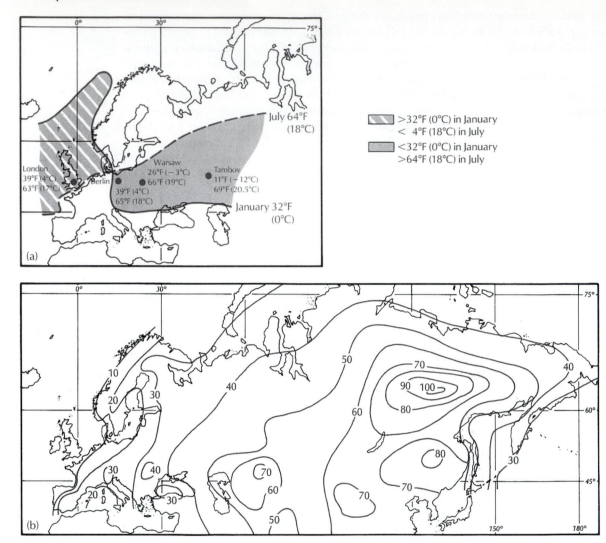

Figure 12.7
The influence of continentality: (a) In January, continental areas are colder than neighboring oceans; in summer, the reverse is true; (b) an index of continentality applied to the Eurasian land mass (from Boucher K., *Global Climates*, © 1975, Hoder & Stoughton, Ltd.).

Winter is the dry season. Dry weather increases Earth cooling during the long nights with the result that mean winter temperatures are usually a few degrees lower than in the humid climates. Extreme low temperatures are 3°C to 5°C (5°F–9°F) cooler. It is this climatic region that experiences the lowest temperatures in the northern hemisphere. In North America, the temperature dropped to −52°C (−62.8°F) at Snag, Yukon, on February 3, 1947. At Tanana, Alaska, a −51°C was recorded in January 1886. The lowest official temperature of the northern hemisphere is −57°C at Verkoyansk, U.S.S.R., and an unofficial −61°C was recorded at Oimekon, U.S.S.R. As a result of less humid conditions and excessive Earth radiation, the average annual range is up to 34°C (67°F) in extreme cases and the absolute range up to 100°C (186°F). A winter characteristic of these areas is rapidly changing temperatures. At Browning, Montana, on January 23–24, 1916, the temperature dropped from 6.7°C to −48.9°C (44°F−−45°F)—55.6°C in 24 hours. Such

QUANTITATIVE EXPRESSION: BOX 12.1

Continentality and Oceanicity

To measure the effect of a continental land mass on climate (i.e., a minimum impact of oceans), climatologists use the concept of **continentality**. As early as 1888, it was suggested that by measuring the average annual range of climate and adjusting for latitude, continentality could be quantified. Later workers used this idea as illustrated by the formula used by Gorzynsky

$$K = [1.7(A/\sin\phi)] - 14$$

where K is continentality, A is annual thermal amplitude (°C), and ϕ is latitude.

Many modifications of this formula have been suggested. The formula proposed by Conrad in 1946 where continentality is given by

$$[1.7A/\sin(\phi + 10)] - 14,$$

Thorshaven, in the Faeroe Islands, has a continentality index of 0 using this formula.

While maritime climates may be expressed by low values of continental indices, oceanicity indices have also been proposed. The first, introduced by Kerner in 1905, is given by

$$O = 100[(T_o - T_a)/A]$$

where O is oceanicity, T_o and T_a are the mean monthly temperatures for October and April, and A is mean annual range of temperature (all in degrees C). Another, developed by Kotilainen in Norway, also includes precipitation.

$$K = N \, dt/100\Delta,$$

where N is precipitation in millimeters per year, dt is the number of days with mean temperature between 0 and 10°C, and Δ is the difference between the warmest and coldest month.

Although the scientific validity of the empiric measures has been questioned, the measures do provide a general comparison of the oceanic and continental influences.

drastic changes are usually the result of the inflow of warm air, but adiabatically heated air will cause equally sharp rises in the temperature.

These regions are subject to the effects of a monsoon. During summer, the excessive heating of land in the northern hemisphere causes the subpolar low to expand equatorward. This forms a low trough or cell in the middle of the continent. Converging air brings moisture to the continent, producing a summer rainy season. During the winter cooling of the land mass, the subpolar trough splits into cells over the oceans and a ridge of high pressure develops between the polar high and the subtropical high. This reduces the probability of precipitation.

Winds are a significant factor in the weather. With the seasonal change in pressure is a seasonal change in wind direction. During the high-sun season, onshore winds prevail. During the winter, there is a switch of 90° to 180° in prevailing wind direction. The degree of wind shift depends to a large extent on the continental position. The more interior the station, the greater the wind shift. Wind velocities are high and persistent. In the United States, Oklahoma City has the highest average wind velocity of any city, with a mean of 22 kilometers per hour. Chinook winds are a winter phenomenon that produces extremely rapid changes in temperatures. They occur in the Great Plains from Alberta to Colorado. On January 22, 1943, Spearfish, South Dakota, observed a 2-minute rise of 27°C from −20°C to 7°C (−4°F–45°F).

Atmospheric humidity and precipitation are seasonal in character, and cloud cover varies with humidity. Precipitation decreases with distance from the equator and the sea. Although it is difficult to delimit annual precipitation, most areas receive between 300 and 900 millimeters (12 and 35 in). Of the annual total, about 75 percent falls during the summer half-year. The annual total is highly variable and subject to long-term cycles as yet unexplained. Several factors are responsible for the variable amount of rainfall. Nearly all rainfall east of the Rocky Mountains comes from mid-latitude cyclones. Winds from the Gulf of Mexico take a curved path across the eastern United States, moving east of the Great Plains. The dry tropical air masses that move northward across the Great Plains develop in Mexico and contain little moisture.

In the Great Plains, the cold fronts are often very steep as they move to the southeast, and extremely severe rainstorms develop along them. The unofficial heaviest 12-hour rain fell at Thrall, Texas, on September 9, 1921: 810 millimeters (32 in) of rain fell. At Holt, Missouri, on June 22, 1947, an unofficial rain of 42-minute duration totaled 300 millimeters (12 in). The frequency of hailstorms is high here as a response to drier air. The largest recorded hailstone fell in Potter, Nebraska, on July 7, 1928. It measured 137.4 millimeters (5.4 in) in diameter.

The winter dry climate is well defined over the Eurasian land mass. Here the largest contrasts between summer and winter weather conditions occur, with the possible exception of the two polar regions. Much of Asia is subject to seasonal reversal of pressure, wind direction, and precipitation.

Summers are hot and humid with intense summer convectional storms. Along the equatorward fringe as much as 70 percent of the annual precipitation occurs in the summer months; toward the interior, this increases to as much as 80 percent. The summer precipitation results from convectional instability, and severe rainstorms are frequent. A significant portion of the annual rainfall may come in a single storm. Total annual rainfall may be as much as a meter along the coastal areas, but this decreases inland to less than 675 millimeters (25 in).

In winter, cold, dry, continental air from the Siberian high-pressure zone brings a majority of clear days, with short periods of cyclonic disturbances. In the interior, winter temperatures drop below −34°C (−30°F). The frost-free growing season is short—less than 100 days in most years, particularly on the northern fringes. Since most of the precipitation occurs in the summer, winter snowfall is not particularly abundant. Snow stays on the ground for as much as 200 days in the northern interior.

The extreme of this climate occurs in northeastern Siberia. Summers can be warm and winters severely cold. At Oimekon, the range in temperature between January and July is 67°C (115°F). The range between the highest and lowest temperature recorded is 104°C (187°F). Winter is at least 7 months long. Snow often remains on the ground until May when temperatures rise rapidly.

Permafrost appears in the coldest parts of the climatic zone. Since there is often little snow to cover the soil, the ground is frozen to a depth of 50 meters (150 ft). In summer, the top few centimeters thaw, leaving a wet, soggy soil and extensive swamps.

THE SUMMER-DRY CLIMATES

One area with a seasonal rainfall regime differs from the rest. In this climate, the rainy season comes in the winter rather than in the summer as illustrated by data in Table 12.5. This type of climate occurs on the western margins of the continents between 30° and 40°. It normally does not spread into the continents very far, and thus is limited in extent. There are three main areas of summer-dry climates. The largest is around the Mediterranean Sea and extends eastward

Table 12.5 Monthly Data for Summer-Dry Climates

	Jan.	Feb.	Mar.	Apr.	May	June	July	Aug.	Sept.	Oct.	Nov.	Dec.	Yr.
Santiago, Chile 33°27' S, 70°40' W: 512 m													
Temp. (°C)	19	19	17	13	11	8	8	9	11	13	16	19	14
Precip. (mm)	3	3	5	13	64	84	76	56	30	13	8	5	360
Los Angeles, CA 33°56' N, 118°23' W: 37 m													
Temp. (°C)	13	14	15	17	18	20	23	23	22	18	17	15	18
Precip. (mm)	78	85	57	30	4	2	T	1	6	10	27	73	373
Rome, Italy 41°48' N, 12°36' E: 131 m													
Temp. (°C)	8	8	10	13	17	22	24	24	21	16	12	9	15
Precip. (mm)	65	65	56	65	51	30	25	17	66	78	96	97	711

into Iran. The second is the west coast of North America from near the Mexico–United States boundary northward into the state of Washington. It exists along this coast between the Pacific Ocean and the Sierra Nevada and Cascade Mountains. The third major area is across Southern Australia. Smaller areas are in the Capetown district of South Africa and in central Chile. The climate is often called the Mediterranean climate for the area associated with the Mediterranean Sea. In total, only about 2 percent of the land area of the earth experiences this type of climate.

The seasonal distribution of precipitation makes this climate distinct from the rest of the earth's climates. Much of the earth's land area receives most precipitation in the summer or spread over the year. In this climate, there is a pronounced winter maximum of precipitation. At Perth, Australia, 85 percent of the annual precipitation occurs in the winter 6 months. At Istanbul, Turkey, some 70 percent of the yearly total falls in the winter half year. San Diego, California, receives 90 percent of its annual total from November to April. Santa Monica, California, has recorded only traces of precipitation in the 3 months of June, July, and August. The 6 months of April through September average only 25 millimeters out of a total for the year of 381 millimeters. The occurrence of the precipitation in winter and the relatively low intensity of the cyclonic precipitation permit more efficient plant growth than is the case in the climates where the rainy season is in summer. Although mean winter temperatures average above freezing, occasional snow falls even along the equatorward margins. It rarely stays on the ground very long as daytime temperatures rise high enough to melt the snow. Total annual precipitation averages less than 750 millimeters (30 in) and tends to increase poleward. Table 12.6 provides the annual rainfall for several cities along the west coast of North America. The data show the rapid increase in annual total that takes place in a poleward direction.

Several climatic and topographic factors give prominence to these climatic regions as resort areas. In summer, the coastal areas are sunny but not as hot as other areas as the cool ocean waters and the sea breeze offset the intensity of the sunshine. Winter temperatures are mild, with a lot of sunny weather and short periods of rain. The subsidence associated with the subtropical high keeps the sky relatively clear so it has an unusually bright blue color. The Mediterranean Sea is noted for its brilliant blue color. This is partly due to sky color and partly due to the low organic content of the water.

Table 12.6 Latitudinal Variation in Annual Precipitation in the Summer-Dry Climates

San Diego	269 mm
Los Angeles	380 mm
San Francisco	510 mm
Portland	900 mm

Like all climates that have a rainy season, these environments have two distinct seasons, each associated with a different air mass and a different part of the general circulation. During summer, these areas are under the influence of stable oceanic subtropical highs, giving them essentially tropical desert weather conditions. In winter months, the anticyclonic circulation moves equatorward, allowing the westerlies to bring moisture to the region.

Mid-Latitude Deserts

The mid-latitude deserts are basically in the interiors of continents, although they merge with tropical deserts on the west sides of the land masses. In North America, there is desert from southern Arizona northward into British Columbia between the Sierra Nevada and Rocky Mountains. In Eurasia, they are in the trans-Eurasia **cordillera** or lie on the flanks of these mountain masses. All of the central Asian countries have areas of deserts within their boundaries. The southern hemisphere has a much smaller area of desert in mid-latitude locations. The basic reason for the small size of the deserts south of the equator is a lack of land area in the latitudes where the deserts are most extensive.

These deserts have the highest percentage of possible sunshine of any of the mid-latitude climates. Arizona averages 85 percent possible sunshine through the year with an average of 94 percent in June and 76 percent in January. Temperatures reflect the clear skies that dominate the weather. Summer temperatures are the highest of the mid-latitude climates (Table 12.7). The summer averages are 5°C to 8°C (9°F–14°F) higher than in humid areas, and the maximum temperatures are also higher. During summer, when the days are long and the sun high in the sky, temperatures will reach above 50°C (122°F). Death Valley in California recorded an official shade temperature of 56.7°C (134°F). This is only 1°C below the highest sea-level temperature ever recorded. Winter temperatures compare with those of the humid climates at the same latitude primarily because the coldest temperatures anywhere in mid-latitudes occur with outbreaks of cold arctic air.

This climatic type is the most difficult to place in the pattern of the general circulation. Several factors contribute to the aridity of these regions. Each is not enough to produce extreme drought, but coupled with other elements results in aridity. Lying between the subpolar low and the subtropical highs, these deserts maintain dry conditions due to low moisture supply and intermittent anticyclonic circulation.

The mid-latitude deserts are not as arid as the tropical deserts. During the winter months, the general circulation shifts equatorward far enough to allow occasional passage of weak cyclones associated with the polar front. Precipitation totals are, of course, very low. Phoenix, Arizona, averages 180 millimeters (7 in) of precipitation per year, and Ellensburg, Washington, 230 millimeters (9 in). There is little seasonal pattern to precipitation. Phoenix receives a trace of snow, and

Table 12.7 Monthly Data for Mid-Latitude Deserts

	Jan.	Feb.	Mar.	Apr.	May	June	July	Aug.	Sept.	Oct.	Nov.	Dec.	Yr.
Yuma, AZ 32°40′ N, 114°36′ W: 62 m													
Temp. (°C)	13	15	19	22	26	31	35	34	31	25	18	14	24
Precip. (mm)	10	9	6	2	T	T	6	13	10	10	3	8	77
Winnemucca, NE 40°50′ N, 117°43′ W: 1312 m													
Temp. (°C)	−3	0	3	8	12	16	22	20	15	9	2	−1	9
Precip. (mm)	27	24	21	21	24	19	7	4	9	21	20	24	219
Alice Springs, NT, Australia 23°38′ S, 133°35′ E: 546 m													
Temp. (°C)	28	27	25	20	16	12	12	14	18	23	25	28	21
Precip. (mm)	27	45	18	10	18	15	14	10	6	25	23	29	250
Ashkhabad, U.S.S.R. 39°45′ N, 37°57′ E: 230 m													
Temp. (°C)	2	5	9	16	23	29	31	29	24	16	8	3	16
Precip. (mm)	22	21	44	38	28	6	2	1	3	11	15	19	210

Ellensburg, Washington, averages 780 millimeters (31 in) of snow each winter. Moisture efficiency is very low throughout these deserts. In some areas, potential evaporation exceeds the precipitation by as much as 10 times.

SUMMARY

Mid-latitude climates have marked differences in temperature through the year in response to changes in the intensity and duration of insolation. Winter is accentuated by the incursion of cold air from the polar regions, and it is winter that distinguishes these climates from the tropics. The convergence of very different air streams results in the development of actively moving low-pressure systems, which often produce severe storms. These changing atmospheric conditions provide a distinct pattern of summer, fall, winter, and spring. Like the Tropics, mid-latitude climates subdivide on the basis of the distribution of precipitation through the year. In mid-latitudes, the wet and dry climates are further subdivided into summer-dry and winter-dry regimes. The fact of having the rainy season in the winter in the one case, and the summer in the other results in different types of climate.

KEY TERMS

Continentality   Diurnal range   Subtropical jet
Cordillera   Permafrost

REVIEW QUESTIONS

1. How does the variation in solar radiation differ though the year between the tropical and mid-latitude climates?

2. In the summer months, why do the 48 contiguous states in the United States receive more total radiation than is received at the equator?

3. Why is the precipitation more efficient for plant growth in the summer dry climates than in the winter dry climates?

4. What impact does the freezing of lakes and soil moisture have on the water balance?

5. What effect does continentality have on temperatures?

6. Why do the coldest winter temperatures in North America occur in the interior of the continent?

7. Why does annual precipitation decrease from north to south along our Pacific Coast?

8. Why does California receive most of its precipitation in the winter months?

9. The Great Plains tend to be very windy. Why does this region have higher average wind velocities than the states east of the Mississippi River?

10. The state of Washington has the fewest thunderstorms of any of the contiguous 48 states. Why is this the case?

C H A P T E R

13

Polar and Highland Climates

*T*he distinguishing feature of the polar climates is cold weather brought about by low levels of solar radiation. The radiation balance differs from that of the Tropics and mid-latitudes in both the diurnal and annual pattern of solar energy. In the Tropics, the major energy change is from day to night, in which daily changes in radiant energy and temperature are greater than from season to season. In mid-latitudes, there are large fluctuations in both the diurnal and annual periods. The more poleward the location, the larger the seasonal fluctuation in radiation and temperature relative to the diurnal changes. Like the Tropics, the polar climates are dominated by a single periodic fluctuation in energy. In the Tropics, the major fluctuation is diurnal, but in polar areas it is the annual fluctuation that is most important (Fig. 13.1).

Every place on the surface of the earth receives nearly the same number of hours of sunlight during the year. However, there is a big difference in the distribution through the year. In the Tropics, the radiation is evenly spread throughout the year, with about 12 hours of insolation each day. At latitudes higher than the Arctic and Antarctic circles, most of the energy comes in one continuous dose during the summer half year. The other half year receives very little energy. At the time of the summer solstice in the northern hemisphere (June 22), the north polar axis tilts toward the sun 23.5°. This is the longest day of the year in the northern hemisphere. On this date, the sun does not drop below the horizon in the zone from 66.5° N to the North Pole. At Murmansk (69° N), the continual daylight of summer lasts for 70 days. At Spitzbergen (80° N), there are 163 days of continuous light, and at the North Pole there are 189 days of continuous daylight.

At the time of the fall equinox, the days are 12 hours long everywhere. The vertical rays of the sun are over the equator, sunset is occurring at the North Pole, and sunrise is at the South Pole. The equinoxes are far more important to the polar regions than are the solstices because they mark major times of change in energy. The equinoxes represent times when solar radiation either becomes significant or drops to near zero.

At the time of the winter solstice, the days in the northern hemisphere are the shortest. The North Pole is in the midst of a single period of darkness lasting 176 days. At Spitzbergen, the night is 150 days long, and at Murmansk continuous darkness lasts for 55 days—from November 26 to January 20.

During the winter months in polar regions, direct solar radiation may be absent. Some radiation, including some in the visible range, continues to find its way

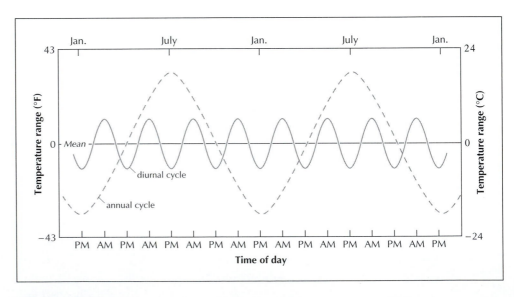

Figure 13.1
Relationship between the annual and diurnal energy cycles in polar climates.

to the surface. Total darkness seldom exists. When the sun is below the horizon, refraction and scattering of the rays bring some sunlight to the surface until the sun is 18° below the horizon (**astronomical twilight**). Thus, at some latitudes, there may be several months with no direct sunlight but almost continuous twilight. The stars, moon, and **auroras** are also sources of light for the poles, so the darkness is not as intense as it might be.

Another important element in the polar climates is that the intensity of radiation is never very high for any length of time compared with mid-latitude or tropical regions. Even on the equatorward margins the sun rarely reaches more than 45° above the horizon. In the summer months, however, although intensity is low, the duration of daylight is quite long, with over 20 hours of daylight near the summer solstice. Although summer insolation persists for many hours a day, temperatures do not get very high. This is because radiation intensity is so low and because much of the energy goes to melt snow and ice.

Although there are elements of similarity between the two polar areas, there are also basic differences that influence the climate of each. There is one important and obvious difference between the Arctic and Antarctic. The Antarctic is a relatively large and high land mass surrounded by water, whereas the Arctic is a sea surrounded by land. The relationship of the land–sea interface reverses at the two poles, and this is very significant to the climate of the two areas. The three climates of the polar regions are the polar wet, wet and dry, and dry (Fig. 13.2). The polar wet and dry climate exists primarily in the northern hemisphere, and the other two climates occur mainly in the southern hemisphere.

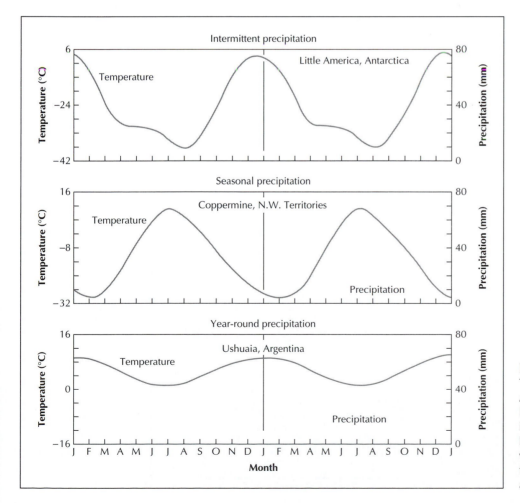

Figure 13.2
The polar environments: All three environments have a low solar energy input year round. The heat from atmospheric water and the sea keeps the average temperature at Ushuaia above freezing.

THE ARCTIC BASIN

The Arctic Ocean consists of all the sea surface north of the Arctic Circle, whether covered by ice or open water (Fig. 13.3). The central part of the Arctic Ocean is frozen most of the year but does have occasional open leads. The central ice pack covers 11.7 million km² (4.5 million mi²) in winter and melts back to 7.8 million km² (3 million mi²) in the warmest months. The ice drifts from east to west. In the process of moving, leads open and close so some unfrozen areas are always present. The ice ranges in age up to 20 years and in thickness up to 5 meters (16.5 ft), with the oldest ice being the thickest. In the summer, the ice melts away from the edge of North America and Eurasia so there may be up to 2 months when there is mainly open water with patches of drifting ice. Because of the heat in the Gulf Stream, the Norwegian Sea and Barents Sea remain open all year at latitudes where the sea is frozen in other areas (Fig. 13.4).

The earth–space energy balance in the polar areas is such that the poles are major energy sinks. There is much more energy emitted by the polar regions than

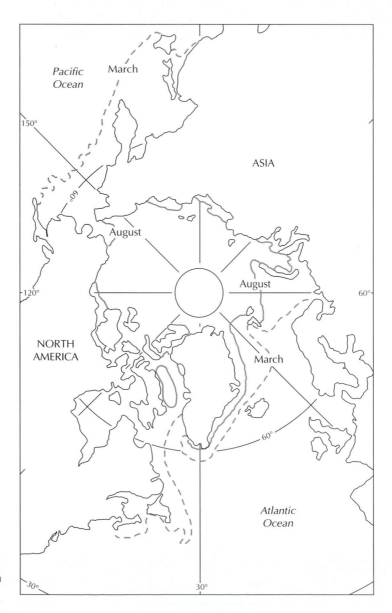

Figure 13.3
Map of the Arctic Basin showing the average location of the ice pack in March and August.

Figure 13.4
Arctic ice pack showing open leads (Peter Arnold, Inc. © Norbert Wu).

received from solar radiation. In the Arctic Basin, radiation reflected and radiated away as long-wave radiation is 60 percent greater than direct solar radiation. For this to happen, a supplementary heat source must make up the difference. This energy source is heat carried poleward by air and water currents. It is also this steady flow of energy poleward that maintains Arctic temperatures at a level much higher than insolation would allow (Fig. 13.5). For the Arctic Basin, the primary source of heat is the inflow from the sea and atmosphere. This energy is nearly double that of solar radiation absorbed at the surface. Warm ocean currents move much of this heat into the Arctic. The water releases heat to the atmosphere by radiation, conduction, and evaporation. The amount of open water compared with ice is of extreme importance. The heat transferred to the atmosphere from a water surface is more than 100 times that of an ice surface.

Sea water acts differently from fresh water as it cools toward the freezing point. In fresh water, the maximum density is at 4°C (39°F), and fusion takes place

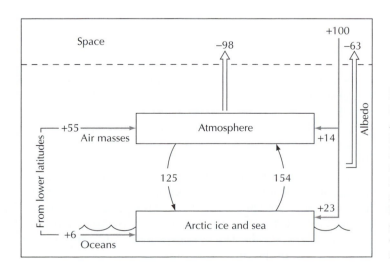

Figure 13.5
Energy balance of the Arctic Basin. Note that because of the influx of heat brought into the arctic basin by the Gulf Stream and atmospheric storms that the radiation loss to space is much greater than the radiant energy received from space.

at 0°C (32°F). In sea water, the maximum density is just near the fusing point. This is about −2°C (28°F) depending on salinity. One result of this phenomenon is that surface water in polar seas may cool up to 6°C (11°F) more than would otherwise be the case before ice forms. The colder polar sea water is more dense than either fresh or warmer sea water. As it chills to near the freezing point, it sinks to the bottom and spreads out, settling in the deeper parts of the ocean basins. This settling cold water helps promote circulation in the ocean, which in turn helps equalize the energy between the equator and poles. This fact also explains why most of the water in the ocean is so cold. The average temperature of water in the sea is about 3.5°C (38°F). Warm surface water in tropical regions and along beaches in mid-latitudes is the exception and not the rule.

The atmosphere over the Arctic Basin never becomes really cold because the water beneath the ice and in open areas does not drop below a temperature of −1.6°C (29°F). The sea, even when frozen, modifies atmospheric temperatures. Whenever the air temperature over the ice drops below −1.6°C (29°F), there is heat transferred through the ice and from the open water. Over much of the Arctic Basin, temperatures over the ice range between −20°C (4°F) and −40°C (−40°F), with the lowest temperature yet recorded near the surface being −50°C (−58°F). The influence of the Arctic Ocean as a heat source is such that there is a very abrupt atmospheric change along the coastline.

Polar Lows and Arctic Hurricanes

It has been known for many years that intense and short-lived low-pressure systems developed along the edge of the continents in the North Pacific and North Atlantic Oceans. Tor Bergeron used the term *extratropical hurricane* in 1954 to describe severe storms in the Arctic. More recently, they have been referred to as **Arctic hurricanes**. These polar lows are small compared with mid-latitude lows. The polar lows are less than 1000 kilometers in diameter, and sometimes a fully developed system is only 300 kilometers across. Wind speeds may be sustained at velocities of 40 knots and there may be gusts to 60 knots. The central pressure drops to 970 millibars or less. The same type of low forms around the Antarctic continent year round.

In some respects, these storms are similar to tropical hurricanes. They are circular, have a relatively clear eye, and sustain high wind velocities. They also have symmetrical spiral bands of deep cumulus clouds. Like tropical cyclones, there is a high-level outward flow of air aloft. Like tropical cyclones, they dissipate rapidly when they reach land. Like tropical hurricanes, the storm reaches to the tropopause. The outflow of moist air aloft produces a broad and thick layer of cirrus clouds.

They differ from tropical cyclones in important respects as well. They develop to full strength in 12 to 24 hours and may have a total life span of only 36 to 48 hours before they strike land. They form under very different conditions than tropical lows. They form in cold air north of the polar front. They travel at speeds up to 30 knots—nearly twice as fast as the average tropical hurricane. They have gale-force winds and heavy snow and sleet. In the northern hemisphere, they form near the ice pack and move eastward toward land. They form where there are extreme differences in the merging air streams along an arctic front and where there is an existing low-pressure disturbance along the edge of the ice pack. This low draws in the cold air from off the ice sheet and relatively warm moist air from the subpolar ocean. Very cold air develops over the pack ice in winter. Temperatures may be as low as −40°C (−40°F). The temperature contrast between the air masses may be more than 40°C (72°F). The sea-surface flux of heat plays a major role in intensifying and maintaining the mature storm.

THE POLAR WET AND DRY CLIMATE

The polar wet and dry climate, found primarily along the shores of the Arctic Ocean, has cold winters, cool summers, and a summer rainfall regime. Specific areas experiencing this climate are the North American Arctic coast, Iceland, Spitzbergen, coastal Greenland, the Arctic coast of Eurasia, and the southern hemisphere islands of McQuarie, Kerguelen, and South Georgia. The land areas with this climate occupy only about 5 percent of the land surface of the earth (Table 13.1).

A sample of five tundra weather stations in the northern hemisphere shows a January average temperature of −30°C (−22°F). Mean annual temperatures are below freezing, but summer temperatures go above freezing at least for a few weeks. The summer period, if measured by above-freezing temperatures, averages 2 to 3 months, but may be as long as 5 months. Mean temperatures in the warmed months seldom go above 5°C (41°F). High temperatures range from 21°C to 26°C (70°F to 79°F).

Pressure gradients are weak in the tundra, so it is not a stormy region. In fact, it is one of the least stormy areas of the earth's surface. High-pressure systems with cold, still, dry air are frequent, particularly in winter. The North American Arctic is even less stormy than other areas. Because of the mountainous character of Alaska, the cyclones of the Pacific Ocean do not move inland very far or very often. Spring and fall are the seasons with the most active weather, as in most high-latitude locations. Winds are light in velocity and variable in direction. The average velocity is less than 8 knots and less than 5 knots in the Canadian tundra.

Atmospheric humidity is high during the summer months, averaging 40 to 80 percent. During the winter it is less, averaging 40 to 60 percent. Annual precipitation averages less than 250 millimeters (10 in) for most stations; more than three fourths of the annual total falls during the summer half year. Winter snowfall is only about 50 millimeters (2 in) of water because the snow is dry and compact. A deep snow cover is absent over most of the tundra as winter snow is not excessive and tends to drift easily. Thus, exposed areas are often free of snow, although the sheltered areas along streams may have a deep snow cover. Cloudiness varies through the year in the tundra. In the summer months, cloud cover may average 80 percent, but during the winter it drops to around 40 percent.

THE ANTARCTIC CONTINENT

The Antarctic Continent (Fig. 13.6) is not only the coldest area on the earth's surface, but also the least-known continent. It is about 14.2 million square kilometers (5.5 million mi²) in size. It is nearly twice the size of the United States and nearly three times greater than the Arctic Basin. The harshness of the environment

Table 13.1   Classification of Polar Climates

Climates with seasonal flux	Köppen type	ET
of mP and mT air masses	Thornthwaite type	E
Climates dominated by cP	Köppen type	EF
air masses	Thornthwaite type	F
Climates dominated by mP	Köppen type	EM
air masses	Thornthwaite type	AE

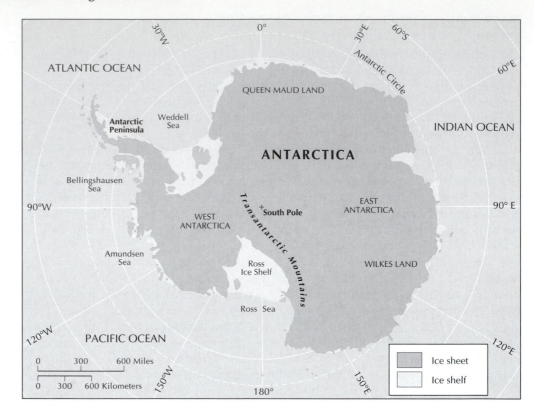

Figure 13.6
The Antarctic Continent (from McKnight T. L. and Hess D., *Physical Geography: A Landscape Appreciation*, © Prentice Hall, Upper Saddle River, N.J.).

around this land mass is such that even the existence of the continent was not known until extremely late in history. Throughout much of historic time, it showed on maps as *Terra Australis Incognito* ("unknown southern land"). Nathaniel Palmer first discovered it on November 17, 1820. The first expedition commissioned by the United States set out in 1838 under Lieutenant Charles Wilks.

The outstanding feature of the Antarctic land mass is, of course, the huge mass of ice surmounting and surrounding the continent. A sheet of ice that is more than a mile thick covers 95 percent of the land mass. In some places, the ice is more than 4000 meters (13,000 ft) thick. This ice sheet is divided into two separate pieces. They are the West Antarctic ice sheet and the East Antarctic ice sheet. The West Antarctic ice sheet is thicker and at higher elevations than the the other. The surface ranges from 2400 meters to over 3500 meters (8000 to over 10,000 ft) in elevation. The eastern ice sheet is firmly attached to the ground beneath, but the western sheet is fairly unstable. Some 86 percent of the planetary ice is on this land mass. The pole of inaccessibility, or the center of the ice mass, is about 4000 meters (13,200 ft) elevation, and it is 670 kilometers (400 mi) away from the geographic pole.

The polar ice cap is not the thickest where it is the coldest. The ice reaches its maximum depth where air temperatures average a little below freezing. The accumulation of ice is quite slow in some areas. Near Byrd Station at a depth of about 300 meters (1000 ft), the ice formed from snow that fell some 1600 years ago. The ice mass moves seaward over much of the continent. The rate of movement of the ice seaward is highly variable, in some cases moving seaward as much as a kilometer per year and in some locations becoming buoyant and extending far out to sea.

There are several locations where the ice sheet moves out onto the sea as a large floating ice shelf. One of these, the Ross Ice Shelf, extends almost 800 kilometers out to sea and covers an area of some 520,000 kilometers^2. This floating ice is as much as 270 meters (890 ft) thick, with roughly 40 meters (130 ft) above water. Icebergs breaking off from this shelf ice are tabular in shape and may be very large. A huge iceberg broke away from the Ross Ice shelf in early October 1987. It was about 154 kilometers long and 36 kilometers wide (92.4 × 21.6 mi). The surface area was about 4750 kilometers^2 (1834 mi^2). It is known to be at least 250 meters (825 ft) thick and may be more than that in some places.

The oldest known drifting iceberg is known as C-2. It was first spotted by satellite in 1978 and broke up in 1990. It was a fairly large iceberg. It was about 30 kilometers in diameter and over 60 meters thick. For over 4 years, it was grounded in 60 meters of water. It traveled nearly around the Antarctic Continent for a distance of about 10,000 kilometers.

The Energy Balance

The Antarctic region is a major energy sink. Almost all of the stations on the continent have a net radiation loss on an annual basis. Most, in fact, have a net radiation loss even in the summer. The winter is 6 months long and there are really only 2 months of summer. The heat loss to space is greatest over the ocean surrounding the land mass. In July in the midst of the southern hemisphere winter, there is radiative heat loss to space over the entire area from the pole halfway to the equator. In January, the area of net loss shrinks southward to as far as 85° in some places.

Antarctic temperatures average 16°C (30°F) below those of the Arctic. There are several reasons for the colder regime. The area is mainly a land mass rather than water. The average elevation is high, with parts of the ice sheet 3000 meters (10,000 ft) above sea level. The South Pole is 3 million miles farther from the sun during its winter than is the Arctic during its winter. This distance reduces solar radiation seven percent below that of the Arctic winter. Temperatures reflect both elevation and proximity to the ocean. Mean annual temperatures range from −19°C (−2°F) to as low as −57°C (−70°F) at Vosktok. Minimum temperatures in the winter drop to −57°C along the coast. They decrease with elevation and distance from open water. Vostok, a Russian Antarctic station located inland at an elevation of 3950 meters (13,000 ft), has recorded the coldest temperature at the surface of the earth, −88.3°C (−127°F). The geographical South Pole is at 3000 meters (10,000 ft) elevation. Table 13.2 provides illustrative data of the temperatures that occur.

Table 13.2 Climatic Data for Amundsen–Scott Station (South Pole)

Mean annual temperature	20 years (1957–76)	−49.3°C
Coldest year	1976	−50.0°C
Coldest day of record	July 22, 1965	−80.6°C
Coldest month of record	August 1976	−65.1°C
Coldest half year	1976 (Apr.–Sept.)	−60.7°C
Warmest day of record	Jan. 12, 1958	−15.0°C

In the Antarctic, the coldest temperatures lag far behind the solstice as a result of the long absence of direct solar radiation. At Vostok the coldest days occur in their spring after the sun has come above the horizon for a day or two. The sun comes above the horizon there on August 22, and the coldest temperature yet observed occurred two days later, on August 24.

In winter, the changes in weather from day to day are brought about by the advection of air masses with differing amounts of moisture and cloud cover. These changing systems are mostly above the inversion.

As stated earlier in the chapter, the intensity of radiation reaching the surface in summer is low. The total amount of radiation received at the surface is large. In fact, stations in the Antarctic hold the record for the most solar energy received in a monthly period. Stations on the Antarctic continent also hold the records for the longest continuous period of direct sunlight, and the most hours of daily sunshine in a month. These aspects of the energy balance do not change the fact that the coldest temperatures on the planet are also found here.

In summers, the temperatures range upward toward the freezing mark but rarely reach it. Over most of the Antarctic, it does not get warm enough in summer to melt the surface snow, although there are a few areas that do shed their snow cover. Temperatures above the ice cannot rise much above the freezing point because the surface does not get above freezing.

The times of the equinoxes are periods of great change in the Antarctic. Around the time of the spring equinox (October), there is a burst of solar energy into the polar atmosphere. It is this burst of energy that begins the mixing process that causes the chlorine compounds to interact with the other gases to remove ozone. At the fall equinox, temperatures drop rapidly over the ice sheet. This results in very strong temperature, pressure, and moisture gradients around the edge of the continent. Intense cyclones develop and move around the perimeter. These are the arctic lows discussed earlier in the chapter. These storms send large quantities of relatively warm moist air inland above the temperature inversion. This influx both retards cooling in the fall and provides heat to the surface throughout the winter.

By far the dominant feature of the Antarctic is the persistent cold. Second to the cold is the prevalence of a temperature inversion above the ice. It exists over the entire continent in winter and much of the continent in summer. The inversion ranges up to 25° in the lower 4 kilometers. On July 2, 1960, there was an inversion of 30°C at Vostok. The amount of the inversion increases away from the coast in winter, reaching 25° or more at the cold pole. The strong inversion is brought about by the high altitude and clear atmospheric conditions, which permit rapid radiation loss from the surface.

Diurnal temperature ranges are small, less than 5°C (0°F), as a result of the small difference in insolation during the day. The absence of the normal pattern of day and night and the high heat capacity of the snow and ice reduce fluctuations in temperature on a diurnal basis. Table 13.3 contains data to show the extreme nature of the temperature regime of the Antarctic.

The energy balance of the Antarctic consists of rapid radiative heat loss from the ice- and snow-covered surface, which is partly offset by advection of heat in the warm moist air from the surrounding ocean. The sea air contains both sensible and latent heat.

The extreme cold of the Antarctic Continent influences the climate far from the ice cap. Cities that lie on the southern margins of the southern hemisphere land mass have average temperatures 3°C cooler than cities at the same latitude in the northern hemisphere. Several of the islands in the South Seas support glaciers, including Kerguelan Island and Heard Island. In South America, mountain glaciers reach the ocean at latitudes nearer the equator than do those in the northern hemisphere.

Table 13.3 World Climatic Extremes Recorded on the
Antarctic Continent

Condition		Record
SOLAR RADIATION		
Longest continuous period of direct sunlight	Dec. 9–12, 1911	60 hr
Highest monthly mean (hours)	Syowa Base (69° S 40° E)	14 hr/day
Highest mean monthly intensity	South Pole (Dec.)	955 ly/day
TEMPERATURE		
Lowest mean monthly diurnal range	South Pole (Dec.)	2°C (3.5°F)
Lowest average temperature (warmest month)	Vostok	−32.5°C (−27.6°F)
Lowest mean annual temperature	Cold Pole (78° S 96° E)	−58°C (−72°F)
Lowest mean monthly maximum	Vostok	−66.5°C (−88°F)
Lowest monthly mean	Plateau Station (Aug.)	−71.7°C (−97°F)
Lowest mean monthly minimum	Vostok	−75°C (−103°F)
Lowest daily maximum	Vostok	−83°C (−117°F)
Lowest absolute temperature	Vostok	−88°C (−126.9°F)
HUMIDITY		
Lowest dew point	South Pole	−101°C (−150°F)
WIND VELOCITY		
Highest annual mean	Cape Denison	19.3 m per sec (43 mph)
Highest monthly mean	Cape Denison (July 1913)	24.5 m per sec (55 mph)

The Windy Continent

Winds are a predominant factor in polar deserts' weather. Westerly winds are common at the surface poleward to around 65°, where they give way to low-level easterly winds that extend on to about 75°. Cyclonic storms develop over the oceans in the westerlies and move around the Antarctic from west to east. They normally do not penetrate far inland, and they account for much of the precipitation and weather along the coast. High-velocity winds that move snow and produce blizzard conditions occur at Byrd Station about 65 percent of the time. They have high enough velocity to produce zero visibility about 30 percent of the time. A slight increase in wind velocity brings a large increase in blowing snow. The amount of snow moved by the wind varies as the third power of the velocity. It is unfortunate for those working in the Antarctic, but the most accessible locations are also the windiest. Mawson's Base at Commonwealth Bay experiences winds that are above gale force (44 kph or 28 mph) more than 340 days a year. At Cape Denison, the mean wind velocity is 19.3 meters per second (43 mph), and during July 1913, it averaged 24.5 meters per second (55 mph).

Strong gravity winds develop and blow off the land masses. These **katabatic winds** are the flow of cold, dense air down a topographic slope due to the force of gravity. They result from the extreme cooling of the air over the ice. A topographic slope of 0.2 percent is about equal to the force of the average pressure gradient over the Antarctic. Where the slopes are steep, the katabatic winds may regularly flow in a direction other than that of the pressure gradient. The strongest winds occur when the pressure gradient and topographic slope are coincident. The katabatic winds are the prevailing winds over much of the Antarctic, particularly where there is a steep drop from the interior highlands to the coast. Where these winds are fairly constant, they will produce formations of waves in the snow and ice surface called **sastrugi**. These frozen waves grow as high as 2 meters and make surface travel extremely difficult. Strong winds, blizzards, and rising temperatures often occur together. This is because the usual temperature inversion disappears, and warmer maritime air will move inland.

Even the interior can be very stormy, particularly in winter. On July 27, 1989, a six-member international team of men and dogs set out to cross Antarctica by dog sleds and skis. It took them 7 months to make the crossing, the first on foot. They encountered (90 mph) winds and temperatures as low as −45°C. One of them got lost in a blizzard one evening only a matter of meters from the camp. He dug himself a little trench in the snow and lay down. Snow soon buried him. When morning came, they started looking for him again. When he heard them calling, he crawled out of the snow, cold, but alive and unharmed. There were only rare interludes without howling wind and blowing snow. Visibility was often zero, in which case they could not travel. Sometimes they were forced to stay put for as much as 3 days at a time. One blizzard lasted for 17 days and another stormy period lasted 60 days. Summers are not so stormy. Temperatures go up as high as −20°F and winds drop to the 20 miles per hour range.

Precipitation

The ice sheet is comparable to the tropical deserts in amount of precipitation. The extremely cold air is incapable of holding much water vapor. Average annual precipitation probably does not exceed 200 millimeters (8 in) in any place on the ice cap. Moisture data are more difficult to get for the polar deserts than are temperature data. Data show that relative humidity, absolute humidity, and cloud cover decrease inland. Relative humidity often drops as low as 1 percent. Humidity and cloud cover are also lower in winter as a result of the stronger subsidence and the surface inversion. The cloud decks that appear over the ice caps produce warmer temperatures as they provide a heat source that radiates heat to the ground. Precipitation may be fairly frequent, but it is light. There is little moisture for precipitation in the atmosphere because of the low temperatures. It probably does not exceed 10 millimeters (0.4 in) at any time. Annual precipitation ranges from as little as 2 inches on the cold plateau to as much as 20 inches at some peninsular locations. Most of the precipitation occurs as snowfall, averaging 300 to 600 millimeters (12–24 in) per year. At Halley Bay, there is snowfall about 200 days per year. Because of the high winds, blizzards and drifting snow are common. Drifting snow also occurs 200 days a year. Blowing snow reduces visibility to less than 1 kilometer about 170 days a year. When there is no wind, loose dry snow will pile up to as much as a meter deep. Some of the precipitation occurs in the form of ice pellets, and some moisture is deposited directly as condensation. Along the coasts, instances of steam fogs drifting inland have occurred when temperatures were as low as −30°C. These steam fogs add to the accumulation of moisture.

THE POLAR WET CLIMATES

Surrounding the Antarctic is the planetary ocean. It is uninterrupted by land masses north to a latitude of 40°S except at the tip of South America. Over this ocean, the atmosphere has year-round high humidity, a high frequency of precipitation, and low temperatures. The only land areas are a series of islands scattered around the Antarctic Continent. Table 13.4 contains data representative of this climate type.

Humidity is very high, averaging over 50 percent all year. Cloud cover is also extensive, averaging 80 percent during the winter and slightly less in the summer. Precipitation frequency is high in all areas, with at least a 25 percent chance of precipitation on any day of the year. The annual total varies, however, as the intensity of precipitation varies with latitude and distance from the ocean. The totals vary from 370 millimeters to 2.92 meters (15–117 in). The annual variation in precipitation is among the lowest found anywhere. Snow is a common form of precipitation. Most snows are very wet and last but a very short time.

The high humidity and cloud cover have a considerable degree of control over insolation and temperature. The summer averages range up to 10°C (50°F), and winter averages are between −6.7°C and 0°C, so the annual range in temperature is not very large. Winter lows are exceptionally warm for a polar climate. The reason for the mild winter temperatures and low annual range is the marine location of these areas. Because the ocean does not freeze, it provides a constant source of heat for the atmosphere. The diurnal ranges are also quite low because of the high humidity. A frost-free season is nonexistent. Frost can occur any day of the year, and these areas average over 100 days annually with frost.

In winter, there is a much greater frequency of storms with high winds. This is the area named "the roaring forties, the furious fifties, and the shrieking sixties" by the sailors of the 16th century. They gave these names to the area when they started sailing around Cape Horn in South America.

This climate is distinctive in several ways. It is the cloudiest of all climates, with most of the area experiencing few clear days. No place can equal the South Atlantic zone for the frequency and severity of storms.

HIGHLAND CLIMATES

The tropical, mid-latitude, and polar systems are distinguished from each other on the basis of periodic aspects of the energy balance. We subdivide each of these

Table 13.4   World Climatic Extremes Recorded in the Polar Wet Climate

Lowest mean annual diurnal temperature range	Heard Island, Indian Ocean (53°10′ S 74°35′ E)	3.3°C (−0.5°–2.8°C) 6°F (31°–37°F)
	Macquarie Island Indian Ocean (54°36′ S 185°45′ E)	3.3°C (2.8°–6.1°F) 6°F (37°–43°F)
Lowest mean annual hours of sunshine	Laurie Island (66°44′ S 44°44′ W)	500 hr
Lowest monthly percentage of possible sunshine	Argentine Island (June) (61°15′ S 64°15′ W)	5%
Highest mean annual relative humidity	Deception Island (63° S 61° W)	90%

latitudinal regimes in turn on the basis of the seasonal pattern in the hydrologic cycle. Highlands, or mountain areas, exist in all of these different climatic regions and hence are subject to the same diurnal and seasonal patterns of solar energy and moisture as the surrounding lowlands. For example, the coast ranges of California are subject to a strong seasonal pattern of energy and a seasonal moisture regime consisting of a dry summer and wet winter.

The primary aspect of the climate of highland areas is the rapid change that takes place with elevation and orientation of the slopes to the sun or prevailing wind. Radiation, temperature, humidity, precipitation, and atmospheric pressure all change rapidly with height. Radiation intensity increases with height because of a steady decrease in thickness of the atmosphere. La Quiaca, Argentina, for instance, has the highest average annual radiation level of any known surface location (Table 13.5). The increase in radiation intensity with height is responsible for the sunburn that skiers or hikers often get in mountain

Table 13.5 World Climatic Extremes Recorded in Highland Climates

Extreme		Amount Recorded
RAINFALL		
Highest monthly mean	Cherrapunji, India (July)	2692 mm (106 in)
Highest monthly total	Cherrapunji, India (July 1861)	9296 mm (366 in)
Highest annual mean	Mt. Waialeale (Kauai, HI)	11.68 m (460 in)
Highest total (4-month period)	Cherrapunji, India (April-July 1861)	18.74 m (737.7 in)
Highest annual total	Cherrapunji, India (Aug. 1, 1860–July 31, 1861)	26.47 m (1042 in)
SNOWFALL		
Highest daily total	Silver Lake, CO (April 14–15, 1921)	1930 mm (76 in)
Highest 6-day total	Thompson Pass, CO (Dec. 26–31, 1955)	4420 mm (174 in)
Highest 12-day total	Norden Summit, CA (Feb. 1–12, 1938)	7722 mm (304 in)
Highest monthly total	Tamarack, CA (Feb. 1911)	9906 mm (390 in)
Greatest depth on ground	Tamarack, CA (Mar. 9, 1911)	11.53 m (454 in)
Highest annual mean	Paradise Ranger Station (Mt. Rainier, WA)	14.78 m (582 in)
Highest annual total	Paradise Ranger Station (Mt. Rainier, WA)	25.83 m (1017 in)
WIND VELOCITY		
Highest 1-hr mean	Mt. Washington, NH	77.3 m per sec (173 mph)
Highest 24-hr mean	Mt. Washington, NH	57.7 m per sec (129 mph)
RADIATION		
Highest mean annual	La Quiaca, Argentina (22° S, 3459 m elev.)	667 ly/day

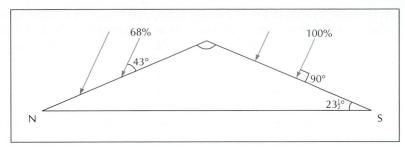

Figure 13.7
Slope orientation and radiation intensity. The figure represents the difference in radiation falling on a south-facing slope versus a north-facing one.

resort areas. There is a sharp increase in ultraviolet radiation as the atmosphere thins. Ambient air temperatures decrease with height above sea level at about 1°C per 100 meters. As the density of the atmosphere decreases, the concentration of water vapor and carbon dioxide decreases. In the absence of these gases, the greenhouse effect of the atmosphere drops sharply, and the ground surface heats and cools rapidly. Higher diurnal temperature ranges result. El Alton, Bolivia, at an elevation of 4,081 meters (13,468 ft), has a mean annual temperature of 9°C (48°F) and a mean daily range of 13°C (23°F).

The microclimate of a site in highlands also depends on slope orientation. Orientation to the sun is significant in determining local heating characteristics. A slope facing the sun has much higher surface temperatures and resultant air temperatures than one facing away from the sun. This is a result of greater intensity of solar radiation (Fig. 13.7).

Relative humidity increases with elevation up to a point. This is because the ability to hold moisture is a function of atmospheric temperature and temperature decreases with height. For this reason, fog is more prevalent at higher elevations (Table 13.6). Moose Peak in Maine averages 1580 hours of fog annually, and Mt. Washington, New Hampshire, averages 318 days per year with fog. The seasonal pattern of precipitation is similar to that of surrounding lowlands. The amount of precipitation changes rapidly with elevation and orientation to the wind. Since highlands provide a mechanical lifting mechanism, precipitation tends to increase up to a point and then decrease again as moisture precipitates out. Orientation of slopes to the prevailing winds is just as significant as elevation in determining local precipitation. Windward slopes

Table 13.6 Comparison of Number of Days per Year with Fog at Neighboring Mountain and Valley Stations

Station	Elevation (ft)	Fog days/year
Mt. Washington, NH	6280	318
Pinkham Notch	2000	28
Pikes Peak, CO	14,140	119
Colorado Springs	6072	14
Mt. Weather, VA	1725	95
Washington, DC	110	11
Zupspitze, Germany	9715	270
Garmish, Germany	2300	10
Taunis, Germany	2627	230
Frankfurt, Germany	360	30

Source: Landsberg H. L., *Physical Climatology*, 2e., (Dubois, PA: Gray Printing Co., 1958), p. 140.

receive up to five times the amount of precipitation that leeward slopes get. The rainiest areas found on the earth are where there is orographic lifting of onshore winds.

Since temperature decreases with elevation, a greater percentage of total precipitation occurs as snow in highland areas. The mean snow line for the earth is about 4550 meters (15,000 ft). In the tropical regions, it rarely goes above 5450 meters (18,000 ft). The snow line is usually higher on the leeward sides of mountain ranges because there is less snow, more insolation, and Foehn winds melt the snow. Mountains in Africa, South America, and Asia all exceed this elevation, and so there are snow-capped mountains even along the equator. Extremely heavy snowfalls are common in some highlands. The 25 meters of snow that fell at Paradise Ranger Station on Mt. Rainier in the winter of 1970–1971 serve as an extreme example.

CHINOOK WINDS

The *chinook* is a down-slope breeze or wind found along the east side of the Cascade and Rocky Mountains of North America. The name comes from the Chinook Indians, who lived along the lower reaches of the Columbia River. The chinook is a relatively warm and dry wind. It develops with the uplift of relatively mild stable air ascending windward slopes of mountains. The wind develops fairly quickly with a sharp rise in temperature and a drop in relative humidity. These winds are most often identified with the sudden changes in temperature brought about by these winds. On January 22, 1943, the temperature rose from −20°C (−4°F) at 7:30 A.M. to 7°C (45°F) at 7:32 A.M. at Spearfish, South Dakota. This was a total rise of 27°C in 2 minutes. These winds occur most often in a narrow zone about 300 kilometers east of the crest of the Rockies from New Mexico north into Canada. They form on the leeward sides of mountain ranges.

The primary source of heat in the chinook is the heat of compression. Chinooks form when there is a strong westerly flow of air across the mountains. When there is a strong wind, a trough or cell of low pressure forms on the east side of the mountains. As the air is stable, the low-pressure trough pulls the wind down the eastern slope of the mountains to its original altitude. The subsiding air heats at the dry adiabatic rate of 10°C/kilometer.

The chinook is strengthened if precipitation takes place in the air stream as it rises over the windward side of the mountains. Condensation adds heat to the air. When the air descends on the lee side of the mountains, it is warmer than at the same height on the west side.

A cloud called a chinook wall cloud forms over the Rockies when a chinook is blowing. The cloud forms and remains over the mountain crests providing a visible sign of the wind system. As the air descends the eastern flanks of the mountains, the clouds evaporate in the warmer air and a clear sky may exist above the chinook.

Chinook winds are gusty and can reach velocities as high as 160 kilometers (100 mi) per hour. The chinook has the capacity to absorb much water. Because they are warm and dry, in the winter they may sublimate large amounts of snow. It is not unusual for these winds to remove 150 millimeters (6 in) of snow a day. They have removed a half meter (19.5 in) of snow in a single day. It is for this reason that these winds are called *snow eaters*.

The same wind occurs in other parts of the world. It is the Foehn wind in the Alps Mountains of Europe. In Argentina, the wind is a **Zonda**.

SUMMARY

Cold is the distinguishing characteristic of the polar climates. Solar radiation of very low intensity is the primary factor in producing the cold in both polar regions. The Antarctic is colder than the Arctic because it is an elevated land mass. The

Arctic Ocean continually supplies heat to the atmosphere even when covered with ice. For this reason, the northern hemisphere cold pole is some distance away from the geographic pole. The major continental ice sheets existing on the planet at present are in Greenland and the Antarctic. Both are related to moderate snowfall and cold temperature. The primary circulation of the atmosphere carries heat into the polar areas to raise the temperature above what it would otherwise be.

Mountain climates are unique in that elevation and orientation to the winds are the primary variables associated with temperature and precipitation amounts. The seasonal distribution of precipitation and insolation is the same in the mountains as in the surrounding lowlands.

KEY TERMS

Arctic hurricanes	Aurora	Sastrugi
Astronomical twilight	Katabatic wind	Zonda

REVIEW QUESTIONS

1. How do the number of hours of sunlight received at the two poles over a period of a year compare with the number of hours of sunlight at the equator?

2. What is the primary factor resulting in the low temperatures at the two poles?

3. What is the fundamental difference in the physical environment at the North Pole from that of the South Pole?

4. Why do air temperatures not get as cold in the Arctic as they do in the Antarctic?

5. Temperatures do not change much during the normal 24-hour period in the Antarctic. Why is this the case?

6. Annual precipitation is low in the Antarctic. It is sometimes referred to as a desert. Why is precipitation so limited over the continent?

7. Laurie Island in the south Atlantic Ocean averages the fewest hours of sunshine per year of any location on Earth. Why does this island have so little sunshine?

8. Mountain ranges are often places of extremes. Paradise Ranger Station on the southwest slope of Mount Rainier averages more snow per year than any site in the United States. What factors contribute to this high amount of snowfall?

9. Why do Chinook winds occur along the Rocky Mountains but not along the Appalachian Mountains?

10. How can there be snow-capped mountains and glaciers along the equator when radiation intensity is so high?

PART III
Past and Future Climates

Frost Fair Upon the Thames © Culver Pictures

CHAPTER 14

Reconstructing the Past

CHAPTER OUTLINE

U sing the instruments of today, minor changes or trends in weather are routinely monitored and analyzed. However, the period for which weather instruments have been available is but a tiny fraction of time in Earth's history. To understand current climates and predict future climates, it is absolutely essential that they be considered in the framework of climatic change over geologic time. To achieve this, the climates of the past must be reconstructed. This process reflects much painstaking, detective-like research because much of the evidence is based on the relationship between climate and other environmental processes and past life. Prior to presenting an account of how climate has varied, this chapter deals first with the way in which climates are deduced.

The reconstruction of climates over time is a fascinating puzzle. Instrumental records of meaningful spatial extent have become available only in very recent times, and information about earlier climates requires the use of **proxy data**— observations of other variables that serve as a substitute or proxy for the actual climatic record. Proxies are paleoclimatological archives. In recent years, the development of highly sophisticated methods of analyzing and dating materials has led to great advances in reconstructing the past. Table 14.1 provides a listing of paleoclimatic data sources, some of which are discussed in the following pages.

EVIDENCE FROM ICE

Study of the processes that modify Earth's surface was the first indication that climates have varied over time, particularly in relation to the existence of ice ages. Perhaps the best-known early researcher in formulating these ideas was Swiss scientist Louis Agassiz, whose work began in the early 1800s.

Beginning his research in Switzerland, Agassiz noted that the valleys had a U-shape rather than the typical V-shape of river valleys. Other researchers had previously noted the same phenomenon and commented on the fact that the U-shaped valley contained a relatively small river that was out of proportion to the valley's size. The other researchers explained this occurrence by referring to the biblical flood so vividly recorded in the Old Testament. It was assumed that the huge valleys must have been carved by much larger streams when the flood occurred.

Agassiz was also impressed by the presence of large boulders set amid assorted finer sediments that had obviously been transported from an area from where they landed. These boulders, called *erratics*, aroused much curiosity, and their presence was again attributed to the great biblical flood.

Other geologists disagreed with this view of a catastrophic cause for the erratics. These geologists maintained that the forces that caused the U-shaped valleys and erratics were the same forces that now operated. They maintained that present processes provided the key to what happened in the past. Accordingly, they believed that the large erratics scattered over many areas of the world had been dumped by ice; basing their ideas on observed facts, they concluded that the erratics had been deposited by icebergs that floated on an extensive sea that formerly covered large areas of Europe. Using the same reasoning, they decided that the assorted finer materials in which erratics occurred had been derived from melting icebergs, and they applied the term *drift* to these assorted sediments.

At first, Agassiz was convinced that the iceberg theory best explained the erratics, but in his work he met other scholars who had different ideas. Scientists like Jean de Charpentier and Ignace Venetz had studied the landscape of the Swiss Alps and were convinced that the glaciers they saw had once been much more extensive. They believed the glaciers were responsible for the valley shapes, the drift, the erratics, and the parallel striations found on hard rock surfaces.

Table 14.1   Paleoclimatic Data Sources and Their Characteristics

Data Source	Variable Measured	Potential Geographical Coverage	Period open to Study (years BP)	Climate Inference
Ocean sediments (cores, accumulation rate of < 2 cm /1000 years)	Isotopic composition of planktonic fossils; benthic fossils; mineralogic composition	Global ocean	1,000,000+	Sea-surface temperature, global ice volume; bottom temperature and bottom-water flux; bottom-water chemistry
Ancient soils	Soil type	Lower and midlatitudes	1,000,000	Temperature, precipitation, drainage
Marine shorelines	Coastal features, reef growth	Stable coasts, oceanic islands	400,000	Sea level, ice volume
Ocean sediments (common deep-sea cores, 2–5 cm/1000 years)	Ash and sand accumulation	Global ocean (outside red clay areas)	200,000	Wind direction
Ocean sediments (common deep-sea cores, 2–5 cm/1000 years)	Fossil plankton composition	Global ocean (outside red clay areas)	200,000	Sea-surface temperature, surface salinity, sea-ice extent
Ocean sediments (common deep-sea cores, 2–5 cm/1000 years)	Isotopic composition of planktonic fossils; benthic fossils; mineralogic composition	Global ocean (above $CaCO_3$ compensation level)	200,000	Surface temperature, global ice volume; bottom temperature and bottom-water flux; bottom-water chemistry
Layered ice cores	Oxygen-isotope concentration (long cores)	Antarctica; Greenland	100,000+	Temperature
Closed-basin lakes	Lake level	Lower and midlatitudes	50,000	Evaporation, runoff, precipitation, temperature
Mountain glaciers	Terminal positions	45° S to 70° N	50,000	Extent of mountain glaciers
Ice sheets	Terminal positions	Midlatitudes High latitudes	25,000 to 1,000,000	Area of ice sheets
Bog or lake sediments	Pollen-type and concentration; mineralogic composition	50° S to 70° N	10,000+ to 200,000	Temperature, precipitation, soil moisture
Ocean sediments (rare cores, > 10 cm/1000 years)	Isotopic composition of planktonic fossils; benthic fossils; mineralogic composition	Along continental margins	10,000+	Surface temperature, global ice volume; bottom temperature and bottom-water flux; bottom-water chemistry
Layered ice cores	Oxygen-isotope concentration, thickness (short cores)	Antarctica; Greenland	10,000+	Temperature, accumulation
Layered lake sediments	Pollen type and concentration (annually layered core)	Midlatitude continents	10,000+	Temperature, precipitation, soil moisture
Tree rings	Ring width anomaly, density, isotopic composition	Midlatitude and high-latitude continents	1000 to 8000	Temperature, runoff, precipitation, soil moisture
Written records	Phenology, weather logs, sailing logs, etc.	Global	1000+	Varied
Archeological records	Varied	Global	10,000+	Varied

Source: After Kutzbach, 1975.

The accumulation of this type of evidence prompted Agassiz to formulate a comprehensive theory of extensive **glaciation**. He suggested that a great ice sheet had extended from the North Pole to the Mediterranean and that the moraines, striations, erratics, and drift seen in Switzerland had resulted from the action of glaciers. The idea of an ice age was born. Today it seems somewhat astonishing that the idea of major glaciation covering much of the earth became accepted only in the 19th century.

Glaciers and Glaciation

The results of ice erosion and deposition are characteristics of large parts of the northern hemisphere. Figure 14.1 shows a typical view of an area that has experienced mountain (Alpine) glaciation. The resulting features—the U-shaped valleys, hanging troughs, aretes, and tarns—are well known and clearly indicative of ice activity.

Although the results of the work of ice are an important interpretive device, glaciers provide a measure of predicting temperature and precipitation conditions. The advance or retreat of glaciers has been used in interpreting climatic change over historic periods. There is evidence that glaciers shrunk about 3000 B.C. and that Alpine snow level was at least 1000 feet higher than today. By 500 B.C., there was a marked re-advance followed by a recession. Between the 17th and 19th centuries, a general resurgence was observed in the Alps and Scandinavia. As indicated by European glaciers in Figure 14.2, the 20th century tended to be a time of glacier retreat, although in some locations glaciers were growing (Fig. 14.2).

Features associated with continental glaciation differ markedly from those of Alpine glaciation. Depositional features such as the position of terminal

Figure 14.1
The town of Revelstoke lies on the floodplain of the Columbia River. The valley was carved by glaciation, and the floor is now filled with sediment (from Photo Researchers, Inc.).

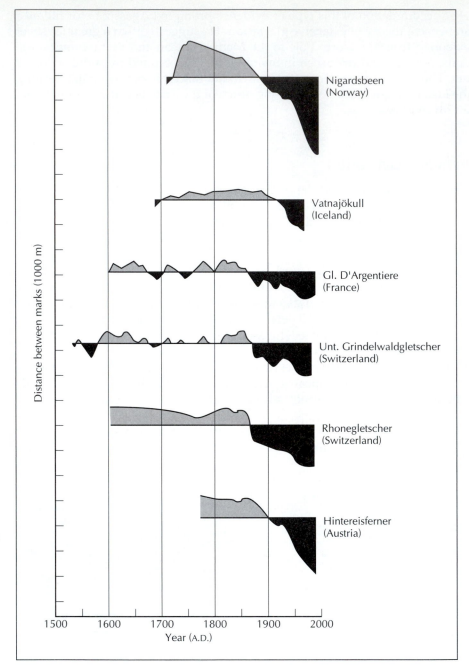

Figure 14.2
Valley glaciers throughout Europe and Iceland have been retreating since the latter part of the 18th century (from Turekian K. K., *Global Environmental Change*, © 1996, Prentice Hall, Upper Saddle River, N.J.).

moraines and the erratic boulders when studied in detail allow reconstruction of the extent and movement of the ice. The erosional features, glacial striations, ice-gouged lakes, and so on provide similar evidence. Using such findings, it is possible to reconstruct conditions that existed in North America during ice advance. Figure 14.3 provides one interpretation.

Ice Sheets and Cores

In the great ice caps of Greenland and Antarctica lies a wealth of climatic information. Ice sheets are formed layer by layer from the snowfall of each year; with time, the snow is compressed into ice often filled with the bubbles of trapped air. By drilling

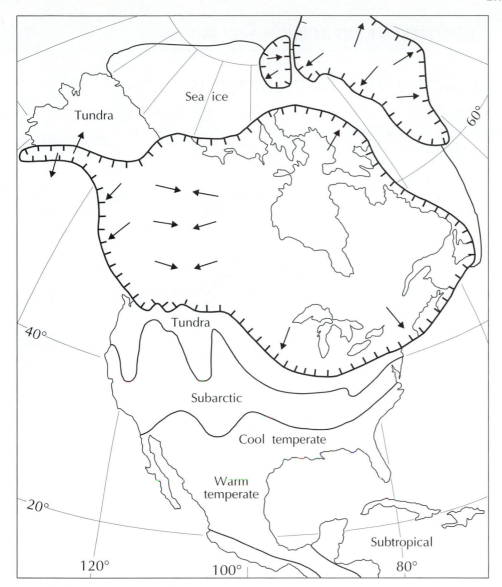

Figure 14.3
The maximum extent of the Pleistocene ice advance in North America. Note the reconstructed region is equatorward of the ice.

into the ice, a great deal can be learned about the atmospheric environment in which original snow formed and was deposited. **Ice cores** are derived by driving hollow tubes into carefully selected ice cap sites where the original ice deposit has been minimally disrupted (Fig. 14.4). The cores are stored at below-freezing conditions for analysis. At the Greenland Ice Sheet Project 2 (GISP 2), a core more than 3050 meters has been obtained. This reflects a record of some 110,000 years. At the Vostok drilling in Antarctica, a core of 3350 meters represents some 160,000 years. The ice cores provide both direct and indirect (proxy) indications of past climates.

The water in ice contains oxygen, and oxygen has atoms whose atomic weights vary from the normal atom (i.e., there are **isotopes** of oxygen, the two most common of which are oxygen 18, the heavy isotope, and the lighter oxygen 16). By analyzing the ratio of one isotope to another in the ice deposit (see Box 14.1), it is possible to determine the existing environmental temperature. From this it is

QUANTITATIVE EXPRESSION: BOX 14.1

Isotopes and Past Climates

Isotopes of an element are those whose atoms have the same number of electrons and protons but have a different number of neutrons. Both oxygen and hydrogen, the elements comprising water/ice, have isotopes. Oxygen commonly exists as oxygen 16 (^{16}O) and the rarer and heavier oxygen 18 (^{18}O). The two naturally occurring isotopes of hydrogen are hydrogen-1 (^{1}H) and hydrogen-2 (^{2}H). The latter, the heavier and rarer isotope, is called deuterium (D).

$^{1}H^{16}O$ is the lighter and more common form of the water molecule. Separation of isotopes (fractionation) is a function of temperature. As noted in the text, water in the oceans contains primarily $^{1}H^{16}O$, which evaporates more readily than the heavier $^{2}H^{18}O$. Depending on the temperature of evaporation and the distance traveled before falling as snow, the ratio of the two isotopes will vary, and this variation is retained in the layers of snow that are deposited. This becomes the ice of permanent ice fields. The ratio may be measured accurately using a mass spectrometer, and results are reported in parts per thousand (‰) using the delta (δ) symbol. Substitution of the derived value of $\delta\,^{18}O‰$ or $\delta\,^{2}H^{18}O$ are used in the following equations:

$$\delta\,^{18}O‰ = \frac{(^{18}O/^{16}O)_{ICE} - (^{18}O/^{16}O)_{SMOW}}{(^{18}O/^{16}O)_{SMOW}} \times 1000. \tag{1}$$

The abbreviation *SMOW* is the Standard Mean Ocean Water, and the equation provides the deviation of the sample from that known quantity.

The same notation is also used to report the relative proportions of light and heavy hydrogen:

$$\delta D = \frac{(^{2}H/^{1}H)_{sample} - (^{2}H/^{1}H)_{SMOW}}{(^{2}H/^{1}H)_{SMOW}} \times 1000. \tag{2}$$

Oxygen isotopes are also used to analyze the tiny shells of foraminifera that are deposited on the ocean floor. In this case, the oxygen is in the calcium carbonate ($CaCO_3$) of the shells. The shells are separated from the mud and silt and are dissolved in acid; the resulting carbon dioxide (CO_2) is analyzed. As in ice cores, the mass spectrometer is used to identify the isotope ratio, and the result is compared to a standard of a powdered sample (supplied by the U.S. National Bureau of Standards). Values are substituted into an equation similar to Eq. 1 using values from the sample and standard.

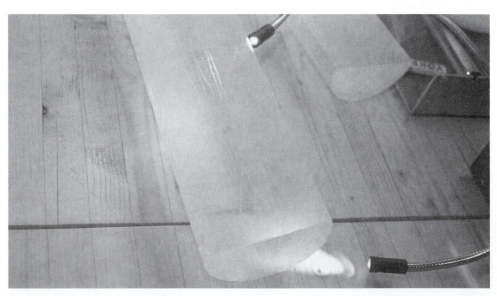

Figure 14.4
Examining ice cores taken from polar ice caps (© Richard Monastersky).

possible to derive a pattern of temperature over time as represented by the layers of ice in the core. Other significant information is gained from the trapped air bubbles; minute samples can be used to determine, for example, the level of carbon dioxide in the atmosphere. Layers of dust may give insight into the storminess of the atmosphere or the extent of volcanic activity.

The snowfall of each year, together with the constituents it contains, is buried by successive annual accumulations. Eventually the weight of overlying snow causes lower layers, below some 80 meters at GISP 2, to turn to ice. Air deposited with the snow is trapped in gas bubbles (or inclusions) in the ice, which represent the composition of air at the time the snow first fell. From these inclusions, it is possible to determine the levels of greenhouse gases over time. Figure 14.5 provides the record of 160,000 years from the Vostok drilling, which, in this case, provides the varying concentrations of carbon dioxide (CO_2). The relationship between the two provides evidence of how global temperatures vary with greenhouse gas concentrations—an important factor in the evaluation of the global warming debate.

Air inclusions in ice cores are used not only to re-create ancient atmospheric chemistry, but also to determine air pollution levels since the Industrial Revolution. Carbon dioxide, methane, and nitrous oxide are some constituents that have been monitored. All show an increase over the last 200 years.

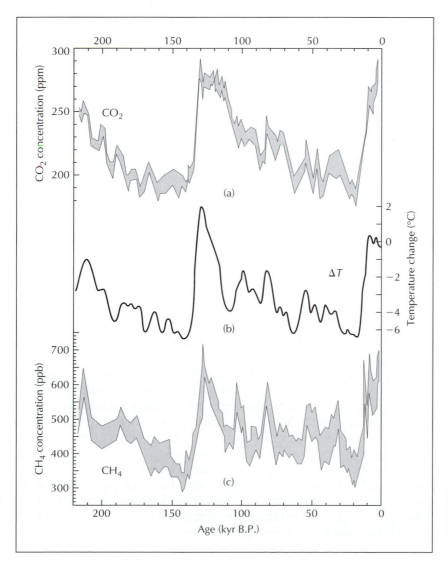

Figure 14.5
The trend of atmospheric carbon dioxide (CO_2), methane (CH_4), and temperature as recorded in the Vostok, Antarctica, ice core. The atmospheric temperature at Vostok is plotted as a deviation from present-day mean temperature (ΔT_{atm}). The different line widths represent the ranges in estimates (from Kump L. R., Kasting J. F. and Crane R. G., *The Earth System*, © 1999, Prentice Hall, Upper Saddle River, N.J.).

The annual layers in ice cores provide considerable information about contaminants in the atmosphere for both ancient and more recent times. For example, radioactivity levels in the ice layers show that large peaks occurred at the time of the Chernobyl disaster in Russia and the times of atomic testing that occurred in the 1950s and 1960s. Volcanic activity may be assessed by examining the annual layers of ice cores, akin to tree ring analysis. The analysis of the core provides the peaks of sulphate levels of volcanic origin. Precise dating of cataclysmic events may be determined. A marked peak occurred some 73,000 years ago when an Indonesian volcano called Toba erupted. Other peaks provide a clear sequence of Iceland volcanic effects.

The improvements in transferrable technology have led to ice cores being derived from places other than Greenland and Antarctica. Of considerable importance are the cores derived from the high mountains of Peru. These provide a guide to temperature changes in low latitudes. Results obtained may be correlated with the temperature changes indicated in tropical marine sediments. Evidence suggests that tropical oceans were much cooler during the ice ages.

Periglacial Evidence

It is not, however, features associated with the ice that supply all geomorphic evidence regarding ice ages. **Periglacial** areas, those close to but not covered by continental ice, experienced a climate quite different from today's. Some areas that are now quite dry experienced much wetter climates as a result of modified circulation patterns. Many inland basins, for example, have been occupied by large lakes as a result of the higher precipitation. Such pluvial lakes, so named because they resulted from increased precipitation in earlier times, have been widely identified.

The western part of the United States, particularly in the Great Basin area, shows fine examples of the extent of such lakes. In this area, it is estimated that glacial lakes Bonneville and Lahontan were enormous. At its maximum, Lake Bonneville occupied some 50,000 kilometers—an area approximately the size of Lake Michigan. Evidence of the extent of this great lake is found in the present Bonneville salt flats and in the *strand lines*—the shore areas indicative of the former level of the lake. Such lakes would necessarily modify the drainage systems, and the formation of a lake and its eventual overflow might lead to a totally new drainage direction.

ANCIENT SEDIMENTS

Sediments deposited during geologic time offer evidence of the climatic environment in which they were formed. One example is offered by **evaporite** (salt) deposits. Salt deposits are formed when, on a long-term basis, evaporation exceeds precipitation in an area where water flows in from other areas. Water evaporates to leave the salts that were formerly in solution as sediments. Eventually these may be buried and turned into sedimentary rock. The large evaporite beds of western United States, Germany, and central Asia are similar deposits being formed at the present time.

Wind-deposited materials also provide a guide to prevailing winds in earlier times. During the last ice age, sand dunes formed around glacial lakes and along shorelines. The dunes' structure and location have been used to estimate local wind conditions. Similarly, pockets of loess were deposited over wide areas. Loess is made up of silt picked up from the edge of the glacier by the wind and scattered over a broad area many miles from the glacier.

Of particular importance in climatic reconstruction are ocean sediments. These provide many kinds of evidence including fossil isotopic composition and plankton accumulation, which are described later in this chapter. Additionally, the sediment contains volcanic ash, which provides information about crustal activity and sand. This may give insight into prevailing wind directions during a past age.

SEA LEVEL CHANGES

Changes in the ocean level can occur when the volume of water or the volume of the ocean basins increases or decreases. Many factors can cause either of these to occur, but most of them—such as the accumulation of sediments on the ocean floor or the extrusion of igneous rocks into the oceans—require many years. Rapid changes mostly result from ice alternately accumulating and melting. If the present Antarctic ice sheet were to melt, some researchers calculate that there would be enough additional water to cause a sea level rise of 60 meters (200 ft). The removal of ice from a large land area would, however, be accompanied by an upward readjustment of the land. The ocean floors would sink farther into the crust under the weight of additional water. Even allowing for this, melting of the Antarctic ice would still cause a rise of 40 meters (135 ft)—sufficient to flood most of the world's major ports.

The rise and fall of sea level during the Pleistocene is an important guide to glacial and interglacial periods. Submergence and emergence of coastal areas and marine terraces point to the amount of water tied up as ice; using appropriate dating methods, such evidence provides a guide to glacial advance and retreat. Figure 14.6 shows global mean sea level over the last 150 thousand years as derived from a variety of sources.

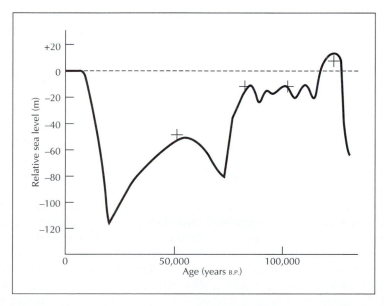

Figure 14.6
Inferred sea-level curve from raised coral terraces and deep-sea core oxygen 18 data (from Turekian K. K., *Global Environmental Change*, © 1996, Prentice Hall, Upper Saddle River, N.J.).

PAST LIFE

Studying species of organisms that lived in the past provides much information about the climate of the time. All plants and animals have a preferred set of physical conditions in which they live. Plant and animal fossils can be used in the reconstruction of past ecologic conditions, including the climates in which they lived.

The Faunal Evidence

Invertebrate fossils are those of creatures without backbones. They are widely used to establish the geologic sequence of rock, and they have proved valuable for reconstruction of past climates. However, caution should be exercised because it is possible to draw biased conclusions from incomplete evidence. This restriction may apply particularly to the invertebrates because many fossil species lived in fresh or salt water; they were not directly affected by the atmospheric climate because the sea, lake, or river in which they lived acted as a buffer to direct exposure.

Although the physiology of fossil animals in relation to modern species is used to determine paleoclimates, much information has recently been gained from the chemistry of invertebrate organisms. Of particular note in this respect is the use of isotopes especially as part of cooperative research.

Climap was one of the important multidisciplinary projects funded to study and understand climates of the past. It investigated ocean atmosphere changes to obtain fundamental knowledge of what produces climatic variations and how the changes might be predicted. Another project, Cohmap (Cooperative Holocene Mapping Project), had similar undertakings, but concentrated on the climates of the last 10,000 years. One of the keys to the Climap research was obtained from cores taken from layers of mud that cover ocean floors.

The layers that exist contain billions of microscopic skeletal remains of plankton. When the creatures die, their remains fall to the ocean depths and are incorporated in the mud layers. Since the tiny organisms are adapted to temperature, each species lives within a certain ocean climate; if that climate changes, they drift away to be replaced by plankton better adapted to the new conditions. Thus, the fossils found in the mud layers provide a record of temperature changes in the oceans—hence, in the atmosphere above the ocean. To obtain a sample of the undisturbed layers, cores up to 30 meters long are taken from the ocean floor. Analysis of the cores is a lengthy task since 20 to 50 species and up to 500 individuals exist in each few centimeters of core.

Dating the time at which the organism was deposited is achieved through *carbon 14* analysis. The basis of this technique is that skeletons contain both ordinary carbon and a minute trace of the isotope carbon 14. The proportion of carbon 14 to carbon 12 remains fixed while the organism is alive. After it dies, the carbon 14 begins to decay; by knowing the ratio of carbon 12 to carbon 14, one can determine the age of the shell.

Fossils with backbones—those of vertebrates—provide important clues to past climates. Much can be interpreted from their fossil distribution and their physiology as related to environment. The great changes in vertebrate life over geologic time result in quite different interpretive methods. For example, the extinction of the dinosaurs in Cretaceous time has promoted much discussion; one view is that progressive and increasing world aridity might have been the cause of their extinction. Alternatively, and as described in the next chapter, an asteroid may have been the forcing factor. Although the physiology of vertebrates is probably the most widely used method for interpreting the ecologic conditions under which they lived, important evidence is also offered by the way in which they are fossilized. For example, some are found right side up in a standing position. To explain their death, it might be assumed that they became bogged down in a swamp environment. By relating such evidence to surrounding deposits and other indicator fossils, it becomes possible to reconstruct the environments in which they lived. The fact that many fossil remains are found close together indicates death of animals through a catastrophe. Obviously, the catastrophe can take various forms—freezing or drought, for example—but by correlation to other past climatic indicators, its nature might be deduced.

The Floral Evidence

Plant distribution provides an important guide to the distribution of climate at the present time; the same is true of paleoclimates. Identification of vegetation patterns and their changes over time is widely used to interpret past climates. Often the evidence is used in relation to other environmental features. For example, it has already been noted that the extent of mountain glaciers varies. A change will be reflected in mountain vegetation, particularly the elevation of the tree line on the mountain. This feature can be traced over time and has been used to evaluate climatic trends in selected areas.

The physiology of plants, like that of animals, again provides much information. The development of drip leaves in plants is indicative of their existence under very moist conditions; the fossil remains of plants with thick, fleshy leaves are probably indicative of arid or semiarid climates. Figure 14.7 provides another method to differentiate the climate under which plants grow.

The interpretation of the climate of the Carboniferous, a time of prolific vegetation that gave rise to great thicknesses of coal measures, provides examples of this use of plant physiology. Many of the fossil plants in the Carboniferous appear to be related to the horsetail and the club mosses, both representative of a marsh or swamp environment. Such an interpretation is endorsed by fossils that suggest plants with layered roots, such as those found in modern bog plants and by many minor structures that appear to indicate that some of the plants actually floated on water. Trees lacked a development of growth rings, indicative of a climate without marked seasonal differences; the dominance of trees over herbaceous plants would further indicate a swamp environment. In all, the representative vegetation suggests a warm, moist climate that favored a luxurious, if wet, plant cover.

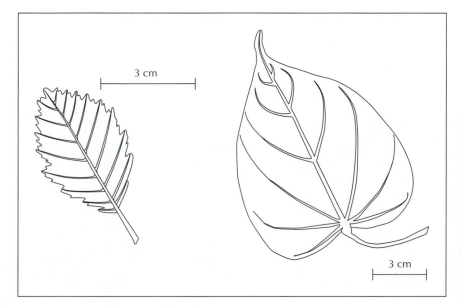

3 cm

3 cm

Figure 14.7
Degree of serration of leaves is related to temperature. The smaller leaf is characteristic of plants in temperate climates, and the larger smooth-edged leaf is typical of tropic plants (from Turekian K., *Global Environmental Change*, © 1996, Prentice Hall, Upper Saddle River, N.J.).

Similar evidence has been used to help reconstruct the climates of late Paleozoic and Mesozoic rocks. More recent deposits can be interpreted by pollen analysis; for much more recent plant distributions, tree ring analysis is used.

The study of pollen grains, or spores, is termed **palynology**. Its success depends on the fact that many plants produce pollen grains in great numbers (e.g., a single green sorrel may produce 393 million grains; a single plant of rye, 21 million grains) and that pollen is widely distributed in the area in which plants are found. Most important, the outer wall of the pollen grain is one of the most durable organic substances known. Even when heated to high temperatures or treated with acid, it is not visibly changed. This is important because pollen possesses morphological characteristics that allow identification of groups above the species level.

For pollen to be of value in interpreting the past distribution of vegetation, and hence inferring the climate that occurred, it is necessary to obtain a layered sequence of the pollen. As shown in Figure 14.8, this often occurs in ancient lakes or peat bogs where seasonal pollen deposits would be covered by sediments. Cores taken would show a sequenced pattern.

Pollens from the cores are identified and a frequency distribution of plant types derived. Thus, a high proportion of spruce pollen in the lower core might give way to oak pollen at higher levels. This variation would indicate that a vegetation change had occurred over time and that the difference could be related to a passage from cool to warmer climatic conditions. Even a relatively crude classification of pollen type (e.g., those from trees compared with those from other plants) could provide a rough guide to changing climatic conditions. The change from pollen associated with the nontree climates of the cold tundra to tree climates might indicate an amelioration of climatic conditions. In-depth statistical counts obviously provide more detail. Much work on this has been completed in Scandinavia, where the first palynological stratigraphy was devised.

Despite the important progress using this method, it does have shortcomings. A vegetation cover only attains maturity after a fairly lengthy period of time, and it is quite feasible that the vegetation established through pollen analysis represents a successional stage that is not totally representative of the prevailing climate. In some areas, the vegetation cover is mixed; thus, it becomes difficult to establish any dominant type that can be related to climate. It has been pointed out that from Neolithic times, people have interfered extensively with the forest cover, and human-induced changes might give misleading results.

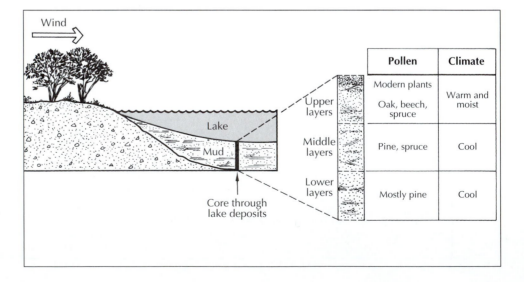

Figure 14.8
Simplified diagram showing the method of reconstructing past climates using pollen analysis.

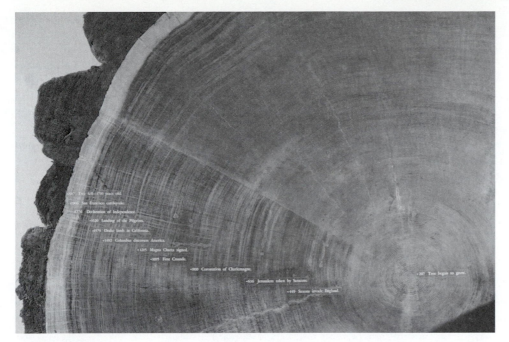

Figure 14.9
Tree ring analysis of the trunk of a 1710-year-old Sequoia (from Photo Researchers, Inc. © Tom McHugh).

Tree ring study, or **dendrochronology**, was pioneered by A.E. Douglas and his colleagues at the University of Arizona. Initial studies were used in an attempt to relate seasonal growth of trees to sunspot cycles, and a great deal of significant work was completed. In the quest for ancient living trees, the *Pinus aristata* was found to be 4000 years old. Analysis of such ancient trees permitted reconstruction of climates of the American southwest during various settlement periods.

Tree ring analysis depends on the fact that growth rings record significant events that happened during the tree's life history. Growth rings are formed in the xylem wood of trees (Fig. 14.9). Early in the season, the xylem cells are smaller and darker. The abrupt change from light to dark-colored rings delineates the annual increments of growth. Study of these rings, their size, and variations provides information about the varying environmental conditions to which the tree was subjected. The method is, of course, most valuable in determining conditions that existed in a relatively recent part of geologic history and is widely used in archeological research.

EVIDENCE FROM THE HISTORICAL PERIOD

Many researchers have used historical records to establish climatic changes that have occurred during the brief existence of humans on Earth. Their findings have allowed fairly detailed reconstruction of climates over the past 6000 years. Medieval chronicles contain many references to prevailing weather conditions. Unfortunately, although not surprisingly, these pertain to exceptional weather events rather than day-to-day conditions. These events include such unusual phenomena as the freezing of the Tiber River in the 9th century and the formation of ice on the Nile River. Although these sources do not supply a continuous record, we can construct an overall view of the usual conditions by assessing the number of times given events are recorded. The freezing of the Thames River provides one such example. Between 800 and 1500, only one or two freezings per century were recorded. In the 16th century, the river froze at least four times, in the next century it froze eight times, and six freezing periods were recorded in the 18th century. One can only suppose that progressive cooling increased the frequency. Guides to the conditions that existed are to be found in artwork and etchings of the periods.

A good example of the use of historic data for climate reconstruction comes from Iceland, where the following sources have been used:

1864 to present: actual meteorological instruments

1781–1845: a reconstruction of weather conditions as derived from the relative severity and frequency of drift ice in the vicinity of Iceland

1591–1780: historical records combined with incomplete drift ice data

900–1590: information from Icelandic sagas indicating times of severe weather and related famines

Such a complete record as this can be used as a base guide to the climate of the entire North Atlantic area. It is indicative of the methods that may be combined to complete a climatic record.

CLIMATIC CHANGE OVER GEOLOGIC TIME

Table 14.2 shows the geologic time divisions, naming the eras and periods, and identifies the times at which Earth was gripped by ice ages. Perhaps the most significant idea expressed in the table is the fact that average global climates have been much warmer than they are today throughout much of geologic time. Periodically, the warmth has been interrupted by times of cooling ice ages.

Table 14.2   Geologic Time Divisions

Era	Period	Beginning (mbp)*	Ice Ages
Cenozoic	Quaternary	2–3	Pleistocene ice age
	Tertiary	65	
Mesozoic	Cretaceous	135	
	Jurassic	190	
	Triassic	225	
Paleozoic	Permian	280	Ice age at approximately 300 mbp
	Carboniferous	345	
	Devonian	400	
	Silurian	440	
	Ordovician	500	Ice age at approximately 450–430 mbp
	Cambrian	570	
Precambrian		> 570	Ice age at approximately 850–600 mbp

*mbp = millions of years before present.

Rocks of the Precambrian period, extending back to the origin of the earth, provide but few details of the climates that existed, and only the late Precambrian times can be reconstructed with any degree of confidence. It is known, however, that during late Precambrian, much of the earth was glaciated, perhaps for the time span 950 million years before present (mbp) to 650 mbp. The Paleozoic (570–225 mbp) represents a time when fossil and sedimentary evidence became more widespread and more details of the Paleozoic environments may be derived. Organisms and rocks suggest that the Cambrian and early Ordovician were largely warm, although there may have been a glaciation toward the end of the Ordovician. The withdrawal of ice and deglaciation at the end of this event made the climates of the Silurian and Devonian similar to those of today. The Pennsylvanian/Mississippian periods were dominated by widespread humid climates with intermittent glaciation. Considerable evidence exists to indicate a major glaciation in the late Paleozoic.

The Mesozoic (225–65 mbp) was essentially a time of widespread warmth and aridity. Climates of the Triassic appear similar to those of the upper Paleozoic—cool and humid—but they gave way to a long period of warmth, especially marked during the Cretaceous. This time of Earth history saw the world in its "greenhouse mode," when climate was predominantly warm, polar ice caps nonexistent, and sea level high. The change from this to an eventual "icehouse mode" may not have been smooth, but rather episodic.

During the early Tertiary (65–22.5 mbp), the warm temperatures of the Mesozoic began to decline. Long episodes of relatively warm climates were punctuated by abrupt drops in temperature. During this time, the first glaciers since the Paleozoic began to form in Antarctica. By the later Tertiary (22.5–2 mbp), wide temperature swings occurred until the Pliocene, when the downward swings produced glaciations such as those associated with the Pleistocene.

The climates of the **Pleistocene** consisted of glacial and interglacial times, with polar ice advancing and retreating from numerous sources. The most recent full glacial, known as the Wisconsin in North America, lasted from perhaps 30,000 to 12,000 years before the present. The coldest temperature was 4°C to 6°C (7°F to 11°F) lower than present and occurred about 18,000 years ago. Huge ice sheets extended as far south as 50° N in Scandinavia and 40° N in North America. Frigid polar water extended in the North Atlantic to 45° N.

CLIMATE SINCE THE ICE RETREAT

The period from 18,000 to 5500 years ago corresponds to the deglaciation of the earth. By 12,000 years ago, only scattered areas of ice sheets remained in western North America, with the main ice sheet confined to eastern Canada.

About 10,200 years ago, a strange event occurred that affected Scandinavia and Scotland particularly: The margins of the remaining ice sheet expanded and some small ice sheets reappeared. This time, known as the Younger Dryas (named for a small flower found in cold climates), did not last long, and shortly after climatic conditons in the northern hemisphere resembled those of the present day. But the rapid temperature decline of the Younger Dryas illustrates that climatic change need not be the long, deliberate process that most people think of it as. Figure 14.10 shows the pattern of temperature change over the past 18,000 years.

After the cooling associated with the Younger Dryas, the global climate warmed. By 7000 years ago, conditions had improved such that only remnants of ice remained. The warm period peaked about 5500 years ago, and most ice disappeared, leaving only the Greenland Ice Sheet and the Arctic Ice that we have today. All

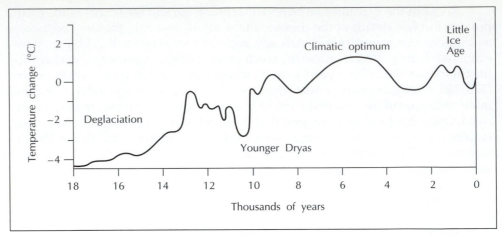

Figure 14.10
Patterns of surface temperatures of the last 18,000 years based largely on Greenland isotope temperatures. Vertical axis provides temperature change from current global average temperature.

evidence points to this being a time when the mean atmospheric temperature of the mid-latitudes was about 2.5°C (4.5°F) above that of the present. This time has been described as the **Climatic Optimum**—a term originally applied to Scandinavia when temperatures were warm enough to favor more varied flora and fauna.

The Last 1000 Years

The time extending from about 950 to 1250 is known as the Little Climatic Optimum (Fig. 14.11). Evidence of agriculture and other indicators has been used to reconstruct the climates that existed at the time Greenland was settled by the Vikings. Under the leadership of Eric the Red, the Vikings passed from Iceland, which they had settled in the 9th century, to Greenland. Although an icy land, it supported sufficient vegetation (dwarf willow, birch, bush berries, pasture land) for settlement. Two colonies were established, and farming was begun. The outposts thrived, and regular communications were established with Iceland.

Between 1250 and 1450 A.D., climate deteriorated over wide areas. Iceland's population declined, and grains, which had been grown in the 10th century, were no longer produced there. Greenland was practically isolated from outside contact, with extensive drift ice preventing ships from reaching the settlements. In Europe, storminess resulted in the formation of the Zuider Zee, and the excessively wet, damp conditions led to a high incidence of the horrifying disease St. Anthony's Fire (ergotism).

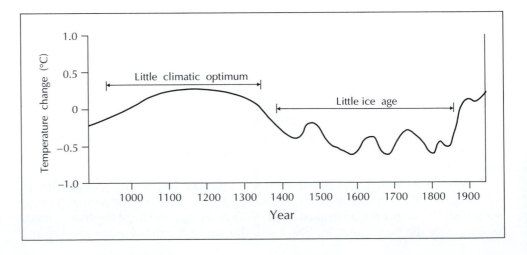

Figure 14.11
Variations from current average surface temperature over the last 1000 years. The Little Climatic Optimum was followed by a deterioration to the Little Ice Age.

Conditions continued to worsen, and from about 1450 to 1880 the time known as the **Little Ice Age** occurred. During this period, glaciers enlarged to the extent that some Alpine villages were overwhelmed by ice. At the same time, very cold winters led to the freezing of rivers and lakes that are seldom deeply frozen today. Figure 14.12 shows an example of the River Thames in London during the 17th century. The ice was so thick that ice fairs were held on the frozen water.

The Little Ice Age marked the end of the Norse settlements in Greenland that had begun in the 10th century. In fact, in 1492, the Pope complained that none of his bishops had visited the Greenland outpost for 80 years because of ice in the northern seas. (He was not aware that the settlements were gone.) By 1516, the settlements had practically been forgotten, and in 1540 a voyager reported seeing signs of the settlements, but no signs of habitation. The settlers had perished.

Whether this was due to the deteriorating climate or invasion by other groups is not known, although a Danish archeological expedition to the sites in 1921 found evidence that deteriorating climate must have played a role in the population's demise. Graves were found in permafrost that had formed since the time of burial. Tree roots entangled in the coffins indicated that the graves were not originally in frozen ground and that the permafrost had moved progressively higher. Examination of skeletons showed that food supplies had been insufficient; most remains were deformed or dwarfed, and evidence of rickets and malnutrition was clear. All the evidence points to a climate that grew progressively cooler, leading eventually to the settlers' isolation and extinction.

Later in the cold spell, the colonies in eastern United States suffered as well. The soldiers of the American Revolution suffered in the cold weather, although the unusual ice sometimes served as a useful tool. British troops, for example, were able to slide their cannons across a frozen river from Manhattan to Staten Island.

From North America comes a well-known account of life during the last years of the Little Ice Age—a description of the year 1816, known as "the year without a summer." The year began with excessively low temperatures across much of the eastern seaboard. But as spring came, the weather seemed to be cool, but not excessively so. In May, however, the temperatures plunged; Indiana had snow or sleet for 17 days, which killed off seedlings before they had a chance to grow. The cold

Figure 14.12
An Ice Fair on the River Thames in the 17th century illustrated the severity of some winters during that time. (© Culver Pictures).

weather continued in June, when snow again fell, devastating any remaining budding crops totally. No crops grew north of a line between the Ohio and Potomac rivers, and returns were scanty south of this line. In the pioneer areas of Indiana and Illinois, the lack of crops meant that the settlers had to rely on fishing and hunting for their food. Reports suggest that raccoons, groundhogs, and the easily trapped passenger pigeons were a major source of food. The settlers also collected many edible wild plants, which proved hardier than cultivated crops.

The image of the period shown by artists of the time is very different from that of today. Paintings of scenery and activities in the Low Countries show winter scenes in which ice and snow are central to the theme. One famous painting by Pieter Brueghel the Elder, shown in Figure 14.13, provides an excellent example of this image.

Fortunately, by the end of the 19th century, the instrumental record shows that the climate was again improving. A reconstructed record of temperature is shown in Figure 14.14. If we consider 1950 to 1980 to be the baseline period, we see that the years prior to that were cool but warming while those after were even warmer. It is this latter trend that raises the specter of global warming.

Figure 14.13
The Hunters in the Snow by Pieter Brueghel the Elder (Kunshistorisches Museum, Vienna).

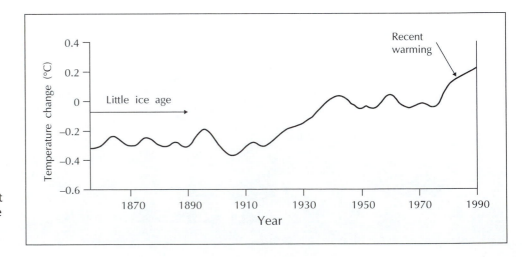

Figure 14.14
Globally average temperatures since about 1860 shown as a change from the 1951–1980 average.

SUMMARY

Reconstructing climates for the periods prior to instrumentation relies on proxy data. Early evidence of past climates was derived from studying ice, but newer methods can more accurately describe a given climatic event. The interpretation of ice cores taken from ice caps permits oxygen isotope analysis to give insight into the prevailing temperature at the time when ice formed. Surface evidence—including sediments, periglacial activity, and changes in sea level—also provides significant proxy data.

Both faunal and floral life of the past is used in climatic reconstruction. Invertebrates, especially those derived in deep sea cores, are of particular significance. Dinosaurs are perhaps the most spectacular of the vertebrates studied, and climatic change may well be significant in explaining their extinction. Floral evidence relies on the physiology of plants, the study of pollen, and tree ring analysis. Floods, droughts, human migration, sailing records, ice cover, and a host of other sources of evidence provide additional clues to climate.

A review of climate over time shows that a number of ice ages have occurred throughout geologic history. The most recent of these, the Pleistocene, was responsible for widespread evidence of glaciation. Deglaciation began about 18,000 years ago, and both warm and cool periods have occurred since that time. During the last 1000 years, a long cool time, known as the Little Ice Age, was a dominant feature. The 20th century also had both warm and cool times, with the 1980s being an exceptionally warm period.

KEY TERMS

Climatic Optimum	Ice cores	Palynology
Dendrochronology	Isotopes	Periglacial
Glaciation	Little Ice Age	Proxy data

REVIEW QUESTIONS

1. What are proxy data?
2. What types of evidence led Louis Agassiz to believe that parts of the earth had been glaciated?
3. What has happened to many European glaciers in the last 500 years?
4. What are isotopes and why are they of value in climate reconstruction?
5. How does the area around the Great Salt Lake in Utah provide information about climate change?
6. What could cause sea level to fall?
7. How can the physiology of plants help us understand past climate?
8. What is palynology?
9. How can the study of tree rings explain past climates?
10. What are the major eras of geologic time?
11. What is the Climatic Optimum? Is it a good name?
12. What type of evidence exists for identification of the Little Ice Age?

CHAPTER
15

Natural Causes of Climatic Change

CHAPTER OUTLINE

Short-Term Changes

Variation in Solar Irradiance

Sunspot Activity

Variation in Atmospheric Dust

Human-Induced Changes in Earth's Surface

Long-Term Climatic Changes

Earth–Sun Relationships

Distribution of Continents

Variation in the Oceans

Extraterrestrial Impacts

Other Theories

Summary

Key Terms

Review Questions

*F*rom all the available evidence, it is clear the climate has changed over time. Given this fact, it is natural that the possible cause of such change has been the basis of much study. The question of what causes climates to change has not been completely answered. This, however, is not because of a lack of theories. In the 100 years or so since the magnitude of climatic change was realized, it has been estimated that, for every year that has passed, a new theory has been postulated. Not all of these have been satisfactory for some have failed to account for the two basic ingredients of a viable explanation. They must (1) explain the onset of ice ages through geologic time (i.e., account for long periods of warmth with cooler interruptions), and (2) account for the warming and cooling periods that occur within an ice age. Not all theories can do this, especially in quantitative terms.

The search for the explanation of climatic change has become increasingly important in recent years. Human activities have resulted in a modification of both the atmosphere and the earth's surface to such a degree that the changes are now an integral part of explaining climatic variations. Given this fact, this chapter is concerned with what may be termed the *natural causes* of climate change. This discussion includes theories that explain changes that occurred long before human habitation of the earth. The role of human activity in inadvertently modifying climate is examined in subsequent chapters. Obviously, human-induced changes and natural changes are not independent for natural changes are still taking place while people are modifying the system.

The basic reason for climatic changes on Earth is essentially very simple. Change is related to the flows of energy into and out of the system and the ways in which energy is exchanged within the earth–ocean–atmosphere system.

SHORT-TERM CHANGES

There are some processes that affect the energy balance for relatively short periods of time. In this case, we are defining *short-term* as changes in weather or climate for periods measured from individual to thousands of years. Of course, some of these changes could be classified as interannual variations, such as discussed in chapter 7. However, there is a tendency for these types of changes to persist for years.

Variation in Solar Irradiance

In most theories of climatic change, it is assumed that the output of energy from the sun is constant or nearly so. However, if it is considered that a fluctuation of less than 10 percent in output from the sun could explain all of the climatic changes that have occurred on Earth, then it is natural that theories of climatic change have used changes in the sun as their basis.

There are two main approaches that have been considered. The first visualizes an actual change in the radiating temperature of the sun over long time periods. The second considers shorter times and deals with periodic phenomena—specifically sunspots.

The solar energy received at the earth's surface can change due to the amount of energy given off by the sun, changes in the transparency of the atmosphere, or changes in the distance between earth and the sun. There is little doubt that solar irradiance affects weather. The average temperature of the sun is near 5438°C, but varies slightly, and hence the energy output varies. The actual variation in solar irradiance is small and difficult to measure. Most measurements are made from within the earth's atmosphere. The turbidity of the atmosphere affects the measurements. Until recently, measured differences in the energy were less than the accuracy limits of the instruments used to record the variations. Beginning in January 1977, the temperature of the sun's surface fell 11°C in a single year. Kitt Peak National Observatory near Tucson, Arizona, measured the drop. This is the first time the observatory measured such a change, but measurements only began in 1975.

During the 1980s, the brightness of the sun faded. The irradiance of the sun declined 0.07 percent from 1981 to 1984. Satellites outside the earth's atmosphere made these measurements. A change as small as 0.1 percent for a decade or more might change the earth's climate in a measurable fashion. In the fall of 1986, the number of sunspots declined. As sunspots decline, so do other elements at the sun's surface. The net effect is less irradiance and less energy reaching Earth. Computer models show that a drop in solar irradiance from 1 to 2 percent would bring about conditions similar to those of the Little Ice Age. Snow and ice would spread over high latitudes in the northern hemisphere. A decline of 2 percent for 50 years would be enough to cause renewed glaciation. A drop of 5 percent should be adequate to bring about a major glaciation of the earth.

Sunspot Activity

Scientists have suggested for a long time that sunspots are responsible for changes in weather patterns and climatic cycles. Detailed analysis of the sun's outer surface, or **photosphere**, shows dark, circular areas known as sunspots. These are areas in the surface where temperatures drop some 1400°C lower than surrounding areas. Intense magnetic fields are associated with them. Sunspots were seen and recorded as early as 28 B.C. in eastern Asia. They have been studied intensively since the invention of the telescope shortly after 1600 A.D. The number of sunspots occurring at any one time varies from as few as 5 or 6 to as many as 100.

Both long- and short-term fluctuations occur in sunspot activity. Two periods of major deviation from normal existed in the past 1000 years. An unusually high number of sunspots formed in a 200-year period around 1180 A.D. Between 1645 and 1715 A.D., no sunspots occurred for years at a time. The total reported for the entire 70 years was less than what usually occur in a single year. There was minimal auroral activity, and the solar corona was less visible. This period, known as the Maunder Minimum, is the only such period in historic times. It was also a period of exceptional cold known as the Little Ice Age. The **Maunder Minimum** appears in both historical records and studies of tree ring growth using carbon-14 dating methods. There is now a record of growth rates of trees dating back 8000 years. Each period of weak sunspot activity correlates with periods of cold as shown by tree ring growth and glacial advances.

There are some signs that sunspots follow an 11-year cycle, and multiples of that cycle appear at 22 and 33 years. Still other cycles appear at periods as short as 5.5 years and as long as 90.4 years. The key to sunspot cycles is that magnetic fields on the surface of the sun reverse their magnetic polarity every 22 years—an interval twice the apparent sunspot cycle.

Repeated studies trying to correlate rainfall with the fluctuation in sunspot cycles have not yet produced statistically significant results. As Tannehill (1947) stated over 50 years ago, "Some drought years have come close to the top of the sunspots (1893 and 1917), some near the bottom (1901 and 1933), some with increasing spots (1925 and 1936), and some with decreasing spots (1910 and 1930). No matter how we select the years or how we group them, we see no obvious relation to sunspots." Attempts to correlate weather with sunspots continue. Karin Labitzke of the Free University of Berlin and Harry van Loon of the National Center for Atmospheric Research found a high correlation between a phenomenon known as the quasibiennial oscillation in the stratosphere over the equator, temperatures above the North Pole at a height of 23 kilometers, and sunspots (1987). There is a lot of controversy about the relationship between sunspots and weather because there is no clear physical connection between them.

Variation in Atmospheric Dust

The amount of energy available at the earth's surface depends on the extent to which the energy is modified as it flows through the atmosphere. Changes in transparency of the atmosphere result from changes in dust content of the atmosphere, cloud cover, and ozone content of the upper atmosphere. A reduction in radiation absorbed by the earth by as little as 1 percent can produce a change in surface temperatures by as much as 1.2°C to 1.5°C.

The amount of fine ash injected into the atmosphere from major volcanic eruptions is sometimes very large. This dust absorbs and scatters a significant portion of solar radiation. Although most of the scattered and absorbed energy eventually reaches the ground, a small part goes back into space without affecting temperatures in the lower atmosphere. Maass and Schneider (1977) examined temperature records for 42 weather stations scattered over the earth, each with a temperature record at least 85 years long. They correlated these data with levels of atmospheric dust and concluded that stratospheric temperatures have increased as a result of the injection of volcanic aerosols. They also found there is a drop in annual temperatures at the earth's surface following major volcanic events. Some individual cases support these findings and others do not. Mt. Asoma in Japan erupted in 1783 and cold years followed from 1784 to 1786. The famous year without a summer in 1816 came the year after the massive eruption of Mt. Tamboro in the Dutch East Indies. So much dust blew into the air that almost total darkness existed for 3 days at distances up to 500 kilometers from the mountain. The explosion of the volcano on the island of Krakatoa in the East Indies blasted some 53 cubic kilometers of solid debris more than 30 kilometers into the atmosphere. Winds in the upper atmosphere distributed the dust over the planet. For the next 3 years, Montpelier Observatory in France recorded a 10 percent drop in the intensity of solar radiation. Unusually cold years from 1884 to 1886 accompanied the drop in insolation. Mt. Katmai in Alaska erupted in 1912, ejecting 21 cubic kilometers of rock. Observations at Mount Wilson, California, and at Bassour, Algeria, show a 20 percent drop in solar radiation during the following months. It is perhaps worth mentioning that unusually cool weather occurred for a month before Mt. Katmai erupted. Mt. Agung in Bali erupted in 1963, and observations at Mauna Loa, Hawaii, show that the receipt of direct solar energy dropped sharply by nearly 2 percent for a period afterward. However, since much of the scattered and absorbed radiation eventually reached the lower atmosphere, total radiation reaching the surface dropped by only 0.5 percent.

Other volcanic eruptions, including some greater than those already mentioned, produced no such cooling effects on the weather. For example, in 1835, Mount Cosequina in Nicaragua blew 49 cubic kilometers of ash into the atmosphere, but there were no atmospheric after effects reported. Other notable eruptions have also occurred without any clear effect on weather.

The 1982 eruption of El Chichon in Mexico blew nearly 10 times as much ash and gas into the atmosphere as Mount St. Helens in 1980. It was the largest volume of rock ejected since the eruption of Mount Katmai in Alaska in 1912. The impact of these two eruptions on the atmosphere was very different. The blast from Mount St. Helens went laterally, and most of the gases and ash stayed in the troposphere where it rapidly precipitated out. Perhaps the most significant aspect of El Chichon is that the main eruption was vertical. As a result, most of the sulfuric gases and sulfuric acid aerosols went into the stratosphere, where they remain for much longer periods. The effect of this cloud of debris on insolation is easier to determine than its effects on actual surface temperatures. The reduction in solar radiation at the surface could have reduced Earth temperatures by about 0.25°C during 1983. However, no actual temperature drop was measured.

The major difference, then, between volcanoes that can influence world temperatures and those that merely have a local effect is penetration of the stratosphere by eruptions that produce large amounts of sulfur dioxide. This gas joins with water vapor in the atmosphere to form tiny droplets of sulfuric acid, which can remain in the atmosphere for several years after an eruption. The sulfuric acid droplets reflect sunlight back to space to produce a cooling effect on surface temperatures.

A number of researchers have attempted to derive measures of the role of volcanoes in climatic change. One, the **Volcanic Explosivity Index** (VEI), is based on the fact that very explosive volcanoes will penetrate the stratosphere and be more effective than those that do not. However, as noted, the impact of the eruption is also contingent on the amount of sulfur dioxide ejected. The Dust Veil Index (DVI), as the name suggests, is a measure of turbidity in the atmosphere—a factor influenced by volcanic activity.

Although there is clear evidence to show that volcanic activity can have an impact on global climate, it is no easy task relating it to climatic changes of the past. Large eruptions such as Krakatoa (Indonesia, 1883), Katmai (Alaska, 1912), and Tambora (Indonesia, 1815) can be shown to have a short-term effect on energy flows. To account for major glacial periods, it is necessary to assume a long-term residence of volcanic dust in the atmosphere. One hypothesis is that, during active times of Earth-building forces, continued volcanic activity over long time periods would have an extended effect. Once glaciation was initiated, the role of volcanic dust would be secondary to changing surface albedos.

Human-Induced Changes in Earth's Surface

In the short period that people have inhabited the earth in terms of geologic time, they have brought about massive changes in the environment. These changes have had a significant impact on the earth's climate. To examine the impacts, it is convenient to deal with changes that modify the earth's surface and those that directly impact the atmosphere.

Humans have been altering the environment since they first controlled fire, domesticated animals, and originated agriculture. Modifications began in an early epoch, when the hunters and gatherers used fire to make hunting easier and to drive game during the hunt. Records from early explorers of Africa refer to massive fires, which probably represented the annual burning over of grazing areas south of the Sahara. In their visits to the Americas, European explorers noted that Native Americans used fire to improve hunting grounds and catch game.

The result of these activities was the deforestation of large areas of the world. In the tropical realm, it may be that the savanna grasslands are a response to deforestation by fire; in temperate regions, grasslands in North America and eastern Europe, prairie and steppe, may at least be a partial response to burning of woodlands that once existed.

With the development of agriculture, deforestation became even more extensive. Once extensive forests in China, the Mediterranean basin, western and central Europe, and North America were cleared for farming. The extent of the change is illustrated by the fact that 50 percent of central Europe was converted from forest to farmland over the last 1000 years. But deforestation was not the only extensive alteration. The misuse of marginal lands led to overuse and the eventual desertlike environment, and it started the process of desertification. Desertification has occurred in India, Africa, and South America.

The advent of a technological society, such as that in which we now live, created further changes. Destruction of the environment in the quest for raw materials, creation of artificial lakes, generation of energy, expansion of farmlands, urbanization, and other processes have significantly changed the face of the earth.

The result of the cumulative changes is their modification on the energy interchange that occurs at the earth's surface. It was pointed out that the surface

Table 15.1   Changes in Albedo That Occur with Land Use

Land Type Changed	Earth's Surface (%)	Change in Albedo
Savanna to desert	1.8	0.16 to 0.35
Temperate forest to field, grassland	1.6	0.12 to 0.15
Tropical forest to field, savanna	1.4	0.07 to 0.16
Salinization, field to salt flat	0.1	0.1 to 0.25

climate that occurs is a function of the energy that arrives at the surface and the way it is utilized. Of considerable importance in this respect is the amount of energy that is reflected from the surface—energy that does not enter the heat balance of the system. The amount of energy reflected depends on the albedo of the surface, and changes in surface cover over time have appreciably altered values. Table 15.1 provides examples of how much albedo may have changed as a result of the human impacts outlined before. It is noted that urbanization (estimated at 0.2 percent of the earth's surface) is not included in the prior list. It is omitted because its role in surface modification is quite variable, and it may lead to a lowering of albedo.

The eventual result of the long-term changes listed is a reduction of surface temperatures. The worldwide temperature decrease resulting from the change has been estimated at about 1°K; this value, of course, is open to question because albedo changes lead to other modifications (e.g., cloud cover and dust) that also play a role in determining the earth's average temperature. Despite this, it is clear that surface change has had a significant impact on Earth's climate as a whole and most certainly on the local areas where changes are the greatest. Tables 15.2 and 15.3 illustrate additional human induced changes.

LONG-TERM CLIMATIC CHANGES

Long-term climate changes are those changes that may persist for as long as millions of years. These processes are thus extremely slow when we consider them in the context of a human lifespan.

Earth–Sun Relationships

Variations in the earth's motion around the sun explain diurnal and seasonal differences in the amount of solar energy arriving at the surface. However, the angle of the earth's axis and the distance from Earth to the sun are not constant values as they vary over time. The actual orbital variations that occur are discussed in the following sections.

The obliquity of the ecliptic.   This term refers to the angle of the axis in relation to the plane in which the earth revolves around the sun. At the present time, the angle is 66.5°, which gives an obliquity angle of 23.5°. This angle is not constant; on a cycle of a period of about 41,000 years, the angle varies some 1.5° about a mean

Table 15.2   Atmospheric Trace Gases That Affect the Energy Balance and Are of Significance to Global Climatic Change

	Carbon Dioxide CO$_2$	Methane CH$_4$	Nitrous Oxide N$_2$O	Chlorofluoro-carbons CFCs	Tropospheric Ozone O$_3$	Carbon Monoxide CO	Water Vapor H$_2$O
Greenhouse role	Heating	Heating	Heating	Heating	Heating	None	Heats in air, cools in clouds
Effect on stratospheric ozone layer	Can increase or decrease	Can increase or decrease	Can increase or decrease	Decrease	None	None	Decrease
Principal anthropogenic sources	Fossil fuels; deforestation	Rice culture; cattle; fossil fuels; biomass burning	Fertilizer; land use conversion	Refrigerants; aerosols; industrial processes	Hydrocarbons (with NOx); biomass burning	Fossil fuels; biomass burning	Land conversion; irrigation
Principal natural sources	Balanced in nature	Wetlands	Soils, tropical forests	None	Hydrocarbons	Hydrocarbon oxidation	Evapotranspiration
Atmospheric lifetime	50–200 yr*	10 yr	150 yr	60–100 yr	Weeks to months	Months	Days
Present atmospheric concentration in parts per billion by volume at surface	353,000	1720	310	CFC-11: 0.28 CFC-12: 0.48	20–40[†]	100[†]	3000–6000 in strato-sphere
Preindustrial concentration (1750–1800) at surface	280,000	790	288	0	10	40–80	Unknown
Present annual rate of increase	0.5%	0.9%	0.3%	4%	0.5–2.0%[†]	0.7–1.0%[†]	Unknown
Relative contribution to the anthropogenic greenhouse effect	60%	15%	5%	12%	8%	None	Unknown

Source: EarthQuest, Spring 1991, Vol. 5, No. 1. Office for Interdisciplinary Earth Studies. Boulder, CO: University Corporation for Atmospheric Research. Reprinted by permission.

*The lifetime of CO$_2$ in the atmosphere is calculated in two ways: one, which is relatively short (10 years), is the residence time of a single molecule before dissociation; more relevant to global change is the longer period that includes the residence time in the atmosphere-ocean system: the time that a CO$_2$ molecule derived from fossil fuel remains in the atmosphere-ocean system before being sequestered as terrestrial humus or deep-sea sediment. The latter is of greater value in calculating future scenarios related to greenhouse warming, since it reflects the relaxation time between a cessation of all industrialized CO$_2$ emissions and the expected detection of a global decrease in the pressure of CO$_2$.

[†]Northern Hemisphere.

Table 15.3 National Emissions Estimates for 1985 (Millions of Metric Tons)

Source Category	Particulates	Sulfur Oxides	Nitrogen Oxides	Volatile Organics	Carbon Monoxide	Lead*
Transportation						
Highway vehicles	1.1	0.5	7.1	6.0	40.7	14.5
Aircraft	0.1	0.0	0.1	0.2	1.1	–
Railroads	0.0	0.1	0.5	0.1	0.2	–
Vessels	0.0	0.2	0.2	0.4	1.4	–
Other off-highway	0.1	0.1	1.0	0.4	4.1	0.9
Transportation total	1.3	0.9	8.9	7.1	47.5	15.4
Stationary source fuel combustion						
Electric utilities	0.6	14.2	6.8	0.0	0.3	0.1
Industrial	0.3	2.2	2.9	0.1	0.6	0.4
Commercial institutional	0.0	0.4	0.2	0.0	0.0	0.0
Residential	1.2	0.2	0.4	2.4	7.1	0.0
Fuel combustion total	2.1	17.0	10.3	2.5	8.0	0.5
Industrial processes	2.7	2.9	0.6	8.6	4.6	2.3
Solid waste disposal						
Incineration	0.1	0.0	0.0	0.3	1.1	–
Open burning	0.2	0.0	0.1	0.3	0.9	–
Solid waste total	0.3	0.0	0.1	0.6	2.0	2.8
Miscellaneous						
Forest fire	0.7	0.0	0.1	0.6	4.7	–
Other burning	0.1	0.0	0.0	0.1	0.6	–
Miscellaneous organic solvent	0.0	0.0	0.0	1.6	0.0	–
Miscellaneous total	0.8	0.0	0.1	2.3	5.3	–
Total of all sources	7.2	20.8	20.0	21.1	67.4	21.0

Source: National Air Pollutant Emission Estimates, 1940–85, U.S. Environmental Protection Agency Publication No. EPA450/4-86-018, 1987.

*Thousands of metric tons.

23.1° (Fig. 15.1a). The effects of changing obliquity are illustrated in Figure 15.1b. The top diagram shows the present position and the lower some hypothetical cases. An **obliquity** of 0° would lead to equal lengths of day and night over the globe and result in a lack of seasonal changes, which would cause well-defined climatic zonation. Another extreme is an angle of obliquity of 54°. Such an angle would produce great extremes in the lengths of summer and winter days and nights. For example, at the December solstice position shown in the diagram, much of the northern hemisphere would have 24 hours of darkness. Extreme temperature differences would occur from summer to winter. Although the actual changes in the angle of obliquity are not as large as these examples, they are sufficient to cause distinctive changes in the distribution of Earth's climates.

Earth's orbital eccentricity. The earth moves around the sun in an elliptical orbit; the **eccentricity** of the orbit is derived by comparing the path to that of a true circle (Fig. 15.2a.) Currently the orbit is relatively close to a circle: Its eccentricity, measured by the method shown in Figure 15.2b, is 0.017. Over the past million years, this value has changed from almost circular (e = 0.001) to an extreme value (e = 0.054). This change influences the amount of solar radiation intercepted by the earth and also modifies the dates at which the solstices and equinoxes occur. This factor is used to derive the precession of the equinoxes.

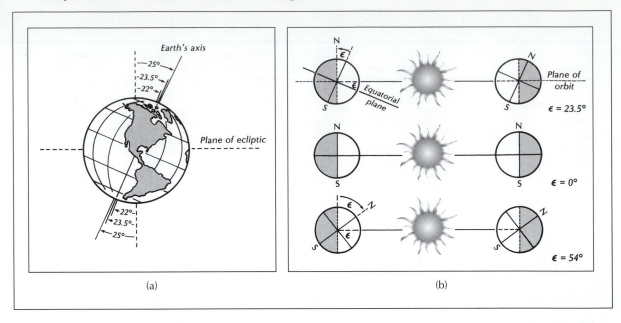

Figure 15.1
Obliquity of the ecliptic: (a) The amount of tilt of the earth's axis changes between 22°
and 25° over a 41,000-year period. The present value is 23.5° (from *Weatherwise*, *35*(3),
1982, a publication of the Helen Dwight Reid Educational Foundation); (b) In the three
cases shown (present angle, 0°, and 54°), the earth's climate would change appreciably.

Figure 15.2
Changes in earth's orbit
around the sun: (a) Over
time, the orbit changes
from elliptical to almost
circular (from *Weatherwise*,
35(3), 1982, a publication
of the Helen Dwight Reid
Educational Foundation);
(b) Eccentricity of the
orbit. The sun is one focus
(F_1) of the elliptical orbit.
The distance from the
center (C) to aphelion or
perihelion is given by half
the major axis, α. The
distance from C to F_1 is
termed *linear eccentricity*
(le), which is used to
determine eccentricity (e),
which is given by
e = le/α.

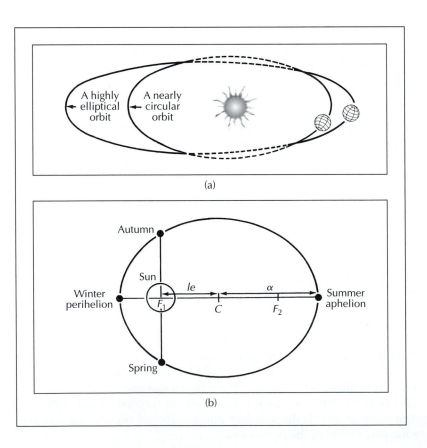

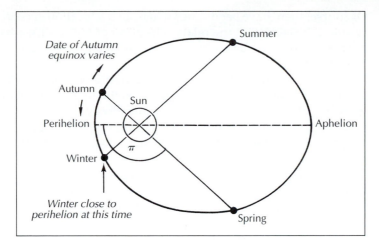

Figure 15.3
Precession of the equinoxes occurs as dates of solstices and equinoxes move along the orbit.

Precession of the equinoxes. Because of varying Earth motions, the days on which the earth reaches its closet and most distant point from the sun (**perihelion** and **aphelion**, respectively) change over time (Fig. 15.3). At present, perihelion occurs during the northern hemisphere winter. Ten thousand years from now, the date of perihelion will pass to the northern hemisphere summer season. The present situation will be reversed, and lowest input of solar radiation will correspond to winter in the northern hemisphere, the land hemisphere, in which extreme cold temperatures are attained.

The dates at which these events occur can be calculated. It is possible to reconstruct the times at which the various cycles of change reinforce one another to produce very high or very low radiation values within a hemisphere. In fact, long before the advent of high-speed computers, the Yugoslavian scientist, Milutin Milankovitch, actually derived values going back thousands of years. Although his derived values provide the basis for understanding the relationships, more recent workers have constructed models of the cycles and their relative weight in influencing climate. Currently, many climatologists think that these orbital variations (**Milankovitch cycles**) are at the base of climatic change and have postulated future conditions from the research findings. Using the orbital variation theory, it is suggested that a long-term cooling trend, which began some 6000 years ago, will continue.

Distribution of Continents

Modern geophysical research has provided a unifying theory to the old idea of continental drift—the concept of plate tectonics. There is now little doubt that the present positions of the world land masses are but a transitory location in the long-term evolution of the continents and oceans. Given this, one set of theories of climatic change deals with the relative location of continental land masses in relation to the position of the poles and the equator.

When the continental distribution during the last two great earth-cooling periods is reconstructed, an interesting pattern results. Reconstructed maps for the Permo-Carboniferous glaciation (250 million years ago) and those of the most recent Pleistocene glaciation show that in both cases there is a concentration of land masses in the polar realms. The presence of large land masses in polar areas is much more conducive to glacial formation because land masses lack the heat storage and transfer mechanisms of oceans. It has been suggested that a primary requirement for the formation of great ice caps is the polar location of continents.

Of course, there have been times when such a location occurred but no glaciation resulted. It seems that the relative location of the continents may well be an important factor in ice age occurrences, but actual ice formation must rest on other causal factors.

Intimately related to the idea of moving continents, although often treated as a separate theory of climatic change, is the role of mountain building and continental uplift. As an explanation of this theory, one only has to think of the formation of permanent snow on Mt. Kilimanjaro, a mountain located astride the equator. With increasing height of land masses, the potential for ice formation is greatly increased.

Geologists have long noted the relationship between times of extensive mountain-building periods and ice ages. For example, both the Permo-Carboniferous and recent ice ages were preceded by extensive mountain-building periods. There is, however, a lag time of millions of years between the mountain building and the onset of the glaciation. To this criticism must be added the fact that some mountain-building periods—for example, the Caledonian period in Europe 370 to 450 million years ago—did not give rise to ice ages.

Despite these criticisms, it is generally agreed that mountain building, or at least the presence of high mountains, certainly contributes to an optimum condition for ice formation. The idea that mountain building can influence climatic change is reinforced by modern research that shows that mountain ranges do influence upper air circulation patterns by changing upper air vorticity. It is known that, on the leeward side of the mountains, a cyclonic circulation—positive vorticity—is induced in air flow.

Variation in the Oceans

Modern research in climatology is paying increasing attention to the role of the oceans in the climatic system. Variations in sea-surface temperatures have been linked to changing circulation patterns and weather anomalies. In terms of climatic change, the oceans have received attention, providing in some cases the basis of entire theories of climatic change. Some of the ways in which the oceans influence the prevailing climates on Earth include the following:

1. As a factor in the relative elevation of land: A drop in sea level would increase the heights of the continents and enlarge land masses in area.
2. As a heat storage mechanism: The oceans are less variable than the continents, and changes in the relative temperature of oceanic waters would influence world climates. Variations in the oceans could occur because of changes in salinity, evaporation rates, and relative solar penetration.
3. As a mobile medium: Ocean water plays a significant role in the redistribution of energy over the earth's surface. Ocean currents transport large amounts of heat, and any changes in their relative extent would have extended results. One example of this role is illustrated in Figure 15.4, which shows paleogeographic maps of oceanic circulation during the warm mid-Tertiary and the circulation that existed in the cold conditions of the late Pleistocene. The essential difference is that a zonal circulation of water occurred because of the equatorial passage of the Tertiary. This closed by Pleistocene times, and the whole circulation pattern was modified to encourage ice formation.

Although the oceans are highly significant in the nature of climate that occurs over the earth's surface, many researchers think that their role in climatic change may be a secondary effect. That is, the oceans react to other changes such as a modified atmospheric circulation; they do not create them. It has been suggested that the atmosphere leads and the ocean follows.

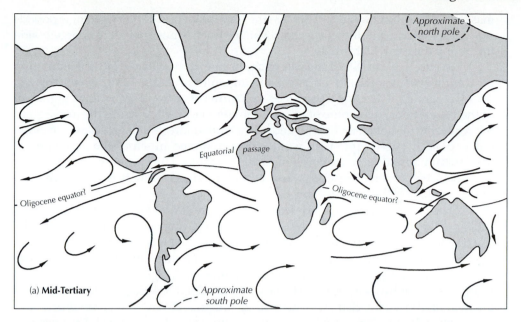

Figure 15.4

World oceanic circulation in the (a) Middle Tertiary, and (b) late Pleistocene. Note how the world encircling equatorial passage was closed by Pleistocene times, whereas a limited north–south exchange occurred when ice occupied large areas (from *The Encyclopedia of Atmospheric Sciences and Astrogeology*, Fairbridge R. W., ed., © 1967, Dowden, Hutchinson & Ross, Stroudsburg).

Extraterrestrial Impacts

At intervals throughout the history of Earth, exceptionally large objects from space have struck the earth. We need only look at the surface of the moon to get some idea of what our planet would look like were it not for weathering and erosion, which continually erase the evidence of impacts. When these bolides have struck Earth, some have altered the climate tremendously for short periods and to some extent for periods of thousands of years. One such example is that associated with K-T boundary mass extinction. The boundary between the Cretaceous and Tertiary geologic periods is approximately 65 million years ago. Much evidence now points to a large bolide impacting the earth at this time.

This impact was either the sole cause or one of the causes for the elimination of over half of all species of organisms living on Earth at the time. It certainly did alter the climate. The impact produced firestorms over the earth, destroying much of the vegetation. At the time of impact, the tremendous heat from the impact and the scattering of molten meteorites and gases ignited forests and grasslands over the entire planet. These same fires would have removed much of the atmospheric oxygen and added large amounts of carbon dioxide. The fires also contributed a huge cloud of soot to the atmosphere. Most large land animals would have been killed quickly by the heat and fire.

The mass of material blown into the atmosphere as the object struck was perhaps 100 times the mass of the meteorite. A huge dust cloud was raised that blocked out the sun for months. The soot from the fires would have further reduced the amount of sunlight reaching the surface halting or greatly reducing photosynthesis. The reduction in photosynthesis had a lethal effect on marine organisms. A high percentage of those existing at the time also became extinct. The sustained cloud of solid particles would have greatly reduced solar radiation to the ground and caused temperatures to drop.

Other drastic changes in atmospheric chemistry also occurred. Precipitation would have turned into acid rain and snow as the highly sulfurous particles combined with water particles in the atmosphere. Undoubtedly the stratospheric ozone layer would have disappeared for some time.

The combination of fire, lack of sunlight, and cold would have destroyed much of the vegetation that constituted the food supply of the larger animals such as the dinosaurs. It was in fact mainly the largest animals that were eliminated. Most were probably killed by the initial blast and firestorm, but those left alive would have succumbed to the loss of food supply.

Following the months of cold, the temperature rebounded to levels higher than before the impact, adding to the stress on living organisms. The impact greatly disturbed the global system for some time, and much of Earth's biota simply was killed outright or could not adjust to the changes. The event affected seasonal variability of climate for nearly a million years. The Milankovitch cycles, which were discussed earlier in this chapter, were greatly strengthened by the event. The amplitude of the cycles jumped substantially following the event and gradually returned to their previous levels after hundreds of thousands of years. The greater seasonal variation in the global climate was most likely due to the greatly reduced seasonal variation in carbon dioxide. Currently, there is more carbon dioxide in winter, helping retain heat, and less in summer. This is due to seasonal changes in photosynthesis. Since Earth's vegetation was largely destroyed during the event, there was little change thereafter in the seasonal level of carbon dioxide to dampen the seasonal variation in heat loss from the ground.

OTHER THEORIES

The theories presented herein are a partial representation of those that have been suggested. Other researchers have introduced ideas ranging from the possible influence of the periodic passage of the earth through an interstellar dust cloud to variations in atmospheric water vapor caused by both natural and human activities. Despite all of these ideas, there is no single theory that can account for all of the observed events; it is evident that earth's climates result from a spectrum of causal elements. This is well shown in Figure 15.5, which presents in diagrammatic form the components of systems that could be modified to cause a significant change in the earth's climate.

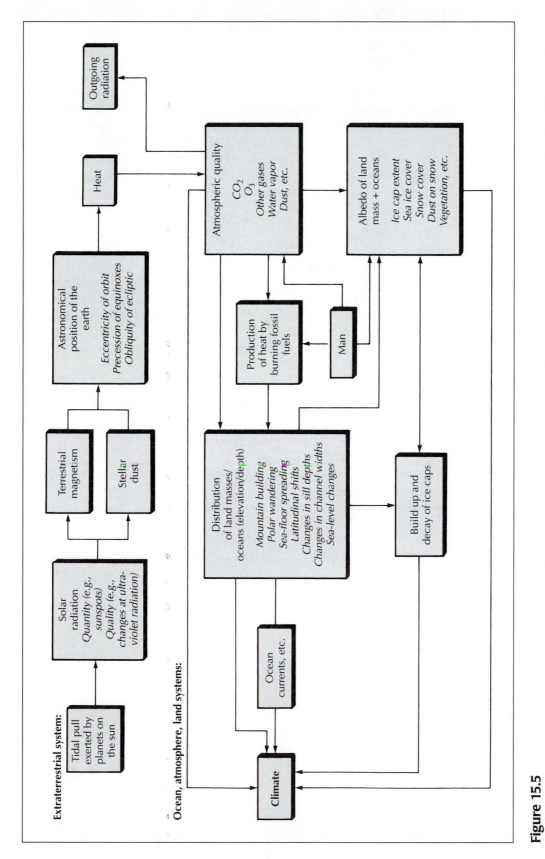

Figure 15.5
A schematic representation of some of the possible influences causing climatic change (from Goudie A. S., *Environmental Change*, © 1977, Oxford, Oxford University Press).

SUMMARY

What causes climates to change is a fascinating question. Many theories have been postulated, but no single one can totally satisfy all necessary requirements. Although there is little doubt that changes in Earth–sun relationships may be a basic cause of long-term climatic change on Earth, it appears that the effects must be considered in conjunction with other factors. The problem of explaining the change is further complicated because human activity is now a major factor in the modification of climate. To explain what is happening, the so-called natural causes must be considered together with the impacts of humans. This is well demonstrated by the extensive surface changes created by human activity and the addition of carbon dioxide to the atmosphere. In cities, the human impact is so great that a new set of climatic conditions is created.

KEY TERMS

Aphelion	Milankovich cycles	Photosphere
Eccentricity	Obliquity	Volcanic Explosivity Index
Maunder minimum	Perihelion	

REVIEW QUESTIONS

1. At the present time, what is the angle of the obliquity of the ecliptic?

2. As the eccentricity of Earth's orbit about the sun increases, what affect does this have on the seasons?

3. The time of year when Earth is closest to the sun is now in the northern hemisphere winter. As perihelion moves to the northern hemisphere summer, what effect will this have on our seasons?

4. What relationship does the Maunder minimum have with the Little Ice Age?

5. Recent large volcanic eruptions appear to have affected global climate for a few years. How have these volcanic eruptions affected climate?

6. What is the mechanism by which volcanic eruptions affect climate?

7. In the past when large amounts of land have been located near the poles, what have been the global climatic conditions?

8. Evidence of climatic changes is found in association with impacts of large objects from space. What has been the effect of these impacts on global climate?

9. What mechanisms appear to have operated to change global temperatures with the impact event at the K-T boundary?

CHAPTER 16

The Warming of Planet Earth

CHAPTER OUTLINE

*I*n previous chapters, we have shown that Earth's temperature has always been changing. Whether we examine data over millions of years or from year to year, it is clear that temperature changes with time. The rate and amount of change vary with time and space, but change is continuous. The present time is no exception.

EVIDENCE FOR GLOBAL WARMING

By the year 2000, many scientists and scientific organizations concurred that Earth was warming at an unusual rate. Evidence of the warming is present in many different forms. The most obvious and mathematically supportable evidence is in the temperature records. Analysis of temperature data gathered over the planet indicates rapidly rising temperatures over the past 140 years. Table 16.1 provides some data for the 20th century that illustrate the rise, particularly toward the end of the record. For example, the 20th century was the warmest century in the last millennium, the 1990s was the warmest decade of the century, and 1998 was the warmest year of the century. Temperatures in the world ocean are also increasing to some depth, not just at the surface.

There are many temperature-dependent phenomena that also indicate the earth is warming. They include the following:

- Earth's mountain glaciers are melting (Table 16.2)
- Antarctica's ice sheets are breaking up
- Sea level is rising
- The temperature of the global ocean is rising
- Northern hemisphere permafrost is melting
- Arctic pack ice is thinning and retreating
- The tree line in mountain ranges is moving upward
- Many tropical diseases are spreading toward the poles and to higher elevations in the tropics.

Changes in the Temperature Record

The year 1880 marks the beginning of the historical record for which there are enough data to provide credible information. Critical study of the data shows that probably not even 0.5°C warming has actually taken place. For example, urbanization has increased the temperature at about a third of the weather stations with long records. In addition, most of the earth is covered with an ocean for

---

Table 16.1   Global Temperature Data

- 1998 was the warmest year of record
- Seven of the 10 warmest years of record occurred since 1990
- The 1990s were the warmest decade on record
- The 1980s were the second warmest decade on record
- The 10 warmest years of record occurred since 1983
- The mean temperature of Earth increased about 1°F (0.4°C) in the 20th century
- The 20th century was the warmest century of the millennium

Table 16.2   Melting of Earth's Ice

- The edge of the West Antarctic ice sheet is shrinking at the rate of 400 feet each year.
- The Larsen ice shelf on Antarctica disintegrated in January 1995.
- Much of Antarctica's Larsen B and Wilkes ice shelves disintegrated in 1998–1999.
- The average elevation of glaciers in the Southern Alps of New Zealand moved upward 300 feet in the 20th century.
- In the Tien Shan Mountains of China, glacial ice shrank nearly 25% in the past 40 years.
- In the Caucasus Mountains of Russia, half of all glacial ice melted away in the past 40 years.
- Garhwal Himalayas of India. Glaciers are rapidly retreating.
- Andes Mountains of Peru. Glacial retreat increases sevenfold from 1978 to 1995.
- Bering Sea. Area of sea ice shrunk 5% in past 40 years.
- Arctic Ocean. Area of sea ice shrunk by 14,000 square miles since 1978.
- The largest glacier on Mt. Kenya almost completely melted away in the 20th century.
- The Bering Glacier in Alaska is retreating.
- The glaciers in Glacier National Park, Monatana, are melting rapidly.
- Glaciers in the Alps Mountains of Europe shrank by about 50% in the 20th century.
- Gruben and Aletch glaciers are among those rapidly melting.

which there are little data. Global circulation models predict the most warming for the region around the Arctic Basin. Climatic data show that average surface temperatures increased over the past century from 0.7°F to 1.4°F (0.4°C–0.8°C). Most of the increases took place in the 1920s and 1990s. The 1990s were the warmest decade on record. For instance, 1998 was the warmest single year on record. Both Earth- and space-based data support the most recent global changes in temperature (Figs. 16.1 and 16.2).

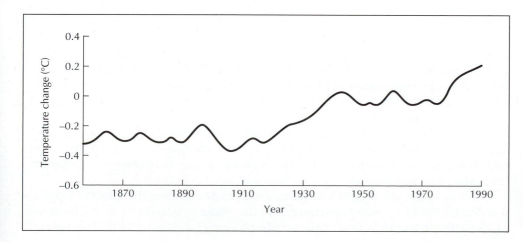

Figure 16.1
Globally averaged temperatures since about 1860 shown as change from the 1951–1980 average (from Hidore J. J., *Global Environmental Change,* © 1996, Prentice Hall, Upper Saddle River, N.J.).

Figure 16.2
Combined land, sea, and air temperatures from 1861 to 1992. The temperatures are expressed as departures from the mean of the climatic normals for 1951–1980 (from Hidore J. J., *Global Environmental Change*, © 1996, Prentice Hall, Upper Saddle River, N.J.).

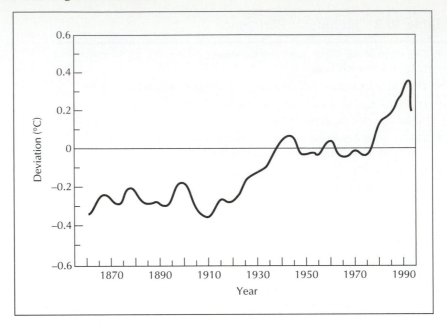

Other temperature-related data also indicate planetary warming. There has been a reduction in the change in temperature from day to night. Most of this reduction is due to nights staying warmer. This is a function of the night air trapping more of Earth's radiation at night. In parts of northeastern United States, the warmer nights have increased the growing season by 11 days since 1970. The length of snow cover has shortened considerably in some parts of the northern hemisphere. In the same seasonal pattern, spring snow melt is taking place earlier in the spring. This in turn affects the annual distribution of rivers and streams in those areas that usually have some snow cover in the winter. It is also fairly evident that the warming may be changing the distribution of El Niño and La Niña events. The frequency of El Niño events has increased in the past 25 years.

Changes in Plant and Animal Life

There are also detectable changes in Earth's animal life. The population of Adelie penguins on the Antarctic dropped 40 percent in the last quarter of the 20th century. The hypothesis is that the warmer temperatures make it harder for the penguins to find food and breed. Off the coast of California, the sea surface has warmed as much as 1.5°C (3°F) since 1951. This has led to an 80 percent decline in zooplankton, which is a basic component of the food chain. Coral reefs are one of the richest ecosystems on the planet, and they are dying at an unprecedented rate. More than 30 nations have reported losses to offshore reefs. The dying of the reefs is attributed to a process caused by bleaching, which occurs when sea temperatures become abnormally high. The bleaching is actually the result of the death of microscopic algae that both color and feed the coral.

Rates of natural migration and adaptation of species and plant communities appear to be much slower than the rate at which climate is changing. Thus, the population size and geographical range of many species may change as temperatures and rainfall change.

Many of America's national parks are already showing the effects of global warming. One of the most spectacular of the parks is Glacier Park in Montana. Model estimates now indicate that the glaciers may be gone by 2030. The natural habitats of many species of plants and animals are being changed by the rapid warming. In the Arctic National Wildlife Refuge in Alaska, the melting of permanently frozen ground is changing the migratory patterns of the caribou. Rising sea levels inundating Maryland's Blackwater National Wildlife Refuge are driving away many species of birds. In Mount Rainier and Olympic National Parks in the state of Washington, plants normally found at low elevations are appearing in the alpine meadows. In Rocky Mountain National Park, there is an unusual growth spurt of spruce trees. The growth rate is normally held down by low temperatures. Millions of sea birds have disappeared off our west coast as offshore waters become warmer. Continued warming will affect the nature of more than a fourth of our national parks. Among the most threatened parks are the Arctic National Wildlife Refuge, the Everglades, Great Smoky Mountains, Haleakala volcano on Hawaii, and Yellowstone and Yosemite Parks.

Spread of Tropical Diseases

Extremes of heat and cold, and wet and dry conditions have been frequent in recent years. Down through history such extreme events have often been followed by outbreaks of disease. Such has been the case in recent decades. These extremes of weather favor the insect pests, bacteria, protozoa, and viruses that spread the diseases. Tropical diseases have become more prevalent in their traditional zones and have broken out in areas where they were previously unknown. Diseases that have increased in frequency or severity include **malaria**, **cholera**, **yellow fever**, **the plague**, **Dengue fever**, and **hantavirus**. Malaria is perhaps the most commonly known of the tropical diseases. Because of the warmer weather in recent years, malaria has been particularly severe. Malaria is currently the most widespread of all mosquito-borne diseases. Increased global temperatures result in an increased range of the *anopheles* mosquito. A rise of 4°F in global temperatures may expand the geographical area in which malaria is found from the present 42 percent to 60 percent of the planet. Not only does the rising temperature increase the range of the *anopheles* mosquito, but it increases their biting rate. The same rise of 2°C (4°F) would more than double the rate of metabolism, leading them to feed more frequently. Mosquitos in many areas are becoming resistant to the most common pesticides, and so the increased temperatures seem certain to increase the perils of malaria.

Cholera is another disease of warm climates. It too seems to be spreading. A cholera outbreak occurred in 1991 along the coast of Peru. It was introduced by a freight-carrying ship from Asia when the waste water was pumped from the hull. There was already serious pollution of the water. The cholera organism thrived in the unusually warm water and algae bloom. A new strain of cholera appeared in Bangladesh in 1993 following heavy monsoon rains.

Dengue fever is another mosquito-borne disease that appeared in South America in the last decade of the 20th century. The symptoms are excruciating pain in the bones. It existed on the Pacific side of Costa Rica for some time. There were epidemics there in 1993 and 1994. In 1995, unusually warm weather allowed the mosquitos to cross the mountains into the rest of the country. Outbreaks of the disease also occurred in other areas of South America. Thousands died of the epidemic, and nearly 150,000 were infected. The disease was not limited to Central and South America. There was an outbreak in Australia in 1992 following a particularly rainy season that favored the breeding of mosquitos.

An ancient nemesis, the plague, made its appearance in Asia once again. In 1994, there was a long rainy monsoon season in northern India. When the rains subsided, there were 3 months of weather with temperatures around 100°F. Rats, which are the primary homes of the flea carrying the plague, converged on the cities. The result was an epidemic of pneumonic plague—a form that is transmitted from person to person through the air.

A relatively recent disease to appear on the scene is hantavirus. A long drought followed by heavy rainfall in America's southwest led to an outbreak of a very serious form of pulmonary hantavirus. It proved fatal in about half of the identified cases. There is a compound problem caused by the warming. The same elements associated with warming that increase the vitality of the disease reduces human immunity to them by weakening natural defenses. Rising temperatures may lead to increased frequency of some of these diseases in the United States.

Melting of Earth's Glacial Ice

Earth is the only planet that we are sure has water in all three physical states—solid, liquid, and gas. The balance among the three is a delicate one. Any global change in temperature will alter the balance among the different forms. If warming is occurring, there should be signs of ice melting. One critical ice mass that is showing changes is that on Antarctica. It is necessary to establish here that Antarctic glaciers have always been producing icebergs. The **ice shelves** move seaward and large pieces of ice break off from the shelves at irregular intervals. In March 2000, the second largest iceberg ever recorded broke free from the Ross Ice Shelf and floated into the Ross Sea. Within a week of this iceberg's formation, three more large pieces broke free from the ice shelf. The huge iceberg (B-15) is some 11,000 square kilometers in size, being about 300 kilometers in length and 37 kilometers wide. The largest iceberg was one that broke away in 1956. This iceberg may well have resulted from processes unrelated to global warming. Temperatures in the area of the Ross Ice Shelf do not seem to have increased.

However, midwinter temperatures increased 4°C to 5°C (7°F to 9°F) on the western Antarctic Peninsula over the last half of the 20th century. The west Antarctic Peninsula, which extends toward South America, warmed by more than 2.5°C (4.5°F) between 1950 and 2000. *Ice shelves* along this peninsula have either broken up completely or are rapidly melting. Three ice shelves disintegrated in the last decade of the 20th century. Five of the ice shelves off both sides of the peninsula have been melting at an ever-increasing rate. These ice shelves are the Prince Gustav Channel, Larsen A, Larsen inlet, Muller Ice Shelf, and Wordie Ice Shelf. Larsen B and Wilkins shelves had an area of about 21,000 square kilometers in 1997. They lost about 3000 square kilometers of this total in just 2 years from 1998 to 2000. In contrast to the huge pieces breaking from the Ross Ice Shelf, these ice shelves are disintegrating in small pieces, which is usually the case with warming temperatures.

The breakup of these ice shelves will not have a major effect on sea level since they are already floating in water. Since the density of the ice is about 9 percent less than water, the volume of water in the iceberg is almost the same as the volume of water the iceberg displaces. What is of much concern is how the breakup of these ice shelves will affect the glaciers that are driving the ice shelves. It is possible that the removal of these ice shelves will result in more rapid advance and faster melting of the glaciers. This would become a global problem of major proportions. If the West Antarctic Ice Sheet melts, sea level will rise by 6 meters. If both the West Antarctic Ice Sheet and the Ross Ice Sheet melt, sea level would rise by 70 meters. What this would do to the global socioeconomic system one can only guess. That it would add tremendous stress is a certainty. Arctic pack ice is also melting. The floating pack in the Arctic Sea has lost 40 percent of its volume since 1958. The only logical cause is surface warming.

Changing Temperatures and Rising Sea Levels

Earth's ocean warmed rapidly over the last four decades of the 20th century. An extensive study was conducted of ocean temperatures over the last half of the 1900s using data from all over the world. Millions of individual temperature recordings were gathered and combined into a single large group. The analysis of the data shows a clear pattern. From 1955 to 1995, the world ocean warmed by an average of .06°C (1°F) from the surface down to about 3000 meters. The Indian Ocean warmed 0.3°C (0.5°F) down to a depth of 800 meters in a period of 20 years.

The depth to which warming can be detected is also increasing. This warming represents the additional storage of tremendous amounts of heat that would otherwise be stored in the atmosphere. This warming of the ocean is temporarily slowing the rate of atmospheric warming. As this additional heat stored in the surface waters of the ocean is released to the atmosphere, it will cause global air temperatures to rise at least another 0.5°C (0.9°F) over the next century. This is separate from the warming caused by the Greenhouse Effect of the atmosphere. Sea level rose 10 to 25 centimeters (5–10 in) in the 20th century. The rise has mostly been due to expansion of the sea water due to warming. Projections are for sea level to rise as much as a meter by 2100. A rise of even half this much would lead to flooding of 5000 square miles of dry land and another 10,360 square kilometers (4000 square miles) of wetlands.

Fresh water lakes are also being affected. Temperatures have warmed suddenly in Lake Superior. In 1998, temperatures at the surface were 20 degrees above normal and the warmest ever recorded. Instead of the frigid 50s, the temperature was in the 70s. The warming was not just restricted to the very surface. The water was warmer by as much as 15 degrees down as far as 100 feet. The impact of this warming on life in the lake is phenomenal. Warmer temperatures would affect the level of the lake. The warmer the water is, the more evaporation there is from the surface. Evaporation is the primary means by which water is removed from the lake. Lake levels in 1999 were the lowest in 30 years. Lakes Huron and Michigan dropped 2 feet between 1998 and 1999. The lower lake levels will affect shipping and power generation among other economic activities.

Changing temperatures will have a detrimental effect on living organisms in the lake. The warmer water kills the plankton on which higher life forms depend. Already the plankton have descended to lower depths where the water is colder. At greater depths, they grow more slowly due to the lack of sunlight. Less plankton will mean less fish. It also encourages many alien species that are prevalent in the other Great Lakes. The cold temperatures in Lake Superior have kept out some undesirable organisms such as *zebra mussels*. These mussels are a serious problem in the other lakes. There are other species as well that would move into the lake if the temperature structure changes. Some of these are competitive enough to destroy the existing native plants and animals.

Northern Hemisphere Melting of Permafrost

In Alaska, the permafrost is melting. The mean annual temperature of the layer above the frozen subsoil rose 3.4°C (6.3°F) between the late 1980s and 1998 in the area north of the Yukon River. The permafrost warmed between 0.6°C and 1.5°C (1 and 2.7°F) south of the Yukon River. In this region, not all the land is underlain by permafrost. Along the boundary between the areas with and without permafrost, the temperatures within the soil are close to the freeze–thaw temperature. Only a slight increase in temperature can melt permafrost over large areas. Along this boundary, the permafrost is melting downward at a rate of about 3 feet every 10 years. Around the base of Mt. McKinley, large areas of arctic meadow have turned into a mix of ponds containing water-loving plants and dry meadows. Much of the permafrost melting took place between 1980 and 2000. Permafrost is not restricted to North America. Somewhere between 20 and 25 percent of Earth's land area contains permafrost. In China, the permafrost is estimated to be melting northward at the rate of about 1 mile every year.

PROCESSES CONTRIBUTING TO GLOBAL WARMING

The evidence that the planet is warming seems to be incontrovertible. The processes responsible for the warming are being debated. The primary reason for the debate is whether the warming is the result of natural processes or whether they are due to human activity. It is possible it is a combination of both. As of this writing, the only strongly correlated variables involve human activity.

Changing Levels of Atmospheric Carbon Dioxide

Carbon dioxide (CO_2) is one of the many variable gases found in the atmosphere. Atmospheric CO_2 is an active trace gas. It is found in only small quantities compared with the major constant gases of nitrogen and oxygen. It averages less than 1/2 of 1 percent of the atmosphere by volume. The amount of CO_2 in the atmosphere varies with time. There are regular oscillations in the CO_2 content on a daily and seasonal basis. During the day, photosynthesis withdraws CO_2 from the atmosphere and at night **respiration** releases CO_2. Daily fluctuations during the growing season in mid-latitudes are as high as 70 parts per million (ppm).

There is also a seasonal change in CO_2 in mid-latitudes due to variation in the rate of photosynthesis (Fig. 16.3). The CO_2 content rises to a peak in spring in the

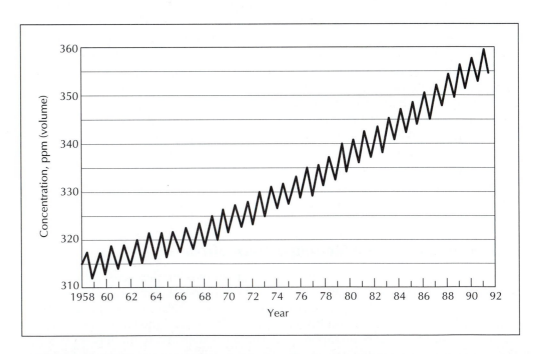

Figure 16.3
Changes in atmospheric CO_2 between 1958 and 1992 at Mauna Loa observatory, Hawaii (from Hidore J. J., *Global Environmental Change*, © 1996, Prentice Hall, Upper Saddle River, N.J.).

northern hemisphere and falls to a minimum in late September or October. The seasonal change in atmospheric CO_2 reflects a very important factor affecting the atmosphere: the metabolism of all living matter. The seasonal change in CO_2 in the atmosphere results from the pulse of photosynthesis. This occurs during the summer in mid-latitudes of both hemispheres. The difference is only about 5 ppm in warm humid areas such as Mauna Loa, Hawaii. It is more than 15 ppm in central Long Island, New York. The difference is less near the equator where the vegetation is green most of the year. It is largest in mid-latitudes where the vegetation is largely deciduous. The difference is also less at higher elevations at all latitudes. These diurnal and seasonal changes in CO_2 have been part of the natural atmosphere over geological time. Irregular changes also take place over longer periods. These are due to changes in volcanic activity, changes in the rate of chemical weathering, and in the volume of the living vegetation on Earth.

There are natural processes that operate to control CO_2 content in the atmosphere. For example, the ocean is a major absorber of atmospheric CO_2. The oceans contain about 50 times more carbon than does the atmosphere. The ocean is an important long-term sink for carbon from the atmosphere. The amount of carbon the ocean takes up and releases depends on a variety of physical, chemical, and biological processes.

There is an exchange of CO_2 between the atmosphere and terrestrial vegetation that is about the same size as that between the atmosphere and the seas. This is about 100 G.T.C. per year. Terrestrial vegetation removes atmospheric CO_2 through photosynthesis and is released when the vegetation is oxidized by either decomposition or burning.

The CO_2 content of the atmosphere dropped steadily through geologic time until about 50 million years ago. The decline slowed after that, and the levels of the present atmosphere have existed for the past 3 to 4 million years. During the last 2 million years, CO_2 varied over a relatively small range—between 200 and 280 ppm. The lowest concentrations were reached during the coldest part of the last glaciation when CO_2 dropped to about 200 ppm. The most likely reason for this drop in CO_2 is that cold water absorbs more CO_2 than does warm water. In the 1000 years preceding 1850, the concentration of CO_2 stayed near 280 ppm.

Data show that the CO_2 content of the atmosphere has been increasing since at least 1850. Just how much the total increase has been is uncertain due to the nature of the early measurements. In 1958, a continuous monitoring station was set up on Mauna Loa volcano on the island of Hawaii. The CO_2 concentration increased from about 280 ppm in 1850 to about 358 ppm in 1996. The increase at Mauna Loa averaged 0.8 ppm per year. It ranged from 0.5 to 1.5 ppm per year. A significant aspect of this change is that the rate of change is increasing. From 1958 to 1968, the increase averaged 0.7 ppm per year. From 1968 to 1978, the average increase was 1.3 ppm per year. A fourth of the total increase occurred from 1967 to 1977. Between 1977 and 1978, the increase was 1.5 ppm for the whole atmosphere and 2.0 ppm over the Antarctic Continent. The CO_2 concentration of 358 ppm in 1996 is greater than at any other time in the past 100,000 years. The concentration is now 30 percent more than in the pre-Industrial era from 1750 to 1800. It is increasing about 1.8 ppm (0.5 percent) annually because of human activities.

Carbon Dioxide and the Theory of Global Warming

The natural Greenhouse Effect of the atmosphere keeps Earth some 33°C (59°F) warmer than it would be otherwise. An increase in CO_2 has considerable potential for changing the earth's energy balance. Carbon dioxide is partially responsible for the Greenhouse Effect of the atmosphere as it is transparent to solar radiation but quite opaque to infrared radiation. Earth's radiation is concentrated in the infrared band from 7 to 20 micrometers. Carbon dioxide is absorbent of radiant energy in the wavelengths in which the earth radiates heat away from the surface, and it is in the 15 to 20 micrometer range where most of the absorption takes place. If CO_2 absorbs Earth radiation and the amount of CO_2 in the atmosphere is increasing, then more Earth energy should be absorbed by the atmosphere. This absorption and reradiation back to the surface shift the energy balance toward increased storage of energy, hence raising the temperature of the earth's surface and atmosphere.

Some scientists suggest the cause for an increase in atmospheric CO_2 and ultimately for planetary warming is from human activities. The main sources for the increased CO_2 are the combination of burning fossil fuels and burning of natural vegetation to clear land for agriculture. Over the past century, fossil fuel use and cement manufacturing released about 200 billion tons of carbon into the atmosphere. Current global emissions of CO_2 from energy use are some 6 gigatons of carbon each year (GiC). By the year 2025, this is expected to increase to between 8 and 15 GiC per year. For years beyond 2025, estimates are highly variable due to uncertainties about whether the countries with major emissions will take any steps to lower their emissions. Emission levels by 2100 may be about the same as those in 2000 or may be six times as high—around 36 GiC. The United States is by far the greatest contributor of atmospheric CO_2.

In the last century, deforestation may have released as much as an additional 115 billion tons of carbon. Deforestation and burning vegetation affect atmospheric CO_2 in at least three ways. First, it directly releases CO_2 into the atmosphere. Second, these processes reduce the removal of CO_2 from the atmosphere. Removing live vegetation reduces photosynthesis, which removes CO_2. Third, deforestation and burning vegetation may also disturb soil processes that affect the CO_2 exchange with the atmosphere.

Thus, the amount of change in global air temperature depends on the rate at which natural mechanisms lower the CO_2 content. The faster these mechanisms operate, the less the temperature change is likely to be. Some 40 percent of the gas placed in the atmosphere was absorbed either by the earth's biomass or by the ocean. The remainder has accumulated in the atmosphere. Uncertainties in the size of individual sources and sinks of CO_2 severely limit the accuracy of forecasts of future atmospheric concentrations. The natural movement of CO_2 into and out of the oceans and terrestrial vegetation is much larger than from human activity. Of the estimated 315 billion tons of carbon placed in the atmosphere since 1850, only 130 billion tons remain there. From 1981 to 1990, about half of the human-contributed emissions stayed in the atmosphere. One of the big questions currently is: Where has the remaining carbon gone and what is the process that removed it?

Other Contributions to Global Warming

There are other gases that play a part in global warming. **Methane** is a gas found in small amounts in the atmosphere under natural conditions. Methane is a greenhouse gas. It is very absorbent of Earth radiation. Methane is about 20 times more effective than CO_2 in trapping Earth radiation. If more methane is released into the atmosphere, there may be a positive feedback effect, which will lead to still more warming and more methane release. Evidence shows that it has begun to accumulate during the past two centuries. Swiss scientists used gases trapped in ice cores from Greenland and Antarctica to study changes in methane. They found that 100,000 years ago methane was present in the atmosphere at about 500 ppb. Seventy thousand years ago, it was about 650 ppb. By 20,000 years ago, at the height of the last glacial advance, the level of methane dropped to 350 ppb. With Holocene warming, the level of methane climbed to near 650 ppb once again. The concentration stayed near this level for the 3000 years prior to 1800. In the past 200 years, levels have risen 250 percent. In the latter half of the 19th century, methane was present at about 800 ppb. This increased to about 1600 ppb in the mid-1990s. About half of the increase came since 1960. The additional methane comes from livestock raising, rice cultivation, industry, mining, and landfills. In latitudes from about 50° to 60° in the northern hemisphere, much of the land surface is covered by permafrost. Permafrost is ground frozen to considerable depth in winter. In summer, the surface melts, but it does not melt down far enough to thaw all the soil. The permafrost provides a barrier for gases beneath and waters above. There is a lot of methane trapped beneath this frozen soil. Because global warming may be quite large in sub-Arctic regions, rapid melting of the permafrost may take place, and this will release methane into the atmosphere.

The **chlorofluorocarbons** (cfcs) also play an important role in the greenhouse process (chapter 17). Although present in much smaller quantities, the impact is large. Because of the high amount of CO_2 already in the atmosphere, it is effectively blocking Earth radiation. However, additional cfc atoms added to the atmosphere are 10,000 times more effective at absorbing infrared radiation than are additional molecules of CO_2. Cfcs probably account for 25 percent of the greenhouse forcing. A quadrupling of the cfcs might produce an increase in temperature from 0.5°C to 1.0°C (1°F–2°F).

Ozone increases in the upper troposphere may also be contributing to the warming. This is not to be confused with stratospheric ozone found at much higher altitudes. Observations indicate that ozone has been accumulating in the troposphere below 10 kilometers at a rate of about 1 percent per year. Ozone in the troposphere acts much differently to radiation than it does in the stratosphere. In the troposphere, it does not absorb the ultraviolet radiation as it does in the stratosphere. Instead, it acts as a greenhouse gas. One of the factors in the increased ozone in the troposphere is the emission of nitrous oxides by aircraft engines. Ozone is not nearly as important to the Greenhouse Effect of the atmosphere. However, the ozone is some 30 times as effective at absorbing Earth radiation as is CO_2.

Thermal Pollution

Thermal pollution resulting from increasing human use of energy and the inevitable discharge of waste heat into the atmosphere or ocean is another source of heat. Although it is not yet significant on a global scale, this heat source may become appreciable by the middle of the 21st century. If future energy production is concentrated in large nuclear power parks, the natural heat balance may be upset even sooner.

FUTURE CHANGES IN EARTH TEMPERATURE

The task of predicting future abundance of atmospheric CO_2 requires scientific information from many scientific disciplines. We must understand how the CO_2 budget operates today. We also need to know how it responds to changes in climate and other environmental conditions. Present data suggest that CO_2 emissions will grow by about 1.6 percent annually to 2025 and then decline to a growth rate of around 1 percent per year. It is both possible and probable that at some time the atmospheric level will reach 400 ppm. Forecasts made in the year 2000 placed the date of reaching the 400 ppm level at the year 2025. Once the concentration reaches 400 ppm, it will be at the highest level attained in the past million years. This level is significant even compared with geological time. The volume of coal known to exist in reserves is far more than necessary to produce such an increase in CO_2. The amount may well double from the 1980 level in the years between 2050 and 2100. If all the known fossil fuel reserves are burned, the CO_2 content in the atmosphere would triple. Once the level goes up, it will remain there for centuries. There is no rapid means of removing it from the atmosphere (Fig. 16.4).

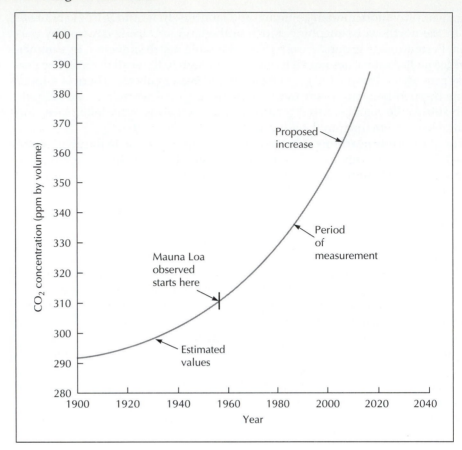

Figure 16.4
Estimated, measured, and projected amount of atmospheric CO_2 from 1900 to 2025.

Natural changes in temperature due to unknown interannual variations are as large as changes predicted due to changes in CO_2. It is a complex process. All the increase in temperature in the past 100 years may be due to natural processes. It is also possible that global warming due to CO_2 forcing is much larger than indicated by mean Earth temperatures. Other processes may be cooling the atmosphere. The earth is now as warm as it has been any time in the past 125,000 years.

The climate has always been changing. For almost all of the 4.6 billion years the planet has been in existence, the climate has changed due to natural processes. This is now no longer true. Climate change in the near future at least will result from human activities. So great has become the global population and its economic processes that it now dominates changes in our atmosphere on a global basis. If the CO_2 content increases, so will global temperatures.

Estimates vary significantly as to how much the temperature will increase. The estimates vary because they come from different mathematical models of atmospheric processes. If CO_2 concentrations double from the pre-Industrial level of 280 ppm, mean world temperatures will likely increase about 1.5°C to 4.5°C (3°F–8°F).

The problem of global warming due to CO_2 is a complex one. Many factors affect atmospheric CO_2. Part of the problem is that other factors cause the temperature to oscillate independently of the CO_2 in the atmosphere. One atmospheric component that affects global temperature is pollutants in the form of particulates. One of these pollutants is sulfur dioxide. Sulfur dioxide particles serve as very small nuclei for cloud particles. Clouds made of very fine particles reflect more solar radiation than clouds made of large particles. Since sulfur dioxide and CO_2 are both products of burning fossil fuels, they work in opposite directions on the energy balance. One leads to warming and the other to cooling. It is possible that increased reflectivity of the atmosphere may be offsetting as much as half of the potential greenhouse warming.

If mean global temperatures increase, the warming will not be the same every place on the globe. The subpolar latitudes will warm two or three times as much as equatorial regions. In the northern hemisphere, there is a zone where the snow cover shifts northward from year to year. In this zone, mean temperatures will increase 4°C if CO_2 content doubles. If other variables do not play a part, the exponential rise in the atmospheric CO_2 content will become significant. Based on the projected increase in CO_2, the mean temperature at the surface of the earth may increase 2.8°C.

Sea levels will continue to rise regardless of measures taken to reduce CO_2 emissions. By 2100, sea level is forecast to rise 15 to 95 centimeters (6–38 in) with a best estimate of 50 centimeters (20 in). Complete melting of the ice caps would change sea level some 70 meters. It is not certain whether warming of the seas would cause the ice caps to grow or shrink. If the water warms slightly, but not enough to raise the temperature of the atmosphere over the ice above freezing, there might be an increase in snowfall and glacial expansion. This would lower sea level. It is uncertain that sea level will continue to rise. In some areas, the rise of coastlines due to crustal decompression from the last ice age is taking place faster than sea level is rising; therefore, the net effect is one of lower sea level (Fig. 16.5).

One set of projections forecasts an increase in global temperature under a business-as-usual scenario. It is assumed that no controls are placed on greenhouse gas emissions. Under these conditions, temperatures will be about 1°C above present levels (2°C above pre-Industrial levels) by 2025. Before the end of the 21st century, global mean temperature will be 3°C above the current level.

There are several uncertainties in the projected temperature increase. First, there is still an incomplete understanding of the nature of sinks for CO_2, and future additions of CO_2 to the atmosphere will depend largely on the actions and policies of governments in relation to fossil fuel use. Second, the role of clouds in a warmer climate is not yet fully understood. Whether clouds will be more efficient reflectors of short-wave radiation or absorbers of long-wave radiation is not yet clear. Third, the role of the oceans, which influence both the pattern and timing of climatic

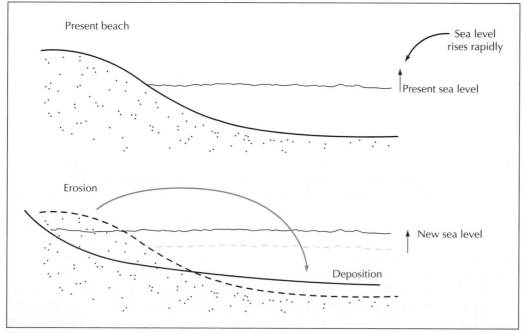

Figure 16.5
Coastal erosion resulting from a rapid rise in sea level.

change, has yet to be fully determined. One thing is certain: Even if major reductions in carbon emissions into the atmosphere are undertaken now, atmospheric warming will continue for several centuries.

To many it would seem that an increase of but a fraction of a degree of temperature is of little consequence in the scheme of a climate where day-to-day changes are in terms of tens of degrees. Of major significance is that the change would lead to a modification of temperature gradients over the earth's surface, and this in turn would lead to a modification of the general circulation pattern. Changes in precipitation and storm tracks are some of the related consequences. Model calculations indicate that evaporation from water surfaces will increase, and this will lead to greater cloud cover and precipitation for the planet as a whole. Much of this additional precipitation will be over the oceans, with some land areas becoming much drier. In the United States, precipitation from 1970 to 2000 was about 5 percent greater than in the prior years of the 20th century. However, the frequency of extreme rainfall events (more than 2 inches per day) over the United States appears to have increased substantially in the 1990s.

Almost all of the debate concerns global warming as a result of the addition of greenhouse gases to the atmosphere. There has been controversy, with some scientists taking a view that the case for such warming and associated change is overstated. Nonetheless, there is a general consensus that global warming is real, with the major areas of disagreement being the interpretation of the rate at which it is occurring and the extent of the impacts of that increase.

Perhaps the best view of the controversy is provided by the final report of the working group of the Intergovernmental Panel on Climatic Change (IPCC), which is sponsored jointly by the World Meteorological Organization and the U.N. Environment Program. The report was based on the scientific assessment of climatic change by several hundred scientists in 25 countries to provide perhaps the most comprehensive view of opinions on the subject. The scientists participating in the study are confident that CO_2 has been responsible for over half of the enhanced Greenhouse Effect in the past and is likely to remain so in the future. The atmospheric concentrations of long-lived gases (CO_2, nitrous oxide, and cfcs) adjust only slowly to changes in emissions. Continued emission of these gases at present rates will commit the atmosphere to increased concentrations for centuries to come. To stabilize concentrations at their present level, emission of long-lived gases would need to be decreased by 60 percent.

Since what we call civilization began on this planet some 6000 years ago, the mean temperature has not varied more than 1°C from the average. The forecast change in temperature from 1.5°C to 4.C° by 2100 has no equal in the recent history of the planet. A 2°C rise in temperature would move the earth to a climatic position similar to that which existed during the climatic optimum 6000 years ago. Because of the new data on the temporary storage of heat in the ocean, some scientists are increasing the forecast to a change of 3°C or more. Changes in the global mean temperature of 2°C or more would have a tremendous impact on global society. The stress of adjustment would be phenomenal. It would alter virtually all aspects of life.

The phenomenon of global warming has become a political issue. Thousands of scientists worldwide agree that warming is occurring. A small, but very vocal number of scientists argue that warming is not occurring. In the United States, the large oil, coal, and automobile companies have maintained a continuous campaign to denounce the reality of global warming and advocate that no action should be taken to reduce carbon emissions. This is in the face of the fact that the United States contributes more carbon to the atmosphere than 150 of the smaller countries of the world combined. Acting on their own, some nations are taking steps to reduce their emission of carbon. Denmark is building some 500 offshore wind turbines to be completed by 2005. They are part of a plan to cut emissions by 50 percent by the year 2030.

SUMMARY

There is a natural storage process operating in the atmosphere that raises the temperature at the earth's surface from below freezing to about 15°C. This storage process is the Greenhouse Effect. It results from some atmospheric gases being fairly transparent to solar radiation but more opaque to Earth radiation. This results in energy flowing back and forth between the surface and the lower atmosphere.

Natural changes in temperature due to unknown interannual variations are as large as changes predicted due to changes in CO_2. It is a complex process. All the increase in temperature in the past 100 years may be due to natural processes. It is also possible that global warming due to CO_2 forcing is much larger than indicated by mean Earth temperatures. Other processes may be cooling the atmosphere. Global temperatures have increased since the middle of the 19th century. Some scientists believe that this increase in temperature is due to the increase in the emission of greenhouse gases such as CO_2 and methane.

KEY TERMS

Chlorofluorocarbons
Cholera
Dengue fever
Hantavirus

Ice shelves
Malaria
Methane
Respiration

The plague
Yellow fever

REVIEW QUESTIONS

1. List four different kinds of evidence that support global warming.
2. Name four national parks in the United States where the natural vegetation is being threatened by global warming.
3. Which tropical disease infects more people than any other and is spreading into areas in which it was previously not found?
4. Name two tropical diseases that have appeared in the United States for the first time since 1980.
5. What is the difference between a glacier and an ice shelf?
6. Is there any evidence that freshwater lakes are increasing in temperature?
7. What is the major sink, or means of CO_2 removal from the atmosphere?
8. How does atmospheric CO_2 act differently on solar radiation and Earth radiation?
9. Deforestation contributes to the accumulation of CO_2 in the atmosphere in three different ways. What are these three processes?
10. How does a molecule of methane compare to a molecule of CO_2 in ability to absorb Earth radiation?

PART IV

Applied Climatology

The Harvesters by Pieter Bruegel the Elder
The Metropolitan Museum of Art, Bogers Fund, 1919. (19.164)
Photograph © 1980 The Metropolitan Museum of Art

CHAPTER
17

The Human Response to Climate

CHAPTER OUTLINE

*T*hroughout history, many scholars have attempted to show that the physical environment is the single most important determinant of human actions and activities. This concept led to a philosophy called *environmental determinism*. This may be defined as the philosophical doctrine that human action is not free but determined by external forces acting on human attributes and behavior.

It was not unnatural that climate, a major component of the environment, gave rise to climatic determinism. In the latter part of the 19th century and up to the 1930s, many books and papers purported to show how environment and climate determined all sorts of activities, ranging from religious beliefs to cultural activities. Deterministic writers often made generalizations, based on limited data, about human–environmental relationships and often ignored contrary evidence. Unfortunately, one outcome of the deterministic viewpoint was that, once it fell into disfavor, it led to a marked decrease in legitimate studies relating human activities to climate. However, the studies of such relationships today are based on better science and rigorous interpretation of data, and the current study of people and the climatic environment deals with many facets of human society. Climate has been related to activities ranging from crime to recreation, from artistic endeavor to the decline of civilizations. This chapter examines but a few of these relationships.

THE PHYSIOLOGICAL RESPONSE

The body temperature of a healthy human varies a little around 98.6°F (37°C). Maintenance of that temperature calls for a balance between the body's heat loss and heat gain. Chemical processes within the human body (**metabolism**) are related to the intake of food, the production of energy (and waste), and the physical status of the body.

Energy production in the body is related to energy gains and losses to and from the surrounding environment. Gains of body heat from the environment occur because of absorption of long-wave radiation and conduction from the surrounding air if air temperature is above skin temperature. Likewise, heat losses occur through radiation, conduction, and, most important, evaporation of moisture from the skin surface (Fig. 17.1). If more heat is lost than is gained, the body temperature must fall. Conversely, if more heat is gained than is lost, then the body temperature must rise. Under conditions of imbalance, a number of physiological responses occur (Table 17.1). The term **hypothermia** refers to conditions that occur when the core body temperature is either reduced or raised beyond a manageable range.

As indicated in Table 17.1, cold conditions cause blood flow to the body periphery (e.g., hands and feet) to be reduced and less heat is lost from the body. Although this reduced outward flow conserves heat and maintains the core temperature of the body, the reduced blood flow to the extremities can have dire effects. Under extreme conditions, frostbite can occur, resulting in frozen tissue and cell destruction.

Shivering is another physiological response to cold conditions. Its function is to increase the metabolic rate. Although this certainly occurs, the shivering function is not a highly efficient process. It brings more blood to the surface layers of the body, but, especially with uncontrolled shivering, increases heat loss by radiation and convection from the body to the surrounding environment. This causes difficulty in breathing, muscular rigidity, unconsciousness, and, eventually, death.

The response of the body to an excessive heat load is variable. An initial effect is dilation of surface blood vessels, which causes a rise in skin temperature. This increased temperature causes sweating, which is a cooling mechanism. The moisture

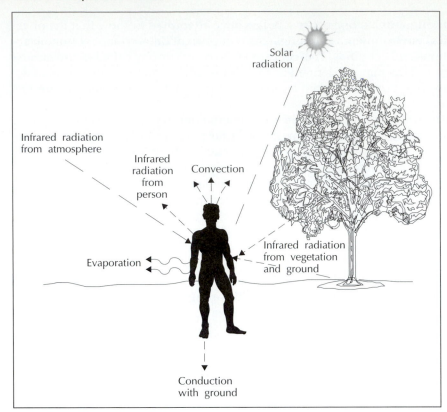

Figure 17.1
The energy exchanges that occur between the human body and the surrounding environment (after Moran and Morgan, 1997).

on the skin surface is evaporated; in the change of state that occurs, heat removed cools the body surface. Such responses are not, however, without danger. Physical exertion in a hot, humid environment can result in heat stroke—a condition caused by cessation of sweating and consequent rise in core body temperature. In a hot, dry environment, excessive sweating causes salt depletion and dehydration. Both heat stroke and dehydration can prove fatal.

BIOMETEOROLOGICAL INDEXES

To better understand the relationships between the human body and a set of climatic conditions, **biometeorological** (or bioclimatological) **indexes** have been developed. These serve to allow prediction of various responses to the sensation of warmth and to assess the physiological strain imposed by combined atmospheric variables. Many indexes have been proposed, and only the most common are considered here.

The human body has specific physiological responses to unusually hot conditions. The efficiency with which the body is cooled often depends on the effectiveness of sweating in cooling the skin surface. At high humidity levels, the rate of evaporation decreases. The result is that one feels sticky and uncomfortable for less evaporation means less cooling and sweat beads on the body. Some indication of how comfortable it may be is given by the dew point of the air; this is the temperature to which air must be cooled for water vapor to condense, and a high dew point (e.g., above 60°F/15.6°C) is indicative of a moist atmosphere. However, to measure this heat/moisture discomfort, a number of indexes have been devised, with the most widely used being **apparent temperature (AT)** and the **heat index (HI)**.

Table 17.1 Summary of Human Responses to Thermal Stress

	To Cold	To Heat
Thermoregulatory responses		
	Constriction of skin blood vessels	Dilation of skin blood vessels
	Concentration of blood	Dilution of blood
	Flexion to reduced exposed body surface	Extension to increase exposed body surface
	Increased muscle tone	Decreased muscle tone
	Shivering	Sweating
	Inclination to increased activity	Inclination to reduced activity
Consequential disturbances		
	Increased urine volume	Decreased urine volume. Thirst and dehydration
	Danger of inadequate blood supply to skin of fingers, toes, and exposed parts leading to frostbite	Difficulty in maintaining blood supply to brain leading to dizziness, nausea, and heat exhaustion. Difficulty in maintaining chloride balance, leading to heat cramps
	Increased hunger	Decreased appetite
Failure of regulation		
	Falling body temperature	Rising body temperature
	Drowsiness	Hear regulating center impaired
	Cessation of heartbeat and respiration	Failure of nervous regulation terminating in cessation of breathing

Source: After Lee 1958.

In deriving the heat stress index, the National Weather Service converted an earlier expression into a multiple regression model to provide an understandable value that is used to alert the public to the hazards of heat waves. The HI incorporates air temperature and relative humidity to determine the apparent temperature. This is what the air temperature feels like for the average person. Due to the design limits of the model, HI is only given when temperatures are 70°F or above.

AT may be derived from a table (Table 17.2a). The ATs have been categorized according to the effect that such temperatures might have on humans (Table 17.2b).

Not all researchers are satisfied with the current system. Some recommend that the use of AT in weather broadcasts be abandoned. The main reason is that the numbers, which are usually near the actual air temperature, can be misleading. Further, there are no specific rules for issuing heat stress advisories, and AT may not be a good indicator of how hot a person feels. Individuals appear to react differently in the same environmental conditions depending on various states of health, mind, and acclimatization. Nonetheless, it does provide a signal for potentially lethal weather conditions.

Just as an index has been derived for heat stress, so has one for the influence of low temperatures on the human body. The term **windchill** was coined by Antarctic explorer Paul A. Siple to describe the cooling power of wind for

Table 17.2(a)   Apparent Temperature Index

Relative Humidity (%)	Air Temperature (°F)										
	70	75	80	85	90	95	100	105	110	115	120
	Apparent Temperature (°F)										
0	64	69	73	78	83	87	91	95	99	103	107
10	65	70	75	80	85	90	95	100	105	111	116
20	66	72	77	82	87	93	99	105	112	120	130
30	67	73	78	84	90	96	104	113	123	135	148
40	68	74	79	86	93	101	110	123	137	151	
50	69	75	81	88	96	107	120	135	150		
60	70	76	82	90	100	114	132	149			
70	70	77	85	93	106	124	144				
80	71	78	86	97	113	136					
90	71	79	88	102	122						
100	72	80	91	108							

Source: National Weather Service, NOAA.

Table 17.2(b)   Hazards Posed by Heat Stress by Range of Apparent Temperature

Category	Apparent Temperature	Heat Syndrome
I	54°C or higher (130°F or higher)	Heatstroke or sunstroke *imminent*.
II	41 to 54°C (105 to 130°F)	Sunstroke, heat cramps, or heat exhaustion *likely*. Heatstroke *possible* with prolonged exposure and physical activity.
III	32 to 41°C (90 to 105°F)	Sunstroke, heat cramps, and heat exhaustion *possible* with prolonged exposure and physical activity.
IV	27 to 32°C (80 to 90°F)	Fatigue *possible* with prolonged exposure and physical activity.

Source: National Weather Service, NOAA.

various combinations of temperatures and wind speeds. Later he provided a method to calculate windchill. This index, developed from experiments conducted in Antarctica, measures the rate of freezing of water at various temperatures and wind speeds.

The original application of the Siple formula was introduced to predict the conditions causing frostbite. A modified windchill formula, which included the effect of clothing, was later introduced. Because a clothed person was considered, the variables of breathing and heat transfer through clothing were incorporated. The model assumes a healthy adult (of height 1.7 m and body surface area of about 1.7 m²) walking outdoors at 1.33 m sec⁻¹ will generate 188 W m⁻² of heat. To maintain thermal equilibrium, the amount of heat loss must not exceed the heat generated. The balance is achieved by wearing an appropriate thickness of clothing. The present windchill advisories are based on the sensation of cold felt by the majority of people. Table 17.3 outlines the system.

Table 17.3 Windchill Equivalent Temperature

Wind Speed (m/sec)	Air Temperature (°C)															
	6	3	0	−3	−6	−9	−12	−15	−18	−21	−24	−27	−30	−33	−36	−39
3	3	−1	−4	−7	−11	−14	−18	−21	−24	−28	−31	−34	−38	−41	−45	−48
6	−2	−6	−10	−14	−18	−22	−26	−30	−34	−38	−42	−46	−50	−54	−58	−62
9	−6	−10	−14	−18	−23	−27	−31	−35	−40	−44	−48	−53	−57	−61	−65	−70
12	−8	−12	−17	−21	−26	−30	−35	−39	−44	−48	−53	−57	−62	−66	−71	−75
15	−9	−14	−18	−23	−27	−32	−37	−41	−46	−51	−55	−60	−65	−69	−74	−79
18	−10	−14	−19	−24	−29	−33	−38	−43	−48	−52	−57	−62	−67	−71	−76	−81
21	−10	−15	−20	−25	−29	−34	−39	−44	−49	−53	−58	−63	−68	−73	−77	−82
24	−10	−15	−20	−25	−30	−35	−39	−44	−49	−54	−59	−63	−68	−73	−78	−83

Wind Speed (mi/hr)	Air Temperature (°F)																		
	45	40	35	30	25	20	15	10	5	0	−5	−10	−15	−20	−25	−30	−35	−40	−45
5	43	37	32	27	20	16	18	6	1	−5	−10	−15	−20	−26	−31	−36	−41	−47	−52
10	34	28	22	16	10	4	−3	−9	−15	−21	−27	−33	−40	−46	−52	−58	−64	−70	−76
15	29	22	16	9	2	−5	−11	−18	−25	−32	−38	−45	−52	−58	−65	−72	−79	−85	−92
20	25	18	11	4	−3	−10	−17	−25	−32	−39	−46	−53	−60	−67	−74	−82	−89	−96	−103
25	23	15	8	0	−7	−15	−22	−29	−37	−44	−52	−59	−66	−74	−81	−89	−96	−104	−111
30	21	13	5	−2	−10	−18	−25	−33	−41	−48	−56	−63	−71	−79	−86	−94	−102	−109	−117
35	19	11	3	−4	−12	−20	−28	−35	−43	−51	−59	−67	−74	−82	−90	−98	−106	−113	−121
40	18	10	2	−6	−14	−22	−29	−37	−45	−53	−61	−69	−77	−85	−93	−101	−108	−116	−124
45	17	9	1	−7	−15	−23	−31	−39	−47	−55	−62	−70	−78	−86	−94	−102	−110	−118	−126

SOLAR RADIATION AND PEOPLE

Solar energy arriving at the top of the atmosphere contains ultraviolet, visible, and infrared radiation. The ultraviolet (UV) radiation has different effects at different wavelengths so three bands are recognized: UV-C, UV-B, and UV-A, from shortest to longest wavelengths. The highly damaging UV-C is absorbed by stratospheric ozone, leaving the other two, together with atmospheric counterradiation, to arrive at the earth's surface.

Radiation determines both the effects of thermoregulation and the photochemical responses that occur in the skin. Radiation absorbed at the body surface has a number of effects. Of prime importance is the production of vitamin D, the vitamin necessary for the prevention of bone disease. Another well-known effect of UV radiation on humans is sunburn. Exposure to UV radiation produces a capillary dilating chemical that immediately induces a reddening of the skin and ultimately, with continued exposure, blistering. Differences in skin types influence the rates of reaction (Table 17.4). A nontanned, White-skinned person will show traces of skin reddening after 3 to 20 minutes of exposure on a clear midsummer day in mid-latitudes. Further exposure leads to a marked photochemical response of the skin. Irradiation at wavelengths of UV-A and UV-B bands produces melanin, which passes into the Malpighian layer. Extensive production over a long period of time leads to skin that appears to age rapidly and, possibly, to skin cancer.

Given that both short- and long-term exposure to UV radiation is a health hazard, the National Oceanic and Atmospheric Administration (NOAA) and the Environmental Protection Agency (EPA) have suggested an **ultraviolet index**. This index, shown in Table 17.5, predicts UV radiation levels for noon standard time for 58 U.S. cities. The UV index is provided with the daily forecast and is derived using satellite and ground-based observations and computer models.

In recent years, psychiatric researchers have been examining the effects and possible treatment of mood disturbances related to seasons. Known as *Seasonal Affective Disorder* (SAD), its most common form is winter depression—a time marked by such things as sadness, decreased physical activity, weight gain, decreased libido, and even interpersonal conflict. There are two main treatments: antidepressant drugs and light therapy. In the latter, the suggested treatment is to expose the patient to all wavelength light using fluorescent bulbs to simulate sunlight for 2 hours each day. It is generally found that this treatment associated with antidepressants is most effective.

Table 17.4 Description of Skin Phototype

Skin Phototypes	Skin Color in Unexposed Area	Tanning History
Never tans/always burns	Pale or milky white; alabaster	Develops red sunburn; painful swelling; skin peels
Sometime tans/usually burns	Very light brown; sometimes freckles	Usually burns; pinkish or red coloring appears; can gradually develop light brown tan
Usually tans/sometimes burns	Light tan, brown, or olive; distinctly pigmented	Infrequently burns; shows moderately rapid tanning response
Always tans/rarely burns	Brown, dark brown, or black	Rarely burns; shows very rapid tanning response

Table 17.5 Range of "Minute to Burn" for Different Index Values

Exposure Categories	Index Values	Minutes to Burn for Most Susceptible Skin Phototype	Minutes to Burn for Least Susceptible Skin Phototype
Minimal	0–2	30	> 120
Low	3	20	90
	4	15	75
Moderate	5	12	60
	6	10	50
High	7	8.5	40
	8	7.5	35
	9	7	33
Very High	10	6	30
	11	5.5	27
	12	5	25
	13	< 5	23
	14	4	21
	15	< 4	20

CLIMATE AND HEALTH

Climatotherapy

The idea of "escaping the weather" has been used by people—at least those who are wealthy enough—for many years. Summer palaces, Mediterranean yachts, and Florida homes all testify to this. The utilization of prevailing climate as therapeutic is called **climatotherapy**. Although travel for climate is perhaps an affluent societal vogue, there is little doubt that it points to the living stresses that given climates can evoke. Intense heat makes great demands on the circulatory system, and it makes good sense for a person suffering from such problems to avoid regions of stress. Cold, damp weather can, as noted later, cause depression and circulatory problems.

Diseases of the respiratory system are worsened in places with high airborne particle counts. Asthma, bronchitis, and tuberculosis cannot be adequately treated in such environments. Unfortunately, some areas that were formerly used for treatment (e.g., the American southwest) now experience enough air pollution to negate the beneficial effects. There is, in fact, a substantial amount of literature on such facts as the climatic conditions most suitable for sanatoria specializing in the treatment of specific diseases.

Morbidity and Mortality

Although many common diseases show a relationship to seasonal and other climatic variations, a direct cause–effect correlation is difficult to obtain. This results from both the problem of differentiating weather-induced illness from other illnesses and from problems of derivation and treatment of data. That is not to say, however, that the relationship cannot be shown; the following cases show that some highly meaningful results can be obtained. Some seasonal variations in respiratory diseases report that the incidence of asthma increases markedly with the onset of cold weather, while there is an indication that respiratory diseases occur much more in winter than in the summer, particularly in cool climates.

It has been shown that mortality rates for the northern and southern hemispheres are out of phase by 6 months. The remarkable feature in Figure 17.2 is the mirror image of the two curves. In England and Wales, the highest number of deaths occur in the winter months—November through March; in the southern hemisphere, the winter deaths occur May through October.

Seasonal variations are increased in the United States largely because deaths from pneumonia and influenza are included in the data. Older persons and very young people are most likely to suffer from these in the winter months. Circulatory system diseases, notably heart failure, show less of a seasonal trend, but are still apparent nonetheless. The atmospheric influence on a person afflicted with heart disease is both direct and indirect. During cold weather, the cooling of the outer parts of the body places greater stress on the heart while increasing blood pressure. For persons with an impaired circulatory system, this proves highly stressful. Stress becomes much greater if physical activity is carried out, and heart failure while shoveling snow is not an uncommon event.

Despite the influence of modern methods on rates and seasonality of mortality, the role of weather is still very important. Figure 17.3 shows the deaths in the United States on a daily basis for a typical year. The upper graph reveals a departure from the winter–summer decline and represents a sudden peaking in the number of deaths. The lower figure presents an enlarged version of the peak and relates it to the daily temperatures at an east coast station—Philadelphia. There is a clear statistical relationship between the number of deaths and the heatwave temperatures. In this case, research shows that the elderly were mostly affected.

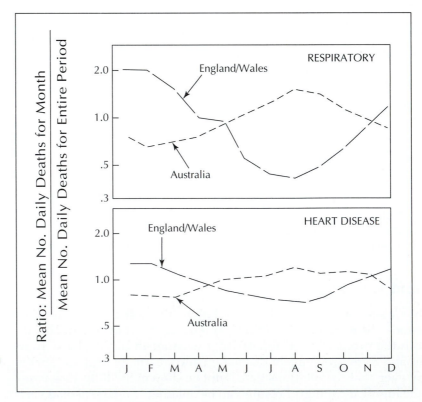

Figure 17.2
Comparisons of seasonal mortality for two causes in England/Wales and Australia.

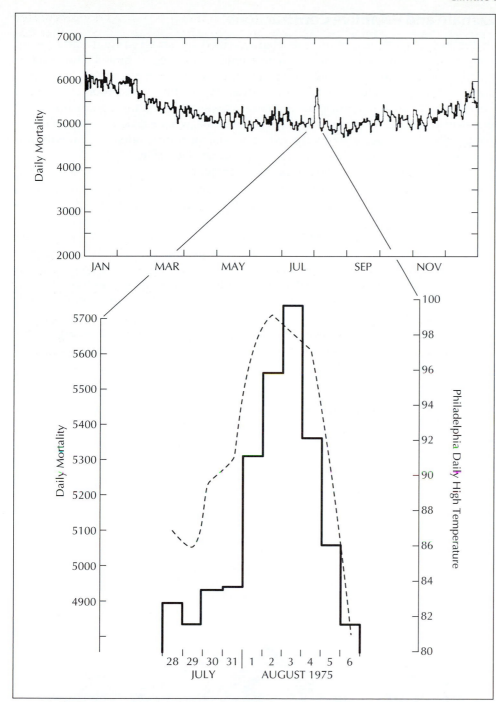

Figure 17.3
Upper graph shows the daily deaths in the United States in a single year. The peak in August of that year corresponded to a deadly heat wave in the northeast. The lower graph shows how deaths peaked in Philadelphia for that period.

CLIMATE AND ARCHITECTURE

Shelter, with food, is one of mainstays of human life on Earth. The nature of shelter required largely depends on the conditions of the environment, with the climate providing one base that determines the type needed. Primitive peoples of the world—primitive in the technological and preliterate sense—using the limited resources at hand often developed shelters that were in harmony with the climatic conditions under which they lived.

Climate and Primitive Constructions

In the hot-wet tropics, climate is characterized by small annual variation of temperature, intense solar radiation, high humidity, and heavy rainfall. Thus, houses should provide maximum ventilation and shade while the walls should be open or supplied with movable shades. Roofs, of necessity, must be waterproof. The profuse vegetation of such areas supplies ample raw material with vines, poles, and bamboo readily available. In the permanent settlements, where agriculture or fishing is the livelihood, houses are constructed on a skeletal frame, often stilted, with impervious thatched roofs that overhang to provide shade (Fig. 17.4).

In the tropical deserts, climatic problems are related to the intense solar radiation and heat during the day and low temperatures at night. Humidity and precipitation are low and present no problem in housing comfort. Some vegetation is available as building material; however, for the most part, mud and straw are the basic materials. These are ideal because their high heat capacity helps maintain an even temperature in the building. The thick walls, with minimum window space, are admirably demonstrated in many desert dwellings. Figure 17.5 shows an example of this and further indicates how grouped dwellings in dry areas are often built close together for maximum shade—a fact well exhibited in the narrow streets of many Arab communities. Modern, expansive boulevards are not so well suited to such conditions.

Savanna areas, located between the desert and the humid tropics, experience seasonal rains of varying intensities. In such locations, homes are frequently dome or cone shaped to facilitate drainage during wet spells. They are constructed of grass, mud, and branches, although animal skins may be used. Such traditional homes are now being replaced over wide areas by cement block types.

The regions that abut the wet and dry zones present special problems of house design. The inhabitants are faced with the dual problem of seasonal climates. In the Mediterranean climatic zone, for example, it is necessary to combine the requirements of buildings of the arid regions with those of places that experience a cool, wet season. In effect, to cope with the cooler wet period, the houses of the Mediterranean climate must be more substantial than those designed for desert

Figure 17.4
The basic design of a traditional house in the wet Tropics. Note the waterproof roof, the stilts, the shaded veranda, and the moveable shutters.

Figure 17.5
Sketch of buildings suited to hot, dry conditions. Note the close grouping of the houses to provide maximum shade.

locations only. Older dwellings in such regions, ranging back to Roman times, are often identified by a central open courtyard. These are shaded during the day because even at the height of summer the sun is never directly overhead in these latitudes. At night they are effective as radiators. The microclimate might be further modified for fountains and pools were often located in the courtyards. During the winter, the substantial homes could be heated to combat cooler periods.

Outside the tropical realm, primitive shelters were necessary to overcome the problem of cold. The best-known construction in this respect is the igloo—a house form that is rarely found today. Using extremely limited resources, the igloo represents a building that is well adapted to hostile environmental conditions. It is hemispherical, which both minimizes the proportion of area exposed and represents a streamlined building capable of withstanding high winds. It is constructed of dry snow blocks that are piled one on the other in an inward spiral. The snow blocks have a low conductivity to help conserve the interior temperatures (Fig. 17.6).

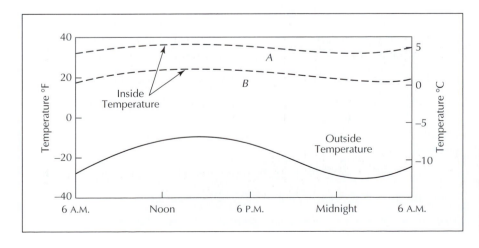

Figure 17.6
Comparison of temperatures inside and outside of an igloo. *A* is air temperature and *B* is floor temperature.

The interior temperature may be as high as 15°C (59°F)—a marked contrast to the temperature prevailing outside. Heat is generated by both oil lamps and body temperatures, and it is not unusual to find animal skins lining the interior of the dwelling. Igloos are winter dwellings. In summer the moderate temperatures and high solar radiation (the sun is in the sky continuously) require dwellings of different characteristics. The available materials—turf, earth, and driftwood—are used to construct sod-roofed dugouts.

Middle-latitude continental climates, with severe winters and warm moist summers, have been countered by some admirable dwellings. Nomadic people of such regions required a home suited to the extremes of climate, but that was portable. Such structures are seen in the yurt (or Kazak) tent of Asia and the familiar Indian tepee. Human ingenuity and comprehension of the surrounding environment have given rise to dwellings of amazing diversity. Equal diversity is also found in modern buildings, but for quite different reasons.

Climate and Modern Structures

The construction of a home or business that uses its background climatic environment as a resource benefits in many ways; some benefits are economic, such as decreased energy costs, and others aesthetic, as exhibited by landscape design. Beginning with site and eventually passing to construction materials, climate can be a significant component of building construction.

Of basic importance in the layout design of a building or building cluster is consideration of the effects of solar radiation. Direct solar radiation on an exterior surface and that which is passed into the structure via openings is, for the most part, a marked asset in cold climates and should be used to full advantage; in hot regions the effects of direct radiation should be avoided as much as possible. Since the sun's path in the sky, and resulting radiation intensity, varies from location to location over the year, an understanding of earth–sun relationships is essential to formulating a rational design plan.

It was shown in chapter 2 that the position of the sun in the sky at any given latitude can be approximated using a simple representation. As in Figure 17.7, the apparent movement of the sun in the sky can be illustrated by considering the observer (O) located at a given latitude (in this case, 4I°N, the latitude of New York City) surrounded by a horizon (H). The large, surrounding circle (S) represented the celestial sphere across which the sun can be assumed to pass. At the equinoxes, the sun will rise in the due east, attain its highest elevation at noon, and set in the due west. The angle of the noon sun is given by

$$90° − latitude \ (in \ this \ case, \ 90° − 41° = 49°).$$

At any other time of the year, the vertical sun is overhead at noon at a latitude between the Tropics. This, the solar declination, may be derived in a number of ways, one of which uses the **analemma**. This is illustrated in Figure 17.8 with an appropriate explanation. Accordingly, a general equation to describe this is

$$Solar \ altitude = 90 − (latitude ± declination \ of \ sun),$$

where the minus sign is used when the sun and latitude are in the same hemisphere and the positive when they are in different hemispheres. So at the winter solstice, when the sun is overhead 23.5°S, the altitude of the noon sun is given by

$$Altitude = 90° − (latitude + 23.5°) \ which \ for \ 41° \ N \ is \ 90 − (41 + 23.5) = 25.5°;$$

when the sun is overhead (Cancer, 23.5°N), in summer

$$Altitude = 90° − (latitude − 23.5°) \ so \ that \ at \ 41° \ N \ is \ 90° − (41° − 23.5°) = 72.5°.$$

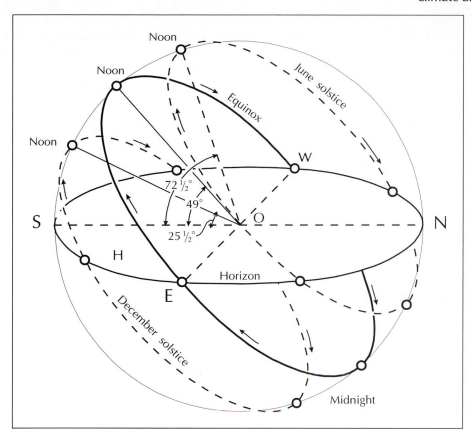

Figure 17.7
Apparent path of the sun in the sky at New York City. To the earth-bound observer, the earth's surface is a flat, horizontal disk. The sun, moon, and stars appear to travel on the inner surface of a hemispheric dome (after Strahler, 1969).

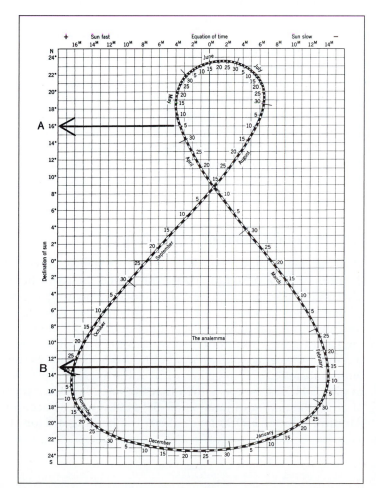

Figure 17.8
The analemma provides the declination of the sun for each day of the year. The examples show that (a) on May 5th, the sun at noon is vertically overhead 16°N, and (b) on February 15th, the declination of the sun is 13°S.

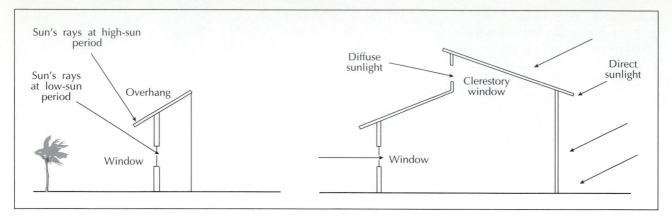

Figure 17.9
Correct use of overhangs (a) can allow penetration of sunlight when needed yet exclude it at other periods, and (b) shows the principle of the clerestory. In this, scattered and reflected light is admitted while direct sunlight is excluded on opaque walls on the sunlit side of the building.

To fully consider the solar radiation load on a structure, it is also necessary to consider the changing lengths of day and night over the period of a year and azimuth, the horizontal angle of the sun measured clockwise a north–south reference line. With such facts, an architect is able to plan to best utilize the solar radiation load. Such information is used to determine shading requirements for, in mid-latitudes, best exposure to summer shade and winter sun. Perhaps the most simple adaptation is the use of an awning, which can be adjusted to the angle of the sun (Fig. 17.9).

The thermal environment in a building is obviously closely related to the solar radiation factor. Within any building, the object is to establish and maintain an internal thermal environment conducive to habitation in terms of the prevailing external climate. The modern approach to maintaining this is to rely, for the most part, on air conditioning and heating systems. Obviously, such systems are needed at times, but overreliance on them has often resulted in unbearable strains on power resources. The amount of energy needed to maintain a required internal temperature in a dwelling depends on both the external climatic conditions and the fabric of which the walls are made. A measure of the former are degree days and of the latter the construction materials used.

Heating and Cooling Degree Days

Heating degree days (HDDs) and **cooling degree days** (CDDs) are used as indicators of the amount of energy consumed for space heating and cooling. The idea of these degree days was formulated by heating engineers and, in the United States, is computed using the Fahrenheit scale. The engineers found that when the outdoor mean daily temperature is below 65°F, most buildings require heating to bring their interior temperatures to 70°F. HDDs are accumulated by subtracting the average daily temperature from 65°F; for example, with a daily high temperature of 70°F and a low of 50°F, the average, 60°F, is subtracted from 65°F to give 5 HDDs. Suppose then that the next day has a mean daily temperature of 58°F; then an additional 7 (65–58) HDDs will be accumulated. By keeping a running total of HDDs throughout the heating season, fuel distributors and power companies can anticipate the general amount of fuel oil to be delivered or power to be generated. Large differences are seen in these accumulated average annual totals. In central North Dakota, for example, some 9,000 HDDs occur, whereas in central Florida HDDs amount to 500. Clearly, heating costs are much greater in the colder northern states.

Cooling degree days (CDDs) are generally computed when the air temperature is 65°F or more, although sometimes a higher base temperature, often 70°F, is used. CDDs are computed in a similar way to HDDs using the accumulated days above the selected base temperature. The higher the value, the greater the amount of energy used for cooling the interior of a structure. The locational energy needs are generally the inverse of the HDDs, with highest values in the United States occurring in the southern states.

Although heating and cooling degrees days were specifically designed for assessing regional energy needs, they also have other uses. For example, by plotting the seasonal degree days over time, it is possible to detect warming and cooling trends or the relative periodicity of cold and warm years.

The Building Fabric

As already noted, both the adobe and igloo seem well suited to the environments in which they occur. These two examples clearly indicate the importance of construction materials under different climatic conditions. In these days of prefabricated houses or uniform building materials, the natural thermal properties of the material are sometimes overlooked. This is in accord with the philosophy that, with artificial heating and cooling methods, most building problems can be overcome by manipulating input of energy. Nothing could be less correct, and application of a few simple principles can provide much insight into optimum thickness and required conductance characteristics suitable for a given climate. Although prefabricated materials can be used in such diverse climatic regions as New England and the American southwest, the manner in which they are arranged needs to be quite different. To understand how a building reacts to a given external climate, it is necessary to estimate a number of thermal characteristics. Basically, the main concern is the flow of heat through the wall or roof in terms of both quantity and time. It is often convenient to express thermal properties as the R value—the heat resistance. It is found that the R value varies considerably. For example, a simple galvanized metal shed may appear to be a potential oven in a hot climate, yet simple modification of the structure and rational use of building materials can mitigate the conditions. The R value for galvanized metal over an open frame is 0.6; if a 1-inch insulating board is used, it becomes 3.6. With the addition of 4 inches of wood shavings under the metal, the R value becomes 10.5.

Once the relationship between building material and heat flow is derived, another important factor must be considered. The transfer of heat through a structure will not occur instantaneously, and there will be a time lag between the maximum outside and maximum inside temperatures. The time lag can vary from about 3 hours for a light wall made of two layers of sheet material separated by an air cavity to at least 8 hours for a compacted Earth wall about 30 centimeters thick. Not only will time lag occur, but there will also be an interior dampening of the extremes found in the exterior climate. As Figure 17.10 shows, the differences can be visualized as two cycles; the time lag causes them to be out of phase while the dampening (or decrement) causes the interior cycle to have less amplitude than that outside. Use of the correct wall materials and thicknesses can result in an ideal situation, where the phase and amplitude are designed to

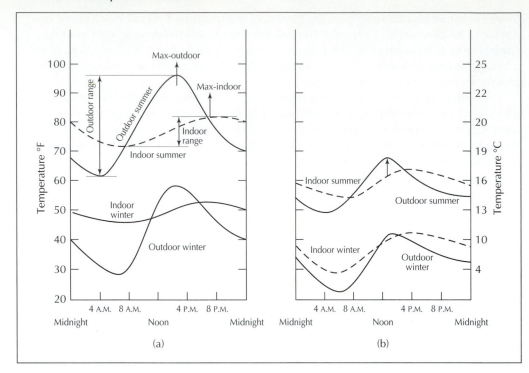

Figure 17.10
Comparison of the inside internal and external temperatures in (a) adobe in hot, dry climates, and (b) a wood structure in mid-latitudes.

meet the needs of the climatic realm in question. In damp climates, building materials also need to act as vapor barriers. Lack of proper use of vapor barriers is seen in peeling surface paint, stains on inside walls of heated buildings, and damaged wood or metal structures. The resistance of material to moisture penetration is given by its vapor resistance.

Layout design also has a marked effect on air motion. Wind in hot, moist climates is an asset and should be exploited as natural air conditioning for it tends to reduce temperatures and excessive humidity. In colder regimes, the main effect of the wind is to induce convective cooling and so lower the temperature sensation. Many of the climatic effects on buildings require both macro- and microanalysis. At the individual home level, ventilation is of considerable significance in climates where external conditions are not extreme. Ventilation maintains inside temperatures near those outside by promoting venting of internal heat (either from that produced inside or from solar loading). Very effective in this regard is cross-ventilation, which uses the natural effect of the pressure variation of the wind in relation to a building. On a windward side, the pressure of the wind against the house leads to a slightly higher pressure; on the leeward side, eddies form and create a slightly lower pressure. Careful placement of building outlets and inlets can use this effect to use the wind to advantage by creating a flow of air through the building.

In many situations, local topographic effects give rise to highly specialized conditions. For example, high winds frequently occur, on the windward sides of slopes in higher latitudes. If such situations are used for building, it then becomes necessary to design the structure in relation to wind breaks or natural topographic barriers. The channeling of wind through valleys also requires special design. The local building response to the cold Mistral winds of the Rhone Valley provides a good example. Small windows, or even no windows at all, are characteristic of the northward-facing sides of buildings where wind breaks are commonly found.

A knowledge of prevailing wind direction is also of value in evaluating potential windborne pollution. Odor and particulate matter are common hazards for the odors generated by some industrial centers can prove quite overwhelming. Many home owners, originally unaware of the problem, have been forced to relocate as a result of such pollution. A similar effect is seen in nonurban areas, but here lessons learned over time are more evident. It would be quite unusual, for example, for a farmer to locate his farmhouse on the downwind side of animal stalls and pigpens.

Regional House Design

Given the foregoing arguments, it is possible to identify the type of house design best suited for a selected regional area. The Department of Housing and Urban Development, together with the Department of Energy and the American Institute of Architects, have attempted such for the United States. Figure 17.11 shows the coterminous United States divided into 16 regions in which climates are best suited for a particular climate-designed home. Like any climate classification, this grouping is somewhat arbitrary and debatable, but it does help identify regional characteristics. Each of the regions is described in Table 17.6.

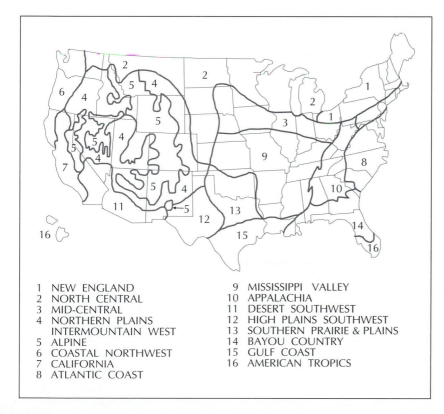

1 NEW ENGLAND	9 MISSISSIPPI VALLEY
2 NORTH CENTRAL	10 APPALACHIA
3 MID-CENTRAL	11 DESERT SOUTHWEST
4 NORTHERN PLAINS	12 HIGH PLAINS SOUTHWEST
INTERMOUNTAIN WEST	13 SOUTHERN PRAIRIE & PLAINS
5 ALPINE	14 BAYOU COUNTRY
6 COASTAL NORTHWEST	15 GULF COAST
7 CALIFORNIA	16 AMERICAN TROPICS
8 ATLANTIC COAST	

Figure 17.11
Regional climate variations identify zones applicable to housing styles (see Table 17.6 for details; after various sources, including Addison, 1994).

Table 17.6 Region, Climates, and House Design

Region	Climate	Design
1. New England	Quick-changing seasons with a harsh winter and cool summers with low-to-moderate humidity.	Use compact shape and control wind and sun effects. Saltbox suits region well.
2. North Central	Cold winters with wide day-to-night (diurnal) temperature swings, summers are also severe; spring and fall are pleasant.	Balance sunshine, wind, and humidity carefully; high-mass, earth-sheltered houses are effective and recall indigenous styles like log or sod cabins.
3. Mid-Central	Temperate, but with fierce wind–chill and humid summers.	Compact shapes with proper solar orientation will thrive.
4. Northern Plains & Intermountain West	Dry with cold, windy, stormy winters; wide diurnal temperature range.	Shading and thermal mass are priorities; evaporative cooling (adding water to dry air to cool it) can help in summer.
5. Alpine	Elevations above 6–7,000 feet are mostly cool to cold with wide diurnal variations.	Simple requirement—keep warm; try to nestle north side into mountain.
6. Coastal Northwest	Mild with only 15-degree seasonal and diurnal ranges; some high humidity.	Few constraintss; compact, small-windowed houses above fog zone are best.
7. California	Chilly, rainy winters; warm-to-hot, dry summers; sunny with wide diurnal range.	Use evaporative cooling (e.g., fountains, pools, and house plants), maximize southern exposure, minimize or eliminate east- and west-facing windows.
8. Atlantic Coast	Relatively temperate with four distinct seasons; pleasant spring and fall; hot, humid summer; winter can be cold and stormy.	Insulate well and maximize ventilation—cover porch, raise first floor, use big windows.
9. Mississippi Valley	Hot, humid summers and cold winters similar to Atlantic Coast.	Shade and enhance ventilation; heavy materials with high thermal mass mitigate seasonal temperature range.

Table 17.6   (continued)

Region	Climate	Design
10. Appalachia	Similar to Atlantic coast, but with wider diurnal range and less humid summers; comfortable days are possible any time of year.	With many climate considerations, houses should be well-insulated, high-mass, shaded, and solar-oriented.
11. Desert Southwest	Extremely hot summers, moderately cold winters, and an extremely wide diurnal temperature range.	Keep sun out; use high-mass, light-colored materials (like adobe); and cool with evaporative techniques.
12. High Plains Southwest	A cousin to the Desert Southwest—sunny, dry, and wide diurnal temperature range.	Wind-swept west Texas has enhanced ventilation and evaporative cooling.
13. Southern Prairie & Plains	Both summers and winters can be uncomfortable; humid and windy summers, but humidity often falls by early evening; significant diurnal temperature range in spring and fall; often sunny in winter.	Single-story ranch houses have more access to earth cooling.
14. Bayou Country	Flat, humid, rainy, with hot summers and some cold winters, though sea breezes and moderate diurnal temperature swings are reliable.	Cool with shading, ventilation, and open layout.
15. Gulf Coast	Mild, sunny winters, but otherwise hot and humid; persistent cloudiness limits diurnal temperature range, and sea breezes add some relief.	Emphasize cross ventilation.
16. American Tropics	Sunny, warm, humid, but comfortable year-round.	Provide shading, ventilation, and light-colored building materials.

(a)

(b)

(c)

(d)

(e)

Figure 17.12
Houses representative of selected regions given in Figure 17.11. (a) The saltbox as may be found in Region 1, (b) the earth-sheltered home may fit well in Region 5, (c) the ranch house, now ubiquitous to all regions, is related to (d) the mission of Region 11, where (e) the abode is also found (from *Weatherwise*, June/July pp. 14–18, 1994. Reprinted with permission of the Helen Dwight Reid Educational Foundation. Published by Heldref Publications, 1319 Eighteenth St., NW, Washington, DC 20036-1802, Copyright © 1994).

Given the descriptions of regional characteristics, the types of buildings found often provide fine examples of design with climate. In Figure 17.12, a selection of houses is shown. That in (a) is the so-called *saltbox* type often found in the New England area. The home shown in (b) is Earth-sheltered and suited for mountain regions located in Region 5. Figure 17.12(d) is the mission-type structure found in the American southwest and in southern California. Its modern equivalent, still suited for that climatic region, is the ranch house shown in (c). This style, however, is very popular and now found all over the United States with no regard to climate. In cold regions, it can be a very expensive home to heat. The optimum structure for the arid southwest, Region 11 for example, is the adobe (e). The small windows and thick walls are as seen in the desert dwelling shown earlier in Figure 17.5.

URBAN CLIMATES

Humans have a remarkable facility to alter the natural environment. Any change results in a modification of the ongoing natural processes at that site. Perhaps the most changes to an environment is the city for the constructed environment of a city creates a totally different climatic realm from that which occurred prior to its founding. Consider the following:

Concrete, asphalt, and glass replace natural vegetation.
Structures of vertical extent replace a largely horizontal interface.
Large amounts of energy are imported and combusted.
Combustion of fossil fuels creates pollution.

These related factors modify the climatic process in the urban environment.

The Modified Processes

Figure 17.13 provides a graphic summary of the modified flows of energy in a city. Of prime importance are the energy characteristics of the asphalt/concrete/glass

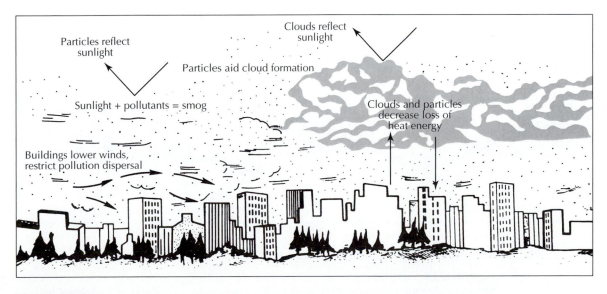

Figure 17.13
Schematic diagram showing the modified flows of energy in an urban environment.

environment as compared with that of the vegetation-covered surfaces of rural areas. For the most part, the city surface has a lower albedo than nonurban areas and greater heat conduction and more heat storage. During the day, the city surfaces absorb heat more readily; after sunset, they become a radiating source that raises night temperatures.

The energy flows are further modified by the geometry of the city buildings (Fig. 17.14a). Walls, roofs, and streets present a much more varied surface to solar radiation than undeveloped areas. Even when the sun is low in the sky, a time when little energy absorption occurs on flat land, vertical city buildings feel the full impact of the sun's rays. In the early morning and late evening, the city is absorbing more energy than surrounding rural areas.

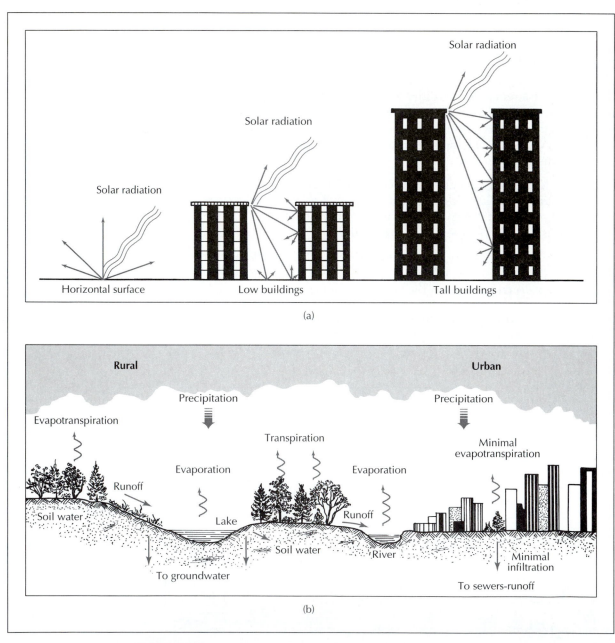

Figure 17.14
Modified climatic processes associated with urbanization: (a) effects of horizontal and vertical surfaces on incoming solar radiation, (b) disposal of precipitation on rural and urban surfaces.

The different surfaces that make up the city modify energy availability by changing the water balance (Fig. 17.14b). Rain falling on urban and nonurban areas is disposed of quite differently. On building-free rural surfaces, some of the water is retained as soil moisture. Plants draw on this source for their needs and eventually return the moisture to that air through transpiration. At the same time, standing water and soil moisture are evaporated. Solar energy is required for both transpiration and evaporation. In the city, pavements and buildings prohibit entry of water into the soil; with limited green areas, transpiration is minimal. Most of the rain water drains off quickly and passes to storm sewers, with the result that water availability for evaporation is greatly diminished. Since both evaporation and transpiration amounts are decreased, the solar energy available for this process provides additional surface heating.

The ratio of energy used for sensible heat flow (which heats the atmosphere) to that of the latent heat flux (the latent energy in water vapor) is given by the Bowen ratio H/LE. In areas such as deserts, the lack of water means that a high Bowen ratio—in the order of 20—occurs. In most vegetated areas, the value is usually less than 1, indicating more transfer by latent than sensible heat. In cities, the ratio is about 4.0, showing that there is a large decrease in latent energy transfer; an increase in sensible heat has resulted. Again, this results in more available energy to heat the city atmosphere.

As areas of concentrated activities, cities are high consumers of energy, and enormous amounts of energy are imported to maintain their functions. In New York City, for example, the amount of heat generated through the burning of fossil fuels in winter is two and one half times the amount of heat energy derived from solar radiation for the same period. This high use of energy means that much *waste heat* from factories, buildings, and transportation systems passes to the atmosphere to add to the warmth of the city.

The actual combustion of fossil fuels results in cities having higher air pollution levels than their rural counterparts. Visible evidence of air pollution over cities is the dust dome that is often produced. This pall of smoke and smog acts in a number of ways to modify the urban climate. Some incoming solar radiation strikes particles in the air and is reflected back to space. Other particles act as a nucleus onto which water vapor condenses to form clouds. That cities tend to be cloudier places than rural areas is illustrated by data from London, England: A comparison of bright sunshine hours showed that the city received 270 hours less sun than that of the surrounding countryside over the period of a year.

The lower input of solar energy because of increased reflection from particles and clouds results in lower temperatures in cities; this lack of energy, however, is more than counterbalanced by other effects. Some of the pollutants absorb rather than reflect energy, and the increased cloudiness reduces loss of long-wave energy to space.

The Observed Results

Of the many changes that occur in the climate of cities, the modified temperature regime is the best known and most closely studied. Cities tend to be warmer than the surrounding nonurban environment. The higher temperatures are best developed at night, when, under stable conditions, a **heat island** is formed. In passing from the center of the city toward the surrounding countryside, temperatures decrease slowly until, in rural areas, they remain about the same. The city is an island of warmth surrounded by cooler air. Figure 17.15a provides a schematic summary of this feature. It is possible to visualize the formation of an urban dome in which the normal lapse rate occurs because of the warmth of the city. This effect is in contrast to rural areas, where, during the night, a low-level

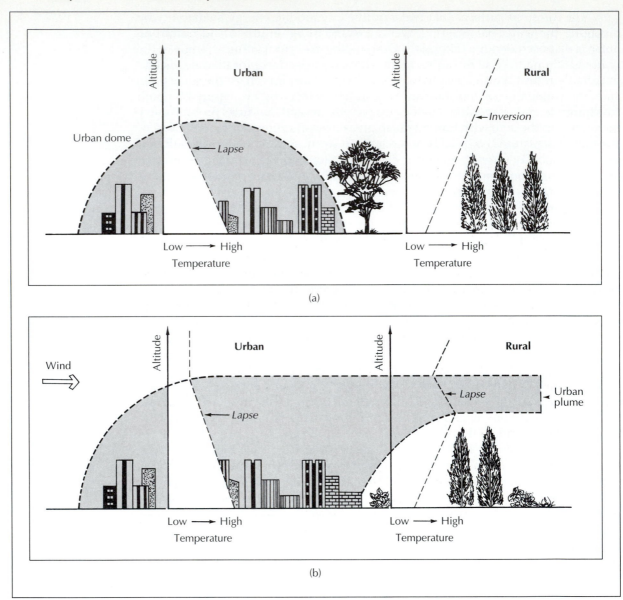

Figure 17.15
Comparison of lapse rates over rural and urban environments: (a) on a still night, normal lapse rate conditions occur in the city and an inversion in the surrounding rural areas; (b) with a slight regional wind, the rural conditions are modified by the creation of an urban plume.

inversion often forms. If a slight regional wind is blowing, the warmer air of the city is swept downwind as an **urban plume**, modifying the rural lapse rate conditions (Fig. 17.15b).

The heat island effect has been measured in many cities around the world. Places of diverse size—from large cities like Tokyo, London, and St. Louis to smaller ones like Corvallis, Oregon, and San Jose, California—show similar patterns. Generally, when the heat island is best developed, temperature differences between city and country are as much as 5.5°C (10°F).

Wind speeds in a city are, on average, lower than those in open surrounding areas because of the increased roughness of the urban fabric. A comparison of highest expected wind speeds in the city and its airport (usually located outside the city in flat, unobstructed land) shows this to be true. In Boston, Massachusetts, the highest airport wind speed is 46 meters per second (102 mph); the value for the city is 32 meters per second (71 mph). The same effect is found in Chicago (city 25 m/s, 56 mph; airport 31 m/s, 70 mph) and Spokane, Washington (city 23 m/s, 52 mph; airport 35 m/s, 78 mph).

The wind in cities may also be modified by the urban heat island. As noted, the heat island is best formed under calm conditions; in such instances, the city may create its own wind pattern. A small pressure gradient results when the warm city air rises and air is replaced by that from surrounding areas. The uplift of air is not high, and a convective cycle results. This process means that the city air is recycled. If pollutants are being added to the air under these conditions, pollution levels can rise dramatically and give rise to a dust dome over the city.

The influence of cities on rainfall is still not completely understood. Generally, however, it is thought that a city gets about 10 percent more rainfall than surrounding rural areas. Intensive studies of this aspect of urban climates are part of the research project called METROMEX—an in-depth study of the conditions in St. Louis. Many significant findings resulted from this project, but the results cannot always be applied to other urban centers in different climatic regimes.

Table 17.7 lists some of the basic climatic modifications that occur within cities. Examples of changes associated with the elements outlined are given, together with

Table 17.7   The Characteristic Mix of Atmospheric Processes over Urban Areas

Radiation and sunshine	Greater scattering of shorter wave radiation by dust but much higher absorption of longer waves owing to surfaces and CO_2. Hence, more diffuse sky radiation with considerable local contrasts owing to variable screening by tall buildings in shaded narrow streets. Reduced visibility arising from industrial haze.

Summary

UV radiation	25 to 90% less
Solar radiation	1 to 25% less
Infrared input	5 to 40% more
Visibility	Reduced

Clouds and fogs	Higher incidence of thicker cloud covers in summer and radiation fogs or smogs in winter because of increased convection and air pollution, respectively. Concentrations of hygroscopic particles accelerate the onset of condensation.

Summary

Cloud	More in and downwind
Fog	Often more

Temperatures	Stronger heat energy retention and release, including fuel combustion, gives significant temperature increases from suburbs into the center of built-up areas creating heat islands. These can be up to 8°C warmer during winter nights. Heating from below increases air mass instability overhead, notably during summer afternoons and evenings. Big local contrasts between sunny and shaded surfaces, especially in the spring.

Summary

Convective heat flux	About 50% greater
Heat storage	About 200% greater
Air temperature	1–3°C/100 years; 1–3°C/annual mean

Pressure and winds	Severe gusting and turbulence around tall buildings causing strong local pressure gradients from windward to leeward walls. Deep narrow streets much calmer unless aligned with prevailing winds to funnel flows along them.

(continues on next page)

Table 17.7   *(continued)*

	Summary	
	Turbulence intensity	10–50% greater
	Wind speed	Decreased 5–30% in strong flow (10 m)
		Increased in heat island weak flows
	Wind direction	Altered 1 to 10°
Humidity	Decreases in relative humidity occur in inner cities owing to lack of available moisture and higher temperatures there. Partly countered in very cold stable conditions by early onset of condensation in low-lying districts and industrial zones.	
	Summary	
	Evaporation	About 50% less
	Humidity	Drier in summer daytime; more moist in winter and summer night
Precipitation	Perceptibly more intense storms particularly during hot summer evenings and nights owing to greater instability and stronger convection above built-up areas. Probably higher incidence of thunder in appropriate locations. Less snowfall and briefer covers even when uncleared.	
	Summary	
	Snow	Less (often turns to rain)
	Total	More, mostly to lee of city
	Thunderstorms	More

Source: After Landsberg, Oke and others.

other parameters. It is an imposing list; it points to the amount of change that can inadvertently occur through modification of the environment by people. As the urban population of the world continues to expand, even more widespread results might be anticipated.

SUMMARY

The study of the relationships between climate and people has a long history. Although it is clear that climate does impact many aspects of human life and activities, the ideas of climatic determinism are not acceptable. However, the human body does react to changes in the surrounding atmosphere, and shivering and sweating provide two ways in which the body reacts to extreme temperatures. Bioclimatological indexes, such as the windchill index and the heat index, have been developed to measure the impacts of combined atmospheric elements on people. The role of climate in human morbidity and mortality has been extensively examined, with the finding that seasonal impact is most clearly seen.

The nature of traditional primitive constructed shelters largely depends on the prevailing climates. Homes in the wet tropics and desert areas provide fine examples. Modern structures are not always designed with climate, with resulting increased energy costs for heating and cooling.

The built environment of the city modifies climate. The urban climate is a result of altered energy flows and imported energy. All major climatic variables are modified, with the best known being the urban heat island—the city being an island of warmth surrounded by a cooler nonurban region.

KEY TERMS

Analemma
Apparent temperature
Biometeorological indexes
Climatotherapy
Cooling degree days

Determinism
Heat index
Heating degree days
Heat island
Hypothermia

Metabolism
Ultraviolet index
Urban plume
Windchill

REVIEW QUESTIONS

1. What is climatic determinism?

2. How is the temperature of the human body maintained?

3. What are the two most widely used bioclimatological indexes? Explain each.

4. Why has it become necessary to derive an ultraviolet index?

5. How does climate influence seasonal mortality?

6. Relate the nature of a traditional home found in the wet Tropics to the prevailing climate.

7. What differences would you find between homes (designed with climate) in New England and the American southwest?

8. What are HDDs and CDDs used for?

9. How does the geometry of a city influence energy flows?

10. Why is the Bowen Ratio less in urban areas than it is in rural areas?

Climate, Agriculture, and Industry

T his chapter relates climate to some economic activities of people. By considering both agriculture and industry, it is clear that only the major facets of each relationship can be discussed within the limits of a single chapter. However, in covering but some of the relationships, it becomes apparent that climate is a significant variable in the economic well-being of people.

AGRICULTURE

From its early origins in the development of civilization, agriculture has evolved into a highly technical field. Through selective breeding and trial-and-error techniques, specialization of cultivated crops took place over the ages, with the most dramatic changes occurring in the 20th century. Genetics, improved fertilizers, increased disease resistance, and the substantial input of energy into farming methods have all led to dramatic increases in productivity. However, the prevailing climate of a region still places distinct limitations on what crops can be grown and, in part, determining the hazards to which the crops are exposed.

Agricultural systems are maintained through human addition of energy for fuel for tractors or food for the farm workers is an energy input as is sunlight. To maximize the benefits of this energy supplement and to obtain the highest yield from all inputs, single crops are grown over wide areas. Such monoculture is well exemplified by the wheat fields of North America. Yet this monoculture is in marked contrast to climax ecosystems, where diversity is a key to the maintenance of the system. Because of lack of biological diversity, crop failure sometimes leads to complete disaster, and most widely grown crops have suffered from disease on grand scales. The boll weevil problem experienced in the old "Cotton Belt" of the southern United States is an appropriate example as is the disastrous potato famine of Ireland and Europe. In the 1840s, the complete failure of the potato crop because of blight led to some 2 million deaths due to starvation, and it is estimated that another 2 million people emigrated.

Because of the weather-related risks involved in agricultural practices, many ways have been devised to ease the vulnerability. Some of the adaptations include:

Risk spreading: Examples—crop insurance schemes and cooperative farming schemes.

Environmental manipulation: Examples—Irrigation and fertilizer application.

Managed diversification: Examples—crop rotation and dual-purpose livestock.

Modified farming system: Examples—modified crop calendar and improved storage.

Modern farming is both technical and highly specialized.

Climate and Crop Distribution

It is not infrequent to find that the limit to the areal extent of a given crop closely corresponds to a climatic limit because crop yields depend highly on climate. It is found that some crops simply will not grow in areas outside the climatic limits, whereas others will grow but will not flower. For example, bananas will not grow in Wisconsin while deciduous fruit trees will grow in the Tropics but do not bear fruit. In agricultural terms, they have a negligible yield. The study of a few widely grown crops provides much insight into the problems and potentials of climate in relation to agriculture. Consider the example of wheat.

Wheat can be produced in quite dry climates through dry-farming methods; the optimum amount of annual precipitation, however, appears to be in the order of 76 centimeters (30 in) because higher amounts of rainfall—particularly in cool conditions—inhibit its growth. The wide distribution of wheat means that yields vary enormously. Some insight into the most productive areas can be obtained by assessing the quality of the wheat, rather than its quantity, because high-quality wheat—those high in protein content—bring the best price to the farmer. The protein content of wheat is highest in the wheat belts of North America and the black-steppe area of the former U.S.S.R. This might be accounted for in a number of ways. Such areas are, of course, technologically advanced and utilize the latest advances in agricultural techniques. They also correspond to the mid-latitude grassland biomes of the world; highest productivity is associated with the biome in which grasses, to which wheat is related, comprise the natural vegetation.

Despite the apparently ideal conditions for growing wheat in the grassland biomes, such areas are not without climatic problems. The continental locations make such areas highly prone, for example, to summer thunderstorms with associated hail. Hailstones can achieve considerable size and, when falling onto standing wheat fields, can cause enormous problems. For example, a single storm in Nebraska destroyed 3 million bushels of standing wheat.

Many agricultural plants are now produced in areas where they do not occur naturally. The banana was native to Southeast Asia, but the main production area is now tropical America. The potato, with its origins in America, is now widely produced in Europe; sugar cane from Southeast Asia is widely produced in tropical America. The list could be extended considerably. In many cases, the relocation of agricultural crops often results in better yields. The reason for this is twofold. First, in the area in which they were native, plants probably became an integral part of the environment in which they were found. They were susceptible to pests and the competition of other plants of that region. On transport to other areas, many of the natural limiting factors were removed. Second, and probably of more consequence, plants transported by people for the express purpose of production are treated with special care and afforded much attention. Were it not for human inputs, such plants might not be able to survive in their new environments.

Perhaps the finest example of a plant that is highly productive in an area remote from its original area is the rubber tree. A tree of the equatorial rainforest (Hevea braziliensis), it was confined to the Amazon basin, its spread perhaps inhibited by the dissimilar climates that occur on either side of the biome. Problems of collecting latex from trees scattered throughout the extensive forest led, through somewhat underhanded means, to its introduction as a plantation crop in Southeast Asia. The intriguing story of how rubber caused a boom in South America and how it was smuggled from there to Kew Gardens, London and, ultimately, to Asia makes fascinating reading.

Clearly climate plays a most significant role in determining what crops are grown in any region. Some of the climatic limitations placed on agriculture are summarized in Table 18.1. While such limitations exist, many ways to overcome climatic problems have been devised, and some crops are now grown in areas because of manipulation of that environment by man.

Extending the Climatic Limits

If a crop is susceptible to **frost**, then the obvious place for it to be grown is where frost does not occur. However, it is sometimes found that some plants grow exceptionally well in areas where only an occasional frost occurs. The farmer is thus faced with a problem; should he take the risk and grow the crop while being aware that high returns are equally balanced against total loss if a frost should occur? In modern agriculture, large investments often require that the end product is guaranteed, and chances should not be taken. To meet this, methods to overcome climatic limitations on crop growth have been implemented. They not only concern the danger of frost, but also the problem of an inadequate water supply.

Table 18.1   Temperature and Precipitation Requirements for Selected Commercial Crops

Crop Type	Temperature	Precipitation	Notes
Cocoa	Since temperatures of between 10 and 15.6°C maybe harmful, the crop cannot be profitably grown in regions where mean maximum of coldest month falls to 13.9°C or where absolute minimum of less than 10°C occurs	Tree not resistant to dry weather so generally restricted to areas where dry season does not exceed 4 months	A tree of humid tropical lowlands—mostly grown within 20° of equator below 457 m
Citrus fruits	Little or no growth where temperature is below 15.6°C. Dormancy in cooler months of subtropical climates. Temperatures slightly below freezing are highly damaging	Requires high soil moisture content. Orchards often irrigated even in fairly moist areas	Can be grown on a variety of soils with high humus content
Coffee	A tropical crop whose temperature requirement varies with species. Generally, optimum temperatures are between 15.6° and 25.6°C	Depending on temperature, optimum amounts vary from 127 to 229 cm per year. The distribution is important with an ideal minimum in the flowering season. Too much water can promote tree disease	Generally does best on a well-drained loam soil. Thus some species are highland varieties. Others do well in lowlands
Cotton	Needs a frost-free growing season of 180 to 200 days. Does not grow below 15.6°C and optimum temperatures are from 21.1 to 22.2°C during the growing season. Four to five months of uniformly high temperatures are beneficial	Can tolerate a wide range in annual precipitation; the distribution during the growing season is of critical importance. Frequent, but light, showers immediately following planting are an attribute	Needs sunshine to ripen the boll in full maturity
Rubber	For optimum growth and yield, a mean maximum of over 24°C. The maximum should not, however, exceed 35°C	Evenly distributed rainfall of more than 178 cm per year. Lengthy dry periods inhibit growth. Good drainage essential	A tree of the equatorial rain forest (Amazonia) transferred to plantations in southeast Asia
Sugar cane	Susceptible to low temperatures. Little or no growth below 10°C while optimum is appreciably higher. Frost very dangerous to young cane	During vegetative growth requires a considerable amount of moisture and is sensitive to drought. In ripening period should be relatively dry to maintain high sucrose level	Often grown in cleared areas formerly occupied by tropical forests
Tea	Optimum temperatures should not fall below 12.8°C nor exceed 32.2°C	Can tolerate high amounts of rainfall (254–381 cm per yr) if rain fairly evenly distributed throughout year	Often grown as an upland crop in tropical areas

Source: Oliver (1973).

Frost Protection. To overcome the problem of frost destruction, it is obviously necessary to be aware of the way in which frost can occur. Radiation frost results from rapid cooling of the air layer above the ground when heat loss through ground radiation causes the ground to be cooler than the air above. This mostly occurs on cool, clear nights when there is little air turbulence. A second type of frost occurs through advection, when cold air flows into a region. Radiation frost is often local, whereas advection frost can cause damage over wide areas.

To protect against frost, farmers have a number of strategies. One that recognizes the nature of radiational frost causes frost-prone plants to be planted away from cold air drainage in valley bottoms, as shown in Figure 18.1a. The cold, clear nights characteristic of radiational cooling form layers of dense cold air near the ground so that mixing the layers may decrease the probability of a frost. This is done by mixing the air, using fans or a heat source (a *smudge pot* as in Figure 18.1b), to create convective currents in the layered air.

Sprinkling and flooding are also used to combat frost. The objective in this method is to reduce excessive cooling and increase thermal conductivity of the ground. Because of latent heat, when water freezes, the temperature of the plants will not fall below freezing as long as the change of state occurs. The method does have limitations; although it retards loss of heat during the night, it also limits heat

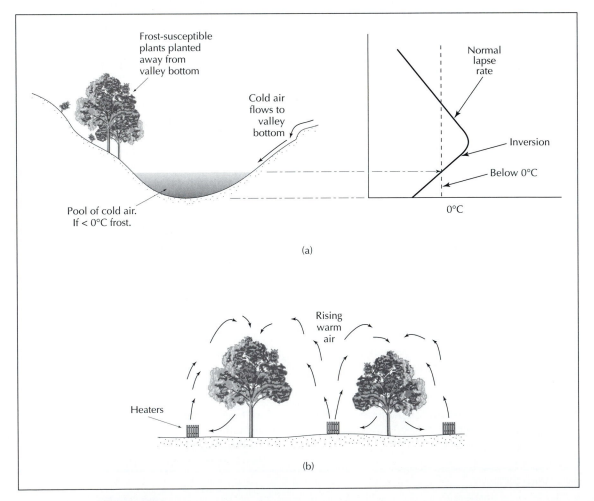

(a)

(b)

Figure 18.1
(a) Cold, dense layers of air close to the ground flow into the valley. Crops susceptible to frost are planted away from the valley floor.
(b) Heaters in an orchard create air motion to decrease the likelihood of frost.

gain during the day, and successive applications of water become less effective. Of the two approaches, sprinkling is probably better, although problems do occur in determining the amount and frequency of water to be added. It has been found that areas not receiving water from the sprinkler are damaged, whereas loss also occurs in those areas receiving limited added water.

Other methods to overcome the frost problem include brushing—adding a protective covering of kraft paper or placing individual covers (hot caps) over plants to reduce nighttime radiation loss.

Irrigation. Water is essential for plant growth. As such, it would seem that agriculture must be restricted to areas in which water is available at the various stages of plant development. This is obviously not the case because people have long since learned that by transporting water to crops agriculture can be successfully carried out. Indeed, irrigation is an integral part of the story of society's development of agriculture, and the great aqueducts and complex water diversion schemes of classical antiquity are well known. Some writers have suggested that the basic cultural aspects of large societies are partly a result of the human control of water.

It is possible to identify three types of regions in which precipitation must be augmented. There are the great deserts of the world in which a perpetual **drought** exists. There are regions in which the distribution of rainfall causes marked deficits of water to occur seasonally. Then there are the usually well-watered areas where, because of periodic droughts, water must be added to maintain the normal production. It is possible then to recognize permanent, seasonal, and periodic drought.

Regardless of the problem of the definition of the drought, the major problem that faces the farmer using irrigation is how much water to add and when to add it. To attain an answer to this question, it is necessary to evaluate the amount of water required by a plant to function at its maximum capacity. It is generally assumed that this amount is equivalent to the potential evapotranspiration. Yet as already noted, there is no single method capable of deriving evapotranspiration with exactitude. Nonetheless, application of an empirically derived formula is probably the most effective way to determine irrigation needs.

Irrigation agriculture has been the subject of many research efforts, and a great deal of literature exists on the topic. These cover such aspects as the methods of irrigation water application, economics of irrigation, possibilities of using seawater or desalinized water, and disease vectors arising from irrigation schemes; the list is a lengthy one and contains much material beyond the scope of climatology.

In terms meaningful from a climatological aspect, however, there are many studies concerning the way in which the creation of irrigation systems modifies the local climates. It will be appreciated that large irrigation schemes mostly occur in dry climates for it is here that irrigation is needed. In many of the schemes, large dams have been built and enormous amounts of water impounded behind them. In effect, people create oases in the middle of dry regions.

Climate and Crop Yields

The purpose of agricultural production is to produce as much foodstuff from a given area as possible. The line between success and failure in farming often depends on the yield per acre in relation to the costs of maintaining that yield. It follows that much research has gone into the determination of optimal conditions to produce the maximum yields, with the light, heat, and moisture conditions being of major importance.

The conversion of radiant energy to chemical energy by plants, **photosynthesis**, is

Carbon dioxide + water + solar energy = hexose sugar + oxygen.

Clearly, sunlight is the basic requirement for the rate at which photosynthesis occurs. Another way in which plants respond is to varying lengths of daylight—**photoperiodism**. Because seasonal climates of the earth are in part a response to differing lengths of day and night over the globe, photoperiodism is an important factor in plant growth and development. Accordingly, plants are to be classed in one of four types:

Long-day plants	Flower only when daylight is greater than 14 hours.
Short-day plants	Flower only when daylight is less than 14 hours.
Day-neutral plants	Bud under any period of illumination.
Intermediate plants	Flower with 12 to 14 hours of daylight but not outside these limits.

This explains, for example, why there are limits to tobacco growth in areas of similar temperature regimes and why chrysanthemums are so colorful during the shorter days of autumn while many other common flowering plants are limited to the longer days of summer.

Just as light plays a role in plant growth and development, so does temperature. The temperature gradient from the Tropics to the polar realm has allowed identification of temperature zones over the earth. For example, it is generally assumed that tropical plants have temperature limits between 20°C and 30°C (68°F–86°F), whereas mid-latitude plants have limits of 15°C to 20°C (59°F–68°F). A number of climatic classifications—that of Köppen, for example—use such temperatures as climatic boundaries. Such are general boundaries for not only is there an upper and lower temperature limit in which a plant will grow, but also a value at which plants develop most rapidly. This is the cardinal temperature—a level not easily defined in general terms for such critical temperatures vary over time with the plant's development.

Such temperatures are gauged using growing degree days (GDD), which estimate the growth and development of plants over the growing season. These are similar to the degree days used by heating and cooling engineers in that their calculation depends on a derived base temperature. The temperature base selected varies on the crop grown. For example, the base temperature to calculate GDDs for wheat, barley, and rye is 40°F; for sunflowers and potatoes it is 45°F; and for corn and soybeans it is 50°F. Interestingly, modified GDDs take into account that growth may be limited when higher temperatures prevail. For example, it is assumed that corn growth is limited once the temperature is above 86°F so that no additional days are accumulated above this value.

From the foregoing account, the role of climate in agriculture stresses but a few cases. So important is the relationship between the two that studies created a specialized discipline called agricultural climatology or **agroclimatology**. This many faceted study has recently been extended to include the potential impact of climate change on farming in different parts of the world. It is a topic about which much more will be written in the years ahead.

INDUSTRY

Industrial activity can be considered in terms of primary, secondary, and tertiary industries. Primary industries are those concerned with the exploitation of raw materials and foodstuffs directly from the physical environment. Agriculture, forestry, and mining fall into this category. The secondary industries utilize resources gained from primary activities for further processing, and activities range from iron and steel works to foodstuff processing. Tertiary industries are the service industries;

they supply the services required by people who might well be employed in any level of industrial activities (e.g., professional services, trading, and tourism). As indicated in the following pages, which provide examples of each of these three levels, climate plays a significant role in industrial activities.

The significance of climate in one of the major primary industries—agriculture—has already been discussed, and it is clear that climate is of outstanding importance in the nature and distribution of agricultural activity. Its role in another primary activity—the extractive industry—is not so clear because Earth resources are not prone to differences in climate over the earth; a mineral or fossil fuel deposit may be the result of paleoclimatic events, but the prevailing climatic conditions have no effect on their distribution. Although at the location level climate is not a significant factor, its impact is certainly felt at the operational level. For example, the iron ore extraction in the subarctic climate of northern Scandinavia has higher costs because of the high wages necessary to attract workers and the costs of heat and light in the long, dark hours of high latitudes.

Industrial Location

It is in the secondary and tertiary activities that the role of climate becomes more visible. The basic factors contributing toward a given industrial location are the historical influence, provision of raw materials, availability of fuel and power resources, supply of labor and market considerations, and transport facilities. Although the word *climate* does not appear in any of these categories, a brief examination of each brings out the role that is played by the climate.

Although the role of climate in the development of some industries is difficult to assess, there are examples in which climate can be shown to play a major part. The rapid growth of industry in southern California illustrates the point for the development of the aircraft industry, which was initially directly related to climate. Early aircraft manufacturers were attracted to the location because its mild winters and light winds (summed up by the term *flying weather*) found favor with early plane manufacturers. A further factor of importance was that the large hangers required in aircraft manufacture did not require expensive heating. This latter point is well demonstrated when the number of heating degree days for California are compared with those of other areas of the United States. As already described, the degree day is used by heating engineers to estimate the amount of fuel likely to be required in given locations. As shown in Figure 18.2, the cumulative number of heating degree days is appreciably less in California than in many other parts of the country. Note, however, that with the advent of air conditioning, cooling costs became significant.

Another interesting example of the development of an industry because of the climatic conditions in southern California is the movie industry. Established at a time when camera equipment was not highly sophisticated, the long hours of sunshine and light proved ideal for outdoor shooting of films. The location was further aided by the myriad of diverse climates and related vegetation associations that occur within a fairly small area around southern California. It was not necessary to travel far to film desert conditions while mountains provided a backdrop for stories associated with everything from Yukon miners to avid Alpinists. As with other industries, this historic inducement was negated by the development of sophisticated equipment. Lightweight cameras and improved transport meant that the situation filmed no longer be simulated; the area about which the film was made could actually be visited. Technology and development of other film-making areas led to the demise of Hollywood as the film capital of the world. Interestingly, and perhaps because of the concentration of expertise and facilities, TV production films remain installed in Hollywood.

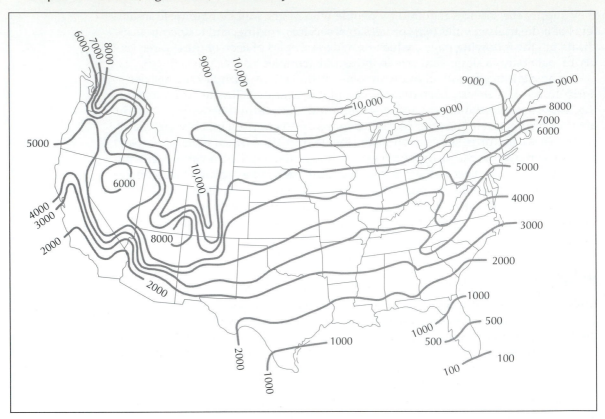

Figure 18.2
The average annual heating degree days (base 65°F) in the United States.

The most obvious influence of the role of climate on industrial location as conditioned by raw materials concerns the industries that are based on agricultural and forest products. In many cases, large industrial processing plants appear where raw materials are found. The meat-packing and grain-processing industries of the American midwest and the location of pulp and paper mills in the forest belt of Canada provide apt examples. Such location is to be expected because the value per unit volume of raw materials is less than that of finished products. As with many primary processing activities, transportation costs are key to industrial location. Of course, there are variations from this pattern, many of them of historic origin.

Although there has been exploitation of natural resources of tropical lands by European powers, there is good reason for the treatment of some products away from their areas of production. An example is shown by the fact that, although West Africa is the major producer of cocoa, western European countries are among the main producers of chocolate and chocolate products. It does not take an expert economist to work out why this is so. The hot, humid conditions of the production areas would necessitate high-cost facilities to stop chocolate products from melting; air-conditioned storage and ships would be required to stop its spoiling.

Energy Production

It has already been noted that present-day climate has little bearing on the location of fossil fuels. However, climate today does influence the location of hydroelectric power generation, solar energy production, and wind energy.

Hydroelectric power production relies on a sustainable flow of water over a gradient. This is most often achieved by construction of dams to create large lakes. To fill and maintain appropriate lake levels, there must be a supply of water, and hence a precipitation source. The source may be quite remote from the power-generation site, as illustrated by the location of the Hoover Dam, which impounds Lake Mead on the Colorado River in a dry region. Much of the precipitation flowing to the lake is from that falling in the Rockies.

The potential for generation of usable energy through solar and wind energy rests largely on the prevailing climate at a location. These sources currently contribute but a tiny fraction of the global energy needs. Yet with a growing world population, environmental concerns, and diminishing fossil fuels, this will change in the years ahead.

Solar energy is used by passive solar collectors in most homes. A south-facing window will permit the sunlight to pass through, be absorbed, and eventually warm the room. In contrast, an active solar collector is often a roof-mounted dark box covered in glass. The heat energy collected in these is transferred to the home by circulating air or fluids through a pipe system and is often used for water heating. In dry, sunny areas, these are quite efficient; for example, it is estimated that in Israel, some 20 percent of all homes have some form of solar device.

The old idea of focusing the sun's rays to produce high temperatures is now used on a large scale. Pioneered by the French at Odeillo, the object is to place rotating mirrors that follow the sun such that they reflect sunlight into a receiving tower. The facility near Barstow in the Mohave Desert uses the concentrated sunlight to heat pressurized water, which is transferred to turbines. Of major importance for the future is the use of the photovoltaic (solar) cell that converts sun's energy directly to electricity.

Wind energy has been used for centuries, as seen in the history of sailing and windmills that were once a common part of the landscape in many areas of the world. A wind-driven water pump remains a common site in the American west. Interest in wind power has been revived in the United States, and the Department of Energy has set up experimental wind farms. That at Altamont Pass in California has 7000 wind turbines and supplies almost 1 percent of California's energy need.

Although the use of the wind has much promise, there are problems relating to its widespread development. Clearly, the location selected should have a good portion of time with a wind blowing; the intermittent nature of the resource would mean that a backup would be required to overcome production shortages. To harness economically the wind also requires many wind machines, and this requires large amounts of space. In densely populated areas, this could create a problem. However, the recent location of a huge facility near Lake Benton in Minnesota suggests that wind power will continue to become more important in the future.

Besides its importance for providing a basic source of power, climate also directly influences the amount of energy required to maintain plants at their optimum working temperatures. Furthermore, it gives rise to many problems in the transmission of power from generating stations to factories (and homes). For any industry to function, it must have reliable and trained employees. In a modern footloose society such as North America, the climate of a location can prove of some significance in attracting workers. The rapid growth of the sun-belt states is indicative of the attractiveness of a warm, almost winter-free environment.

Although the external climatic environment is important, it should be remembered that indoor workers also require a comfortable environment. Table 18.2 shows the suggested optimum indoor operating conditions of selected industries. It shows quite clearly that in many of the highly industrialized areas of the world it is necessary to either heat or cool the factory area and artificially modify the humidity conditions.

Table 18.2   Optimum Indoor Temperatures for Industrial Activity

Industry	Temperature (°F)	R.H (%)
Textile industry		
Cotton	68–77	60
Wool	68–77	70
Silk	71–77	75
Nylon	85	60
Orlon	70	55–60
Food industry		
Milling	65–68	60–80
Flour storage	60	50–60
Bakery	77–81	60–75
Candy	65–68	40–50
Process cheese production	60	90
Miscellaneous industries		
Paper manufacturing	68–75	65
Paper storage	60–70	40–50
Printing	68	50
Drug manufacturing	68–75	60–75
Rubber production	71–76	50–70
Cosmetics manufacturing	68	55–60
Photographic film manufacturing	68	60
Cosmetics storage	50–60	50
Electric equipment manufacturing	70	60–65

Source: From Maunder (1970), after Landsberg and Grundke.

TRANSPORTATION

Air Transport

All phases of air travel are influenced by atmospheric conditions. The effects are felt at all levels of operation—from the construction of a landing strip to the trip en route and the landing and take-off conditions. The significance of atmospheric studies in relation to air travel is illustrated by the simultaneous development of the two. As air transport improved, it became necessary to understand flying weather; as planes flew higher, the altitudinal variations of the atmosphere became a necessary area of study. Development in one led to research in the other. Unfortunately, the importance of atmospheric conditions in air travel is also illustrated by the number of air disasters associated with weather conditions.

Many of the air transportation problems are concerned with conditions that will be encountered at a given time. As such, aviation weather is of more immediate significance than aviation climatology. En route, aircraft face many problems. Most are concerned with meterological conditions and include such factors as icing, in-cloud turbulence, and selection of optimum cruise altitude in relation to aboveground winds.

Icing refers to the formation of ice on lift-producing surfaces (e.g., wings, control surfaces, propellers) so that the smooth flow of air over such airfoils is interrupted. This increases drag and reduces lift to modify the aircraft's flight capabilities. Icing occurs in clouds where air temperature ranges from slightly above freezing to about −20°C (−4°F) when supercooled water droplets exist in the cloud. These droplets will freeze on contact with a surface. In cumuliform clouds, the large size of the droplets allows the formation of clear ice; stratiform clouds result in the formation of rime ice, where supercooled droplets are much smaller. Clear ice is heavier, and thus more hazardous than rime, although most structural ice is often a combination of the two forms.

Aviation personnel are well aware of the dangers of icing, and many forms to combat its effect through anti-icing methods are available. The mechanical boots on the leading edges of wings is a rubber covering that swells and causes ice to break up. Some thermal devices are used to melt ice, but anti-icing fluids are commonly used. For smaller, fixed-wing aircraft, the avoidance of potential icing formations is the best defense.

Turbulence is irregular motion of air over short distances in the atmosphere. Its effect can pass from a slight bumpiness to structural damage of an aircraft. There is no direct measure available for turbulence, so flight forecasters and pilots use their knowledge of meteorology to predict turbulence. Clearly, uneven air flow will occur as a result of surface influences ranging from uneven heating of the ground to air flow over mountains. Wind-shear turbulence results from air moving at different speeds or direction in short distances and can occur anywhere in the atmosphere. One particularly devastating form occurs in the downdraft of thunderstorms. Such downbursts or microbursts have been the cause of aircraft disasters. The introduction of doppler radar at airports greatly reduces the chances of such accidents.

Clear-air turbulence (CAT), which is not normally visible to pilots, poses a severe safety hazard because of the violent motions it induces. Furthermore, aircraft that experience CAT are usually flying at high speeds that will increase the potential for damage or injury. The probability of encountering CAT increases with increasing altitude, being most typical at 9 to 13 kilometers. It is most often associated with the presence of strong vertical wind shears—that is, locations where large vertical gradients of wind speed or direction occur. There are significant temperature differences between the two air masses on either side of the shear. To allow detection of CAT, microwave radar, optical (laser) radar, and infrared sensors have been used, but careful forecasting and pilot knowledge remain the best safety methods.

Air transport is plagued by the problem of fog. Many flight delays, particularly during the cool season, are related to fogged-in airports in a different part of the country. This is particularly problematic when a widespread advection fog forms. Such a problem is also shared by other forms of transportation.

Shipping

The struggle between people and the sea is an age-old one described in literature by many famous writers. Given that most of the problems faced by mariners are related to atmospheric conditions, a study of the marine climate is basic to the shipping transport.

Violent storms at sea are another hazard facing shipping. Probably the most dangerous storms are hurricanes because the effects are felt over thousands of square miles of the oceans. Hurricanes originate over the warm seas in low latitudes. Thereafter, in the northern hemisphere, they take a westward and then a northeastward path. Deriving their energy from the oceans over which they originate, they soon dissipate once they pass over land. However, as well shown in the book/film *The Perfect Storm* (describing a storm off the northeast coast of United States) and described in the song "The Wreck of the Edmund Fitzgerald" (telling of a sinking of an iron ore carrier in the Great Lakes), violent storms can occur wherever there are large bodies of water.

The cause and frequency of fogs have already been discussed in an earlier chapter, but are mentioned here because they are so important to shipping. Many of the fogs are due to advection. The expansion of the infamous fogs off the Newfoundland Banks, for instance, is related to the movement of warm air off the warm waters of the North Atlantic Drift and its sudden chilling as it passes over the colder waters of the Labrador Current. Despite the adoption of modern navigation techniques, accidents still occur in fogs. The distribution of fogs also clearly shows that the coastal deserts of the world are among the foggiest that occur anywhere. It is little wonder that they have been designated by a special symbol (n) in the original Köppen climatic classification.

Successful ocean trade requires that ships can dock in a location that provides their cargo, and an ice-free port is of major importance. The role of climate in relation to this is well shown by considering North Atlantic ports. The 32°F (0°C) isotherm is located almost at 40°N over the northeast United States. In Scandinavia, it is located north of 60°N. Most of the ports located north of the isotherm thus experience subfreezing temperatures for much of the winter and many will be ice-bound. The location of the freezing isotherm is, of course, a partial result of the influence of heat transport of ocean currents. Norway is affected by the warm North Atlantic Drift; waters off Newfoundland are associated with the cold Labrador Current.

The economic benefits of ice-free ports are illustrated by the utilization of Norwegian ports, such as Narvik, for export from Sweden. Swedish ports on the Baltic Sea, an inland sea, are frozen in winter. Iron ore from the northern ore fields in Sweden passes through the ice-free ports of Norway in winter.

Land Transportation

For most persons living in a mid-latitude climate, the significance of climate in relation to land transportation is usually quite clear. Roads covered by snow or ice, or obscured by fog or blowing dust, are events most people face at one time or another. The frequency of these, however, is climate based, and for each hazard a frequency probability may be derived. As an example, one might consider whether a location may have snow during a holiday period. Figure 18.3 shows the probability that there will be 1 inch of snow on the ground in the United States on Christmas Day. Climate, as seen in the influence of latitude and altitude, is clearly evident.

The quality of road transportation is also influenced by climate. The effect of freezing and thawing on roads is widespread, and most temperate regions experience some degree of frost. Frost action on pavements can cause frost heave and settlement that result in loss of compaction, rough surfaces, and deterioration of the surfacing material. The resulting thaw provides both drainage problems and restriction of subsurface drainage.

In frost-susceptible soils, ice may grow in the form of lenses or veins. As the ice crystallizes, water from below is attracted to the freezing mass and increases the volume of ice. The resulting surface raising is termed *frost heave*. The heave is most problematic when it is not uniform, leading to extremely uneven surfaces. This occurs when subgrade material varies, from sand to silt for example, where drains, culverts, or utility lines break the uniformity of the subgrade material. During the thawing period, water released by the topmost melting portion cannot drain through the still frozen lower soils. It therefore tends to move upward to emerge through cracks at the surface and to further weaken the roadbed.

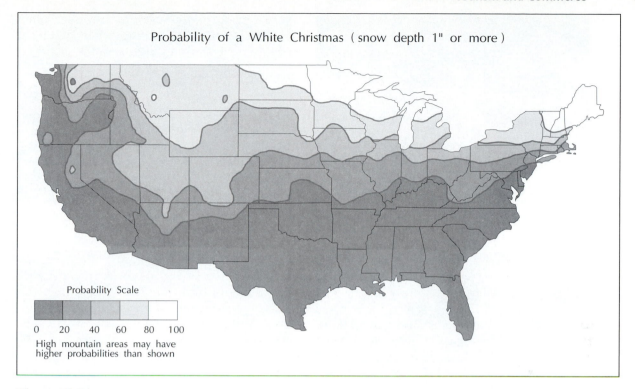

Figure 18.3
The probability of a "White Christmas" when 2.5 centimeters (1 in) or more of snow is on the ground.

TOURISM AND COMMERCE

The tertiary industries cover many sectors of the economic environment. To illustrate how climate plays a role in these, tourism and commerce serve as two good examples.

Tourism

Tourism is the world's fastest growing industry, and some economists have suggested that fairly soon it will become the largest industry in the world. In some locations, it is already the main source of national income; in the Bahamas and Cayman Islands, for example, money from tourism constitutes some 80 percent of the income.

Climate is part of the environment that makes a location a successful tourist center. To be sure, the cultural attraction of places like Paris and London, or the religious centers such as Rome or Mecca, do not rely on the prevailing climate, but the selection of a site to spend a holiday is most often weather related (Fig. 18.4). This is particularly true in climate-dependent tourism, where travel is generated by the reliability of climate as part of the perceived attractiveness of the destination. For the United States, the Caribbean and Hawaiian Islands provide such an outlet, whereas in Europe, the Canary Islands or the Riviera coast are popular with those from more northern realms. In each case, sun and sand are the bases for selection of these and many similar sites.

But there are other activities whose success rests on the climatic regime. This is especially true of specialized sports that require a given set of conditions. Perhaps the most significant of these are those grouped collectively as winter sports. Some locations, such as mountain resorts of the Alps and Rockies, exist in their present

(a)

(b)

(c)

(d)

Figure 18.4
Photograph(s) depicting role of climate in tourism.
(a) from Photo Researchers, Inc. © Agerice Vandystadt
(b) from DRK Photo © Doug Perrine 1988
(c) from Photo Researchers © Jim Corvin
(d) from DRK Photo © Kim Heacox Photography

form because of the way they cater to visitors. Note, however, that the vagaries of winter weather can lead to low snowfall or warm temperatures in some locations. Despite the ability of making snow when conditions are appropriate, resort operators may lose entire winter seasons.

Other sports also require weather conditions that make them favorite locations. **Thermals** are needed for gliding while onshore winds often provide suitable conditions for surfing. Locations that have longer playing seasons for golf have a weather advantage, as do those courses that are open earlier following the winter layoff. Sailing, hot-air ballooning, and hiking also have a climate component in activity location.

Spectator sports in the United States have undergone great changes in recent years, with indoor stadiums providing a comfortable climate-control environment. Such is not always the case, as seen in the storied frigid conditions found in Green Bay and Chicago. It is noteworthy that sites selected for the Super Bowl and College Bowl games are invariably either indoor stadiums or sun-belt cities.

Commerce

Perhaps the most easily observed local impact of weather and climate on commerce is at the local mall or shopping center. The overall regional climate determines the demand for the type of goods sold while weather conditions influence the day-to-day variations in spending.

Retail managers are well aware of the needs of people living in the climate divisions that they serve. The change of seasons influences the types of clothing needed, and stores make the choices available long before the actual seasonal change occurs. For those living in snow areas, the needs of winter—ranging from snow shovels to snowmobiles—are catered to. It is not uncommon to find season-sensitive stores, such as those selling ice cream or fishing supplies, to close during the off season. In those areas where winter is generally little more than a slight decline in temperature, the changes are less radical.

Both good and bad weather can influence retail sales. The weather might actually prevent people from reaching stores, with snowfall being a good example. Similarly, the weather may be such that people simply do not want to go out to a store on a miserable day. Those times when heat waves cause extensive discomfort can lead to extra time spent in shopping malls. In contrast, good weather, such as especially sunny, warm days, can cause people to enjoy the weather by going to the beach or working in their gardens. Less money is spent shopping.

Of the many other commercial activities closely related to climate and weather is insurance, particularly that concerning agriculture. Crop and livestock insurance requires an appreciable amount of climatic information to assess risk. Further thought would identify many others.

SUMMARY

Climate plays a significant part in the economic activities of people and societies. Both modern and traditional agriculture provides many examples for climate places limitations on what crops may be successfully grown at a given location. In many cases, a predominant crop of a region reflects the original natural vegetation cover as indicated by the biome; similarities of climate in world biomes have allowed crops to be produced in areas where they did not exist.

To overcome climatic problems, methods for protection against frost have been introduced. Some of the methods use latent heat, others rely on air mixing, while still others use protective covers over susceptible plants. Irrigation is used to meet the problem of drought; essentially the goal is to add as much water as needed to meet the evapotranspiration requirement. The methods to do this are varied. Both frost protection and irrigation are part of maintaining conditions for the highest crop yield. Sunlight, necessary for photosynthesis, is ultimately the key to success while photoperiodism and temperature limits are a significant part of an agricultural venture's success.

Many examples may be cited to show that climate plays a role in industrial development and location. In some cases, the location may reflect the favorable weather of a location; the movie industry is one such case. In other instances, the location is a function of the natural environment as conditioned by the climate. Agricultural processing serves as an illustration. Similarly, some energy resources are more suitable for some climatic regions than others. Wind and solar energy are certainly applicable examples.

Transportation—whether by air, sea, or land—is susceptible to many problems relating to weather and climate. Turbulence and icing are troublemakers for air travelers, while fog has an adverse effect on all three methods of transport. Storm frequency, a function of climatic location, is obviously of major importance. Examination of other tertiary industries, particularly tourism and commerce, shows that climatic influences play a role in the successful functioning of the economic activities.

KEY TERMS

Clear-air turbulence Icing Photosynthesis
Drought Photoperiodism Thermals

REVIEW QUESTIONS

1. How is the distribution of major wheat growing areas related to climate and natural vegetation?

2. Why is it that some crops introduced into new global areas, to which they are not native, produce good yields?

3. Outline the main methods used to protect crops against frost.

4. Describe the three types of droughts identified in this chapter. What differentiates them?

5. Describe the role of climate in rates of photosynthesis and photoperiodism.

6. How has the climate of California influenced the early location of some industries?

7. What are the major problems associated with the development of solar energy? Wind energy?

8. Describe an example of the economic impact of wind-driven ocean currents.

9. Provide examples of how climate influences sporting activities.

10. How do retail sales reflect the prevailing climate?

CHAPTER 19

Global Changes in Atmospheric Chemistry: Acid Precipitation and Ozone Depletion

CHAPTER OUTLINE

The chemistry of the atmosphere plays a vital role in determining the surface temperature. Although oxygen and nitrogen are by far the major constituents of the atmosphere, many trace elements play an important part in the way in which energy is transferred through the atmosphere. To permit a ready assessment of the changes that are currently occurring, *air pollution* is defined as a change in the concentration of material (or energy) in the air such that the modified content adversely affects the well-being of living things. Air pollution is neither new nor is it necessarily associated with industrial processes. Gases from fetid marshes and smoke from fires are examples of natural air pollution. Even before factory chimneys and automobiles added their output to the atmosphere, pollution was observed. Juan Rodriguez Cabrillo, sailing into San Pedro Bay in 1542, named the Los Angeles Basin the "Bay of Smokes." Prevailing weather conditions caused the smoke from scattered Indian fires to hang as a pall of pollution over the area. As we now know, this was a harbinger of things to come.

It makes no sense in today's world to consider the atmosphere in its natural form for the reason that human activity is rapidly altering atmospheric chemistry. Technological processes inject into the air a wide variety of solids, liquids, and gases collectively called *pollutants*. It is true that the atmosphere is never completely pure under any circumstances. Gases such as sulfur dioxide, hydrogen dioxide, and carbon monoxide are continually released into the air as by-products of natural occurrences such as volcanic activity, decay of vegetation, and range and forest fires. Thus, some **effluents** introduced into the atmosphere are in fact natural constituents.

These materials become pollutants only when they are placed into the atmosphere in abnormally large amounts. The volume of **effluents** placed in the atmosphere has reached the extent where levels of some constituents are increasing beyond natural limits. The Environmental Protection Agency (EPA) released a report in 1989 indicating the magnitude of the industrial injection of chemicals into the atmosphere. In the United States, there are at least 1600 industrial facilities in 46 states that release into the air significant amounts of chemicals suspected of being carcinogenic. Some 125 of these plants release more than 400,000 pounds of chemicals each year. There are some 30 industrial facilities that emit more than 1 million pounds a year each. For instance, 2.7 billion pounds of pollutants were placed into the atmosphere in 1987. Of this amount, 360 million pounds are suspected of being **carcinogenic**.

A phenomenal amount of pollutants is pumped into the atmosphere every year. The potential for ill health resulting from air pollution is well known, and the EPA is responsible for monitoring and enforcing laws pertaining to environmental laws. Air quality standards have been used to produce a Pollution Standards Index (PSI). This is computed for a defined location and uses the value of the pollutant that occurs in the highest concentration in relation to the national air quality standards. If one of the pollutants is at the primary standard, a PSI value of 100 is given. Below that value, the air is considered moderate or good; above the 100 level, the categories pass from unhealthy to hazardous.

ATMOSPHERIC POLLUTANTS

Particulates

In 1994, nearly 23 million Americans, or 9 percent of the total population, lived in areas where **particulates** such as soot and acid aerosols exceed EPA standards. In these areas, the air often is thick and hazy, especially in the summer. Particulates can impair the functioning of the lungs and seriously threaten health. They mainly affect people with chronic respiratory illnesses such as asthma, bronchitis, and emphysema. Lung disease is the nation's third leading cause of death. California has the strictest laws governing particulate air pollution. The EPA standard for particulates is 150 micrograms per cubic meter of air. California has a maximum of 51 micrograms. The primary sources of particulates are diesel trucks and buses, factory and electric utility smokestacks, car exhaust, burning wood, mining, and construction. Small-sized particulates have been linked to disease. Dust-sized particulates of less than 10 microns in diameter have been linked to bronchitis, asthma, pneumonia, and pleurisy in children.

Carbon Monoxide

By volume, the greatest emission from human activity is **carbon monoxide**. It is a colorless, odorless, and tasteless gas. It is formed primarily by incomplete combustion of coal, fuel oil, and gasoline. The largest single source is from the automobile. The gas begins to affect the human body at concentrations of about 100 parts per million (ppm). At this concentration, people develop headaches and may become dizzy. Levels of 100 ppm have been observed in some urban areas, and concentrations of 370 ppm have been recorded inside vehicles trapped in traffic jams. Carbon monoxide has a residence time of several days. Eventually the carbon monoxide combines with oxygen to form carbon dioxide.

Sulfur Compounds

Sulfur oxides are the second most abundant pollutant. Sulfur dioxide (SO_2) is one the major oxides of sulfur. It is a heavy, pungent, colorless gas. It forms from the combination of sulfur from emissions of coal burning industries and atmospheric oxygen. Sulfur dioxide is highly reactive and hence is not cumulative. The maximum residence time is probably 10 days. Much of the compound combines with atmospheric water to form sulfuric acid. Atmospheric sulfuric acid causes the leaves of plants to turn yellow, it dissolves limestone and marble, and it is highly corrosive of iron and steel. It also reduces atmospheric visibility and blocks out sunlight. It is a major irritant to the eyes and respiratory system and is lethal at a few parts per million. In 1985, some 23 million tons of sulfur oxides were emitted into the atmosphere in the United States alone. Seventy percent of the sulfur dioxide (16 million tons) came from burning low-grade coal and petroleum in electric power plants. Sulfur dioxide emissions from electric power plants have declined by about 30 percent since 1975 due to the use of higher-grade coal and cleaner burning plants. Most of the most severe offenders are located in the Ohio River Valley.

Hydrogen sulfide (H_2S) is another sulfur compound that forms in the atmosphere. It forms from organic decay when there is not enough oxygen present to oxidize the organic material. The main sources of hydrogen sulfide are swamps. It has a very bad smell, like rotten eggs, but fortunately has a short residence time. In the atmosphere, it will darken lead in oil-based house paints. It is also responsible for tarnishing copper and silver.

Nitrogen Oxides

Nitrogen and oxygen do not normally interact at standard environmental temperatures. The substantial quantities of nitrogen oxides (NO_x) result from combustion at high temperatures, largely in automobile engines. Nitrogen dioxide (NO_2) is the only widespread pollutant that has a color to it. It is yellow-brown in color and also has a pungent sweet odor. The average residence time is about 3 days. The end product of nitric oxides is nitric acid (HNO_3). Nitrogen oxides (NO_x) are also a major contributor to acid rain and surface ozone. New scrubbers that remove as much as 70 percent of the gas are being developed.

Automobile and truck exhaust yields most of the nitrogen oxides in the United States. Out of 20.5 million tons injected into the atmosphere in 1985, 45 percent came from vehicle exhaust. The distribution of nitrogen oxides is distributed as vehicle traffic is distributed. It is emitted over a much broader area and more evenly than is sulfur dioxide.

Surface Ozone

Ozone is a form of oxygen that contains three atoms of oxygen instead of the usual two. The gas is colorless and odorless except at very high concentrations. It is a major ingredient in smog. Ozone is formed near the ground when pollutants such as unburned petroleum hydrocarbons and nitrogen oxides from automobile exhausts and fossil fuel power plants react in sunlight. The chemical reactions are faster on hot sunny days. It takes several hours after the sun rises for the chemical reactions to reach the level where ozone accumulates. It usually begins to form about 10 A.M. solar time. The ozone that forms in sunlight usually breaks down at night. Thus, the process begins on each day. It does not persist for long periods.

Ground levels of ozone reached their highest levels on record in 1988. About half of the U.S. population lives where ozone exceeds the EPA standard at least part of the time. Ozone is highest in the states east of the Mississippi River, and in California and Texas.

Breathing ozone may cause respiratory problems for those who exercise out of doors. Ozone is an irritant to the lungs and air passages. It is especially irritating to those who engage in vigorous exercise. Some individuals are affected almost immediately if they exercise in air with elevated ozone levels. They may cough or experience chest pain and shortness of breath. To reduce the chance of respiratory irritation, it is best to take precautions when exercising in warm weather when there is a risk of ozone accumulation. It is best to exercise before 10 A.M. Jogging along major traffic thoroughfares is also to be avoided. A general rule is: The more and harder you exercise, the greater the intake of air and therefore of ozone.

ACID PRECIPITATION

Acid rain is a phrase that applies to a process that results in deposition of acid on the surface of the earth. All precipitation is slightly acidic in nature. One index of measuring acidity is the concentration of hydrogen ions (pH). A neutral solution has a pH of 7.0. The lower the pH, the more acidic the water. For each unit the pH drops, the acidity increases by a multiple of 10. Thus, a pH of 6 represents an acidic element 10 times that of a pH of 7. A pH of 5 represents 100 times the acidity of water with a pH of 7.

Natural precipitation has a pH of near 5.6. The term *acid rain* was first used by a British chemist, Angus Smith, in 1858. It refers to precipitation with a pH of less than 5.6. When precipitation has a pH of less than 5.6, it is usually due to the injection of sulphur compounds or nitrogen oxides into the atmosphere. Coal-burning electric power plants, industrial furnaces, and motor vehicles inject large amounts of these chemicals into the atmosphere. In the atmosphere, the chemicals combine with water to form sulfuric acid and nitric acid. These droplets may be transported great distances by wind before they precipitate to the ground.

The acidity of precipitation has increased over North America to the level where the pH is less than 4.6 over most of the continent east of a line from Houston, Texas, to the southern tip of Hudson Bay. In 1980, the pH of precipitation dropped to an average of 4.1 over part of the Ohio River Valley and the Adirondack Mountains. In the Great Smoky Mountains, precipitation with a pH of 3.3 was measured. At Wheeling, West Virginia, in 1980, rainfall with a pH of 1.4 was recorded. Ordinary battery acid has a pH of 1.1.

Much of the attention on the pollution causing acid rain in the northeast of the United States and in Canada has focused on the Ohio River Valley and other areas of the midwest, where there are large concentrations of coal-fired power plants. Nine states produce 52 percent of the sulfur dioxide emissions. They are Georgia, Illinois, Indiana, Kentucky, Missouri, Ohio, Pennsylvania, Tennessee, and West Virginia.

Global Distribution of Acid Rain

Acid rain has become a worldwide problem. In Europe, Norway, Sweden, Denmark, The Netherlands, and West Germany all have a problem with acid rain. In Sweden alone, an estimated 18,000 lakes are more acid than natural rainfall. Great Britain is accused by the continental nations as being the main source of the pollutants. Great Britain has admitted to being a source of sulfur dioxides and nitrogen oxides.

Great Britain has also had its share of **acid precipitation** problems. The worst case of acid mist occurred in 1989. The mist came in over the east coast on September 9, affecting a 1000-square mile area. The area was mainly in Norfolk and Lincolnshire. The mist was estimated to have a pH of 2.0. The mist was so acidic that it corroded aluminum instruments. It damaged thousands of trees. The acid killed the leaves, turning them brown overnight. The source of the sulfuric and nitric acid particles is believed to be automobile traffic on the continent. This incident was worse than the incident at Pitlochry, Scotland, in 1974. The Pitlochry acid mist had a pH level of 2.4—stronger than vinegar.

Precipitation in other parts of the world is also acidic. In the city of Guiyang, concentrations of the sulfate ion are about six times greater than in New York City and 20 to 100 times greater than that over Katherine, Australia. Katherine is considered to be representative of an area little affected by industrial pollution. The higher concentration of sulfates in China is due to the large use of coal as a primary fuel for home cooking, heating, and electricity generation. There are virtually no controls on the use of coal as a fuel. Although concentrations of sulfates are higher in China, there is a lower concentration of nitrates primarily due to fewer automobiles. The nitrate concentration is highest in Beijing where there are the most automobiles.

Impact of Acid Precipitation

The impact of acid rain on aquatic environments, particularly fresh-water lakes, has been clearly established. Aquatic systems are very susceptible to acidification. Fish are very susceptible to acidification. Fish become endangered when the pH drops to about 5.5. Most species of fish stop reproducing at pH levels of between 5.3 and 5.6. Fish are hurt by acidification in a number of ways. As acidity increases, more trace metals are dissolved in the water. Aluminum is one such metal. Young fish are particularly susceptible to increased aluminum concentrations. The aluminum collects in their gills. In trying to get rid of it, the young fish strangle in their own mucus. Above normal acidity also prevents fish from absorbing calcium and sodium. The lack of calcium weakens their bone structure and their skeletons become deformed and are easily damaged. Lack of sodium causes convulsions, which kill the fish.

Fish are also susceptible to acid shock. Acid shock is the sudden introduction of large amounts of acid. It is commonly associated with spring snowmelt in mountain regions. The acid is deposited in the snow crystals and remains on the ground for periods of up to several months. With spring melting, large amounts of acid enter the streams and lakes resulting in a sudden, if temporary, increase in acidity. In northern United States, some winters, such as that of 1993 to 1994, have a lot of snow accumulation, and there may be widespread acid shock in the spring. Other years it may be minimal. In Canada, snow accumulates in most winters and so there are annual episodes of acid shock.

By the time the pH of a lake drops to 5.0, between 30 and 50 percent of the natural biota cease to exist. The most susceptible are the smaller organisms such as mollusks and minnows. Many lakes contain water with a pH of less than 4.5. At 4.5, all fish are gone, and the water supports completely different organisms from normal lake water. High acidity favors the growth of sphagnum mosses and filamentous algae.

The sensitivity of lakes to acidification depends on their natural ability to neutralize the acidic runoff into the lake. Lakes located in areas where the parent rock is igneous and metamorphic containing lots of silicates are most sensitive to acid deposition. The dissolved minerals from these rocks result in acidic runoff. In regions where the parent rock is high in the mineral salts such as calcium, magnesium, and phosphorous, lakes can better tolerate the acid runoff. The reason is that the soil solution tends more toward alkaline and the salts neutralize the acid.

On a global scale, there may be more than 1000 lakes that have become too acidic to support life. There may be several thousand that receive episodes of acid shock. The greatest share of these are in eastern Canada and Scandinavia. Acidification is a particularly severe problem in the Adirondack Mountains of New York and New England. The parent rock in these areas is high in the silicates. In these areas, soils are thin and acidic under natural conditions and so the runoff from the acid rain remains highly acidic. Acidity in lakes in these areas has increased sharply since 1950.

The Impact of Acid Precipitation on Terrestrial Systems

The impact on terrestrial ecosystems is less clear than for aquatic systems. A major area of controversy is whether acid clouds and acid precipitation are damaging world forests. In September 1990, the National Acid Precipitation Assessment Program (NAPAP) released the results of a 10-year study. One of the study's conclusions was that there is now widespread forest damage in North America that can be directly linked to acid rain. Many scientists are convinced that acid rain is the leading cause or at least the catalyst in widespread forest damage in mid-latitudes. Evidence of damage to vegetation in North America is beginning to accumulate.

One area where rapid **dieback** of the forests is occurring is around Mount Mitchell in North Carolina. Dieback is the gradual dying of a tree or trees, either from the crown downward or from the tips of the branches inward toward the trunk. Acid rain and acid fog are factors suggested for the problem. Fog over the mountains has frequently been measured with a pH of between 2.5 and 3.5, and rain with a pH of 2.2 was measured in 1986. The trees affected are a variety of spruce, which is the remnant of a once widespread forest that was logged off more than a century ago. Dieback of this forest has been observed only since 1983. It began at the summit and has progressed down the mountain to lower elevations. Sections of the dead timber can now be seen from the Blue Ridge Parkway that skirts the mountain.

A second area where tree damage has been documented is in New England. Stands of spruce on Whiteface Mountain in New York are dying, and Dr. Hubert Vogelman, a botanist at the University of Vermont, found a 19 percent decline in the number of sugar maple trees over a 20-year period. In 1965, researchers counted 345,493 maple seedlings in a two-acre area on Camels Hump Mountain near Duxbury, Vermont. By 1983, the number had dropped to 53,400. Dr. Vogelman also reported that wood samples show increasingly high concentrations of residues of industrial chemicals and hydrocarbons.

Forests in other parts of the world have been affected. In 1983, it was estimated that one third of the forests in West Germany were damaged from acid rain.

The extent of forests suffering from acid rain in Germany is increasing at a geometric rate. Five percent of the forests were damaged enough to be essentially dead.

It may be that the damage is done through acidification of the soil. When soils become more acidic there are more dissolved metals in the soil water taken in by the tree roots and there is less decomposition of organic matter in the soil.

Acid Rain and Human Health

Dr. David V. Bates, a University of British Columbia physician, found that several years of hospital records indicate admittances increased as atmospheric sulfate levels rose in one urban area of southern Ontario containing some 6 million people. The correlation between admissions for ailments including pneumonia and asthma were significantly related to sulfate levels. His study indicated that 13 percent of the variations in admissions could be explained by changes in sulfate levels. Other researchers suggest that it contributes to emphysema and other respiratory diseases as well, particularly in children.

There may be a health hazard in eating fish taken from streams and lakes with increased acidity. Fish from these waters often have high levels of aluminum, copper, lead, mercury, and zinc. Intake of these metals can affect health. Aluminum may be linked to the onset of Alzheimer's disease. People that die of Alzheimer's disease often have high enough concentrations of aluminum to reduce neural function.

Impact of Acid Rain on Structures

Limestone and marble are soluble in acids. Since many of the major buildings and sculptures are made of limestone and marble, they can be damaged by acid rain. In Great Britain, there is a problem with acid rain in terms of it dissolving the exteriors of major historical buildings. Forty-four of the flying buttresses on Westminister Abbey are needing replacement. These buttresses were rebuilt less than 100 years ago, but the limestone is badly eroded due to the solution by the acid rain. Another famous structure suffering from solution by acidic precipitation is the Taj Mahal in India. The world-reknowned structure, which is built of marble, is rapidly being destroyed. Replacement of damaged panels cannot keep up with the rate of damage. On the Acropolis in Athens stands the Erectheum. It is a small structure away from the Parthenon. The porch roof is supported by six marble statues of maidens carved by the Greek scuptor Phidias in the 5th century B.C. The statues had to be removed because of rapid deterioration due to acid rain. They were replaced in 1977 by fiberglass copies (Fig. 19.1).

STRATOSPHERIC OZONE AND ULTRAVIOLET RADIATION

Ozone is a form of oxygen in which three atoms of oxygen combine to form a single molecule of ozone. Ozone normally is not abundant in the lower atmosphere under natural conditions. It does, however, form in smog by the action of sunlight on oxides of nitrogen and organic compounds. This ozone does not stay in the air for very long. It reacts with other gases in the atmosphere and changes to normal oxygen molecules.

Stratospheric Ozone

Ozone exists in the stratosphere, though the total amount is small. It is concentrated in a layer or layers between altitudes of 12 and 50 kilometers (7 and 30 mi). The ozone is continually formed and then removed. The process that forms ozone is the absorption of ultraviolet radiation in the range from 0.1 to 0.3 microns in length. This absorption

Figure 19.1
Statues of the maidens on the porch of the Erectheum. The Erectheum is one of the structures on the Acropolis in Athens, Greece. These statues are made of fiberglass. The originals were moved to a museum because they were being destroyed by acid precipitation (from Hidore J. J., *Global Environmental Change*, © 1996, Prentice Hall, Upper Saddle River, N.J.).

of radiant energy breaks oxygen molecules apart into single oxygen atoms. Some of the single atoms combine with an oxygen molecule to form ozone. Absorption of additional radiation breaks up the ozone molecules. Most ozone forms over the tropical latitudes. It is here that most solar radiation, and the most intense solar radiation, reaches Earth. The upper atmospheric circulation carries the ozone toward the poles.

The atmosphere absorbs ultraviolet radiation at all altitudes. Single atoms of oxygen absorb the shortest wavelengths (less than 1 micron) at altitudes above 160 kilometers. From 110 to 160 kilometers, oxygen molecules absorb radiation in the range from 0.1 to 0.2 microns in length. Below 110 kilometers, ozone absorbs the longer wavelength ultraviolet radiation. The most ultraviolet radiation is absorbed at heights from 20 to 50 kilometers (Fig. 19.2). Most of the ozone is at these altitudes. Ozone has a broad absorption band peaking at 0.255 microns. The energy absorbed adds to the increase in temperature with height in the stratosphere. The atmosphere, mainly as a result of ozone, absorbs about 98 percent of the ultraviolet radiation reaching Earth. Although the upper atmosphere absorbs most ultraviolet radiation, some reaches the surface. The 2 percent that reaches the surface of the planet is critical to life on Earth.

Chlorofluorocarbons

In 1974, scientists warned there was evidence to suggest that compounds known as chlorofluorocarbons (CFCs) have a depleting effect on stratospheric ozone layers. First synthesized in 1928, these compounds promised to have many uses. They are odorless, nonflammable, nontoxic, and chemically inert. The primary compounds

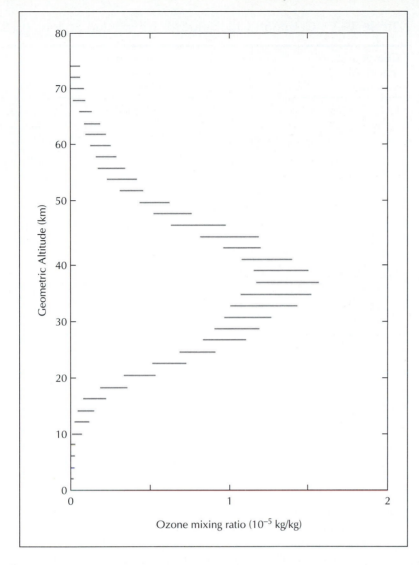

Figure 19.2
The vertical distribution of ozone in the lower 80 kilometers of the atmosphere.

are listed in Table 19.1 along with some of their characteristics. They first came into use in refrigerators in the 1930s. Since World War II, they have been used as propellants in deodorants and hair sprays, in producing plastic foams, and in cleaning electronic parts. The United States, Japan, and Europe produce and consume most of the chemicals. In the United States, per capita use reached 1.1 kilograms (2.4 lbs) per person per year.

Automobile air conditioners use CFC-12. This is among the most damaging of the CFCs. Ninety percent of all new cars in the United States have air conditioners. Automobiles are the largest single source of harmful CFCs. They make up 26.6 percent of the CFCs released into the environment. In 1989, the state of Vermont enacted legislation to outlaw cars with air conditioners using CFCs beginning with the 1993 model year.

These compounds are not natural compounds. They do not react with most products dispersed in spray cans. They are transparent to sunlight in the visible range. They are insoluble in water and are inert to chemical reaction in the lower atmosphere. It is for these reasons that they are valuable compounds. It is for these same reasons that the chemicals are a problem in the stratosphere. The average lifetime of a CFC-11 molecule is between 40 and 80 years. A CFC-12 molecule may last from 80 to 150 years.

Table 19.1 Common Chlorofluorocarbons and Halons (EPA)

Compound (chemical formula)	Ozone Depletion Potential[a]	Atmospheric Lifetime (years)	Major Uses
CFC-11 ($CFCl_3$)	1.0	64	Rigid and flexible foams, refrigeration
CFC-12 (CF_2Cl_2)	1.0	108	Air conditioning, refrigeration, rigid foam
CFC-113 ($C_2F_3Cl_3$)	0.8	88	Solvent
Halon-1211 (CF_3BrCl)	3.0	25	Portable fire extinguishers
Halon-1301 (CF_3Br)	10.0	110	Total flooding fire extinguisher systems
HCFC-22 ($HCClF_2$)	0.05	22	Air conditioning

Source: U.S. Environmental Protection Agency.

[a] Ozone-depleting potentials represent the destructiveness of each compound. They are measured relative to CFC-11, which is given a value of 1.0.

Chlorofluorocarbons rise into the upper atmosphere where they break apart under ultraviolet radiation. The breakdown takes place when the compounds are exposed to radiation of wavelengths of less than 230 nanometers. Ultraviolet radiation of this wavelength or shorter does not reach the troposphere because it is absorbed at altitudes of 20 to 40 kilometers (12–24 mi). This breakdown releases chlorine, which interacts with oxygen atoms to reduce the ozone concentration. The process ends with the chlorine atom once again free in the atmosphere. Each atom of chlorine may persist for years, acting as a catalyst that may remove 100,000 molecules of ozone. The maximum rate of ozone destruction takes place at an altitude of 40 kilometers (25 mi). The final means of disposing of the chlorine is a slow drift downward into the troposphere, where it combines with water molecules and falls out as hydrochloric acid.

Antarctic Ozone Hole

The most disturbing change in atmospheric ozone is that found over the Antarctic Continent called the **ozone hole** (Fig. 19.3). The ozone hole is a loss of stratospheric ozone over Antarctica, which has occurred in September and October since the late 1970s. The hole appears in September when sunlight first reaches the region and ends in October when the general circulation brings final summer warming over Antarctica. During the Antarctic spring, there is a decrease in ozone north from the pole to nearly 45° south latitude.

In August and September 1987, the amount of ozone over the Antarctic reached the lowest level recorded to this date. Nimbus–7 on September 17, 1987, recorded a large area in which the ozone concentration was only about half the surrounding region. That fall the ozone hole, the area of maximum depletion, covered nearly half of the Antarctic Continent.

In the winter over Antarctica, a very large mass of extremely cold, dry air keeps out warmer air surrounding the continent. This cold air gets even colder during the

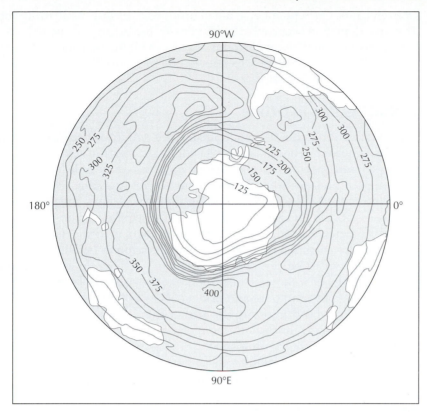

Figure 19.3
The Antarctic ozone hole over the southern hemisphere on September 28, 1992, as measured by the Total Ozone Mapping Spectrometer (TOMS) aboard the Nimbus-7 satellite. Notice that the area of lowest ozone concentration is larger that Antarctica, and that the ozone hole is nearly centered over the South Pole (from Hidore J. J., *Global Environmental Change*, © 1996, Prentice Hall, Upper Saddle River, N.J.).

months when there is no sunlight. Temperatures drop as low as −84°C (−119°F). In the extreme cold, moisture condenses into ice crystals and nitric acid crystals also form. These crystals form very high thin clouds called **polar stratospheric clouds** (PSCs). The cloud crystals play a very important role in the chemistry of the CFCs and ozone depletion. The nitrogen oxide crystals drop out of the stratosphere leaving behind the chlorine and bromine compounds and the ice crystals. Each ice crystal provides a place for accelerated chemical reactions. The chemical processes are more rapid where there is a surface on which the reaction can take place. Ice particles are good surfaces and are some 10 times as efficient as the surface of water droplets. This partially explains the speed with which the process takes place in the Antarctic spring. It also explains why the process is less effective in low latitudes.

The chemical process begins when sunlight appears in the spring. The warming increases the rate of chemical reactions, and chlorine destroys ozone at a rapid rate. The depletion actually first begins near the Antarctic Circle, where sunlight begins to penetrate the stratosphere. It may begin here by mid-August. Spring over the South Pole occurs in September and October. During this time, the ozone level drops until there is no more ozone left or the clouds evaporate. There may be a total loss of up to 60 percent of the ozone in the center of the Antarctic hole. At some altitudes, it is 90 percent. Eventually, air from surrounding regions flows into the area and ozone levels recover. Polar stratospheric clouds disappear with the spring warmup. The same process takes place elsewhere in the atmosphere, but at higher altitudes and at slower rates.

In the spring of 1991, record depletion of ozone occurred in the Antarctic. Record lows in ozone occurred in September. By mid-November, the system broke up and ozone levels recovered. The depletion usually takes place at altitudes of 12 to 22 kilometers. This is the range where most of the polar stratospheric clouds form. In 1991, ozone concentrations dropped to record lows slightly below those of the worst 3 previous years. On August 28, the ozone concentration was about 270

Dobson Units. On September 4, it had dropped to as little as 30 to 40 Dobson Units. This year depletion occurred at levels not affected before. Ozone reductions of nearly 50 percent took place at altitudes of 11 to 13 kilometers and 25 to 30 kilometers. The result of this depletion was a reduction of the total ozone column of 10 to 15 percent more than in past years.

Global Decline in Ozone

Ozone depletion is not as much outside the Antarctic because the stratospheric aerosols are less abundant and consist of liquid sulfuric acid droplets rather than ice. This difference is significant. There is no Arctic ozone hole like that of the Antarctic. Temperatures are warmer, and there is more variable weather in the Arctic, which provide less favorable conditions for the necessary chemical and circulation processes. Ozone levels in the high latitudes of the northern hemisphere have dropped 5 percent since 1971. In 1988, researchers in Thule, Greenland, measured increased concentrations of reactive chlorine compounds. These are the same compounds known to be present over the Antarctic while depletion takes place. Experiments in 1989 showed the presence of the ozone hole and provided detailed measurement of the amount and extent of the depletion.

In the low- and mid-latitude stratosphere, there is greater solar radiation during the winter months and there is a general absence of polar stratosphere clouds. In this part of the atmosphere, the destruction of ozone is due to a combination of chemical processes. Models of the atmosphere show that nitrogen oxides play a leading role in the ozone destruction. Particular nitrogen oxides (N_2O_5 and $ClONO_2$) react on the surface of sulphuric acid solutions, which are similar to stratospheric aerosol particles. Sulphate particles exist throughout the lower stratosphere. They form from biological and volcanic activity. One of the processes is not dependent on extremely cold conditions so it operates much of the time to deplete ozone. The eruption of Mount Pinatubo in June 1991 injected some three times as much sulphur into the stratosphere as did El Chichon in 1982. Since the depletion process normally takes place on sulphuric acid particles, the increase in these particles did not alter the rate of depletion.

New evidence keeps appearing that supports a decline in the global ozone layer. In 1986, Canadian scientists detected a thinning of ozone over the Arctic region. In 1988, NASA established that the global ozone layer was declining faster than expected. By 1990, NASA reported spring losses of ozone in mid-latitudes of the northern hemisphere two to three times greater than before. Also in 1990, British scientists reported an accelerated rate of ozone loss over western Europe. By 1994, there was a 4 to 5 percent decline in stratospheric ozone worldwide.

International Response to Ozone Depletion

Concern over the possible connection between CFCs and ozone loss led to a ban on the use of these compounds as aerosol propellants in the United States effective in 1978. This was part of an EPA ban on all nonessential uses of CFCs. The U.N. Environment Program called a conference in Montreal, Canada, in September 1987 to discuss the possible effects of CFCs on stratospheric ozone. Representatives of more than 30 countries took part in the conference, which drafted a treaty restricting the production of CFCs. The agreement is officially termed the *Montreal protocol*.

Ultraviolet Radiation and Living Organisms

There are two areas of concern about the possible reduction in the ozone layer and an increase in ultraviolet radiation reaching ground level. The first problem concerns public health. The second is the role of CFCs in potential global warming.

Stratospheric ozone filters out ultraviolet radiation in the 280 to 320 nanometer range. This is the high-energy portion of the ultraviolet radiation spectrum known as ultraviolet-B (UVB). This UVB radiation is also very harmful to living organisms. Although the atmosphere blocks most UVB radiation, it does not block all of it. Plants did not flourish on Earth until there was enough atmosphere and ozone to block much of the UVB radiation.

All plants and animals now existing and living in sunlight on Earth have adapted to ultraviolet radiation. Plants vary widely in their tolerance of UVB. Plants that developed in climates with high-intensity sunlight show a variety of defense mechanisms for UVB. Some produce clear or nearly clear pigments that absorb UVB radiation. Marijuana plants produce protective chemicals called *cannabiniods*. These are also the main hallucinogenic ingredients in marijuana. In arid climates, plants develop thick, shiny leaves. Cacti and olive trees are examples.

Although sunlight is essential to most life, there is a limit to how much sunlight is good. One of the effects of ozone depletion is to let more ultraviolet radiation through the atmosphere to the surface. Ultraviolet B can damage DNA. DNA is the genetic code in every living cell. Most living organisms are subject to damage by UVB radiation. Since plants cannot adjust their behavior to changing solar radiation, some are damaged by UVB radiation. The soybean is one such plant. Excessive amounts of UVB slow growth and reduce yields. Soybean yields may drop 1 percent for each 1 percent drop in ozone.

Animals and humans also have adapted to UVB radiation. Nearly 90 percent of marine species living in the surface water surrounding the Antarctic Continent produce some form of chemical sunscreen. Humans manufacture melanin in the skin. This is a pigment that blocks ultraviolet radiation. A summer tan results from increased production of melanin. Persons with very fair skin do not readily manufacture melanin and sunburn very easily.

Ultraviolet-B and Human Health

Exposure to ultraviolet radiation results in aged skin, skin cancer, and a weakened immune system. The main element in the increase is the popular need for a suntan. It is ultraviolet radiation that produces tanning of the skin and also sunburn. The risk of skin cancer is much greater from overexposure as in a sunburn than from steady low doses. A single blistering sunburn in a person 20 to 30 years of age triples the risk of skin cancer.

Melanoma.　One form of skin cancer is **melanoma**. It may start in or near a mole. This involves the cells that give the skin its color and often are a mixture of black or brown, sometimes with red or blue areas. These moles continue to grow and have irregular borders. It is the least common but most lethal form of skin cancer. The fatality rate from melanoma is about 25 percent. It is almost always fatal if it spreads to other parts of the body. Early treatment results in a survival rate of more than 80 percent. The disease is now almost epidemic in the United States. In 1974, there were about 9000 cases diagnosed. Some 27,000 new cases were expected in 1990 (American Cancer Society, 1990). This represents a tripling of the number of cases in 16 years.

The highest incidence of melanoma occurs in individuals that do not tan easily. The incidence is increasing at about 4 percent each year. There is the same rate of increase in other countries where sunbathing and tanning salons are in vogue. Younger and younger persons are diagnosed as having melanoma. When first regularly reported, it was in persons ages 40 or over. By 1990, it was frequent in the age group from 20 to 40. Fatalities in 1990 consisted of about 6300 cases of melanoma and 2500 cases of other kinds of skin cancer.

Basal and Squamous Cell Carcinoma.   Other forms of skin cancer are basal cell carcinoma and squamous cell carcinoma. Basal cell carcinoma is the most common. It is a slow-growing cancer that usually begins with a small, shiny, pearly bump or nodule on the head, neck, or hands. It can bleed, crust over, and then open again. It is not life threatening. Squamous cell carcinoma may start as nodules or red patches with well-defined outlines. It typically develops on the lips, face, or tips of the ears. It can spread to other parts of the body and enlarge. These skin cancers can be removed by simple surgery and are rarely fatal. In 1990, more than 600,000 new cases of nonmelanoma cancer were forecast. It is likely that nearly everyone in the United States over the age of 30 has some skin damage from solar radiation. At current rates, at least one in seven Americans will develop some form of skin cancer.

Reducing the Risk of Skin Cancer.   The risk of getting skin cancer can be reduced with reasonable care. The first rule is to avoid exposure to the midday sun. The most dangerous hours are between 10 A.M. and 2 P.M. local time (11 A.M.–3 P.M. daylight savings time). There is an old saying: "Only mad dogs and Englishmen go out in the noonday sun." If exposure to the sun is necessary, use a sunscreen with a rating of 15 based on Ultraviolet-B radiation. Ultraviolet-A is also harmful to health, but not nearly as much so as Ultraviolet-B. Lotions with a rating of 15 provide protection from both UVA and UVB radiation There is no evidence that sunscreens with a higher rating provide additional protection. Avoid tanning parlors because the radiation is as bad or worse than natural sunlight.

Future Depletion of Ozone

One of the problems associated with ozone depletion is that maximum depletion may not occur until between 2010 and 2020. The Antarctic ozone hole may not fill until as late as 2075. Each CFC molecule has a lifetime of up to 30 years. It takes these molecules 6 to 8 years to rise to the stratosphere. Even if present production of CFCs and related compounds stops, there is a huge quantity of CFCs in old refrigerators, air conditioners, and foam packaging. Production has not stopped. Under present global arrangements, nearly half as much CFCs can be produced in the future as the total produced since the chemicals were first introduced.

Compliance with existing agreements will slow but not stop the accumulation of the chemicals in the stratosphere. Concentrations may grow to as much as 30 times the 1986 levels. Computer models show that at least an 85 percent reduction in CFCs and halon use is needed to stabilize the level of the chemicals in the atmosphere. The chlorine already released will continue to remove ozone for at least a century. If the release of CFCs stops, there will be a lag of several decades before maximum ozone depletion takes place. Only after that time can the rate of removal begin to decline. The chlorine content may continue to rise until it reaches a level as much as six times the 1986 levels. If the release of the compounds continues unabated, a reduction in ozone of about 10 percent will take place with a possible range of 2 to 20 percent.

SUMMARY

Earth's atmosphere is no longer a natural one. That is, it is now greatly altered by human activity. The major gases of the atmosphere are still nitrogen and oxygen, but the variety and amounts of the variable gases have changed. Large amounts of dust, soot, metals, and organic matter are injected into the atmosphere each day.

Large volumes of gases such as carbon monoxide, carbon dioxide, and nitrogen dioxide are also added each day. Many of the gases or their byproducts are harmful to plants and animals.

Sulfur compounds and nitrogen dioxide combine with water droplets in clouds to become acid precipitation. Acid precipitation is harmful to most ecosystems. It is harmful to human health, commerce, and structures. Control of acid precipitation is difficult because the atmosphere is so mobile. The precipitation may fall hundreds or thousands of kilometers from the place where the pollutants are injected into the atmosphere.

Ultraviolet radiation from the sun reaching Earth's atmosphere creates a layer of triatomic oxygen called *ozone*. This process absorbs about 98 percent of the high-energy radiation. One of the many classes of chemical compounds developed by humans are the chlorofluorocarbons. These chemicals escape into the atmosphere and rise into the stratosphere. In the stratosphere, radiation breaks them down in a fashion that releases chlorine. The chlorine removes oxygen ions and reduces the amount of ozone. With a reduction of ozone, there is increased ultraviolet radiation reaching the surface. This may alter the planetary energy balance and also be detrimental to the health of plants and animals.

KEY TERMS

Acid precipitation	Dieback	Ozone hole
Carbon monoxide	Effluents	Particulates
Carcinogens	Melanoma	Polar stratospheric clouds

REVIEW QUESTIONS

1. What is the primary source of carbon monoxide in the atmosphere?

2. A neutral solution has a pH of 7. What is the pH of normal precipitation?

3. Below what level of acidity in fresh water do fish not survive?

4. What locations in the United States have forests suffering from dieback due to atmospheric pollution?

5. How does acid rain affect structures made from marble and limestone?

6. What process creates the stratospheric ozone layer?

7. How do the CFCs that are relatively stable in the troposphere affect the ozone layer in the stratosphere?

8. Why does ozone depletion take place more readily over the two polar regions than in the mid-latitudes and Tropics?

9. Why is Ultraviolet-B radiation damaging to living organisms?

10. What are the three most common forms of skin cancer?

CHAPTER
20

Statistical Analysis Using Climatic Data

CHAPTER OUTLINE

The study of climatology is based on analysis and interpretation of meteorological data collected over many years. To analyze data, an awareness of
basic statistical methods is needed, and anyone seriously concerned with extensive use of climatic data should have a basic knowledge of statistical techniques.
     This chapter offers an overview of some of the statistical methods used in cli-
matology. The coverage is not intended to be comprehensive and uses simple examples of commonly used methods that do not require sophisticated data
manipulation. The concern is with descriptive statistics of data series, distribution
of values, and relationships between variables. Simple time series analysis is included, but no effort is made to cover all useful techniques. Suffice it to say that
climatological applications often require substantial amounts of data, and the
only manner by which to rationally manage them is through statistical analysis.
Note too that in this chapter examples using data from NWS sources retain the
Fahrenheit scale.

MEASURES OF CENTRAL TENDENCY

It is very useful to summarize data by using a single value. The value needs to be
chosen such that it gives a reasonable approximation of what is the normal, where
normal might be defined as a number that locates the center of the measurement distribution (a measure of the **central tendency** of the data set). The most common of
these measures are the mean, median, and mode. Their derivation and application
may be demonstrated using the example of 10-year series of the hours of sunshine
in December at two places in Illinois. Table 20.1 provides the data and calculated
values. (Note that in this and other examples, short-length data series are given to
reduce the necessity of lengthy tables.)
     The **mean** is the sum of the values divided by their number. In short form:

$$\bar{X} = \Sigma X_i/N$$

where $\bar{X}$ is the mean, X_i are the individual values, N is the number of values, and
Σ is sigma—a sign used to show summation.
     At Moline the mean is 113.5, and at Springfield it is 113.4 (Table 20.1a). Two
factors need to be considered in relation to these results. First, the mean in each
case seems to give a greater degree of accuracy than the raw data (i.e., data do not
have decimal places, the mean does). Such accuracy has statistical but not physical
meaning, so that in both cases the mean can be given as 113 hours. The second factor relates to the same derived result. In looking at the raw data, they are clearly different, yet the mean is the same. The mean thus appears to provide an incomplete
description of facts. Later methods to provide information about variance of data
around the mean are given. For now, the other measures of central tendency are
used to show that the data sets are different.
     The **mode** of a data series is that value which occurs with greatest frequen-
cy. In the Moline and Springfield data, no value occurs more than once, so the
mode is estimated by using modal classes or class intervals. The size of the class
interval depends on the size of values in the data set. In this example, we assume
that six classes are to be identified so that the size of the class interval can be estimated by finding the range of values (e.g., in Moline $204 - 61 = 143$) and dividing it by the number of classes required (i.e., $143 \div 6 = 24$). The frequency of
values within the identified model classes is shown in Table 20.1b. Despite the
similar means, the mode of the data sets is different.
     The third measure of central tendency, the **median**, is the middle value (or
the average of the two middle values in an even series) of the data when given in
order of magnitude. For the two stations in question, Moline has a median of 111.5

Table 20.1   Mean, Mode, and Median: Measures of Central Tendency

	December Hours of Sunshine	
Year	Moline	Springfield
1967	86	87
1968	112	111
1969	119	92
1970	149	167
1971	71	74
1972	107	114
1973	115	94
1974	111	106
1975	61	99
1976	204	190
(a) **Mean**	**1135/10 = 113.5**	**1134/10 = 113.4**

Modal Class	Moline f	Springfield f
60–84	= 2	= 1
85–109	= 2	= 5
110–134	= 4	= 2
135–159	= 1	= 0
160–184	= 0	= 1
185–209	= 1	= 1
(b) **Modal Class**	**= 110 to 134**	**= 85 to 109**

(c) **Median**

Moline Ranked Data	61, 71, 86, 107, 111, 112, 119, 149, 208
	Median = (111 + 112)/2 = 111.5
Springfield Ranked Data	74, 87, 92, 94, 106, 111, 114, 167, 190
	Median = (99 + 106)/2 = 102.5

while that for Springfield is 102.5 (Table 20.1c). As in the case of the mode, the median value for Moline is greater than that of Springfield.

The median belongs to a general class of descriptors called the **quantile** (these are sometimes called fractiles)—a value below which lies a given fraction of the data set. For example, the median would separate the data into two quantiles. In climate studies, the quantile often used in precipitation studies is the quartile. To derive this, data are divided into four sets by Q_1, Q_2, and Q_3; 25 percent of the data are below Q_1, 25 percent between Q_1 and Q_2, 25 percent between Q_2 and Q_3, and 25 percent above Q_3. Thus, the rainfall that occurs in a month or season may be compared to previous amounts of data according to the quartile in which it occurs.

The different values determined by measures of central tendency provide important information about the raw data. Figure 20.1 shows the relationships among the mean, mode, and median for three hypothetical distributions of data. In Figure 20.1a, data are evenly distributed in a bell-shaped curve. In such a case, mean = median = mode. Figure 20.1b shows a distribution curve in which the distribution is not symmetrical. It has the property of **skewness**. Since the tail of the distribution is to the left, negative skewness occurs. When this occurs, the measures of central tendency differ, with the mean being the lowest value. With positive

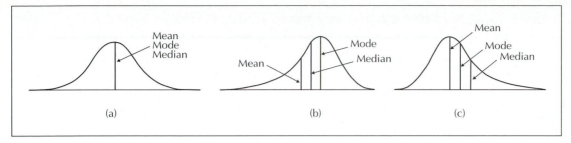

Figure 20.1
Relationships among the mean, mode, and median for three hypothetical distributions of data.

skewness, the third figure, the mean is the largest value and the mode the smallest. In the examples given, the use of only 10 years of data is not climatologically meaningful. However, given the Springfield data, for example, it can be seen that the mean provides the highest value of central tendency and that the limited data are distributed such that positive skewness is illustrated.

Of the measures of central tendency, the mean is the most widely used in climatic analysis. In applications, the use of the mean may lead to dangerous conclusions if used alone. A dry land farmer in central Nebraska might be encouraged to grow corn on looking at the mean precipitation. This ignores the years that have considerably lower than the mean precipitation; wide interannual variability is a feature of the Great Plains climate. As indicated, the mean is an incomplete measure, and measures of variance about the mean provide more complete data description.

MEASURES OF DISPERSION

Measures of central tendency provide no insight into the dispersion of values around the central measure. In climatology, such dispersion is usually expressed using the range or standard deviation. The **range** is obtained by subtracting the smallest individual value from the largest. When presented together with the mean, this provides some insight into variability. For example, the mean annual temperature at Quito, Ecuador, is 15.0°C; at Nashville, Tennessee, it is 15.3°C. From this, it might be construed that Quito and Nashville have similar climates. However, the range at Quito is less than 1°C (15.3°C–14.7°C); at Nashville it is 22.2°C (26.1°C–3.9°C). The two locations experience quite different climatic regimes—that at Quito is a tropical regime modified by altitude, whereas at Nashville the seasonal range reflects the mid-latitude climate regime. The range, although useful, does not provide any clue to how data are distributed. For this, the **standard deviation** is of value.

The standard deviation has many applications in applied climatology. Here it is demonstrated using the dates of the first fall frost at a midwestern U.S. city (Table 20.2). The dates are given in year days (i.e., January 1 is Day 1, January 2 is Day 2, etc. throughout the year). The mean of these dates $(\bar{X})$ is calculated and, for easy derivation of the standard deviation, values for $X_i - \bar{X}$ and $(X_i - \bar{X})^2$ are also given. The first step in the analysis is to derive the variance by obtaining the deviations of the individual values (X_i) from the mean $(\bar{X})$ (second column). The formula for variance (s) is:

$$s^2 = \Sigma(X_i - \bar{X})^2/N - 1,$$

where Σ is the summation sign, $(X_i - \bar{X})^2$ is calculated (third column), and N is the number of observations. Since the standard deviation (s) is merely the square root of the variance, the data can be substituted into

$$s = \sqrt{\Sigma(X_i - \bar{X})^2/N - 1}.$$

Table 20.2 Year Days of First Fall Frost

Year Day X_i	$X_i - \bar{X}$	$(X_i - \bar{X})^2$
291	−8	64
299	0	0
279	−20	400
302	+3	9
280	−19	361
303	+4	12
304	+5	25
307	+8	64
314	+15	225
313		196

$\sum X = 2992$ 1356

$$\bar{X} = \frac{2992}{10} = 299 \qquad (X - \bar{X})^2 = 1356$$

For our example, the standard deviation is 11.6, which to the nearest day $= 12$. This value provides significant information about the distribution of data around the mean:

a. The mean $\pm$ one standard deviation contains approximately 68 percent of the measurements in the series.

b. The mean $\pm$ twice the standard deviation contains approximately 95 percent of the measurements in the series.

Thus, in the example, 95 percent of first fall frosts have occurred between the mean data $\pm$ twice the standard deviation. That is, Day $299 + (2 \times 12) = 299 + 24 = 323$ and $= 299 - 24 = 275$.

Such information is of value to farmers. It suggests that if frost-susceptible plants are harvested by October 2 (Day 275), the chances of frost damage are minimal. The standard deviation has also been used by climatologists to describe abnormal climate conditions:

If standard deviation is beyond	−3	extremely subnormal
is between	−3 and −2	greatly subnormal
	−2 and −1	subnormal
	−1 and +1	normal
	−1 and −2	above normal
	+2 and +3	greatly above normal
If standard deviation is beyond	+3	extremely above normal

The standard deviation is also used to derive the **coefficient of variation** (cv). This is the ratio between standard deviation and mean expressed as a percentage:

$$cv = (s/\bar{x}) \times 100.$$

The cv may be used for applied climatic analysis in a variety of ways for it enables direct comparison of variation for a given data set. A common use is analysis of precipitation variation over a given area for a given time.

FREQUENCY DISTRIBUTION

Frequency distribution is the basic tool for describing and analyzing a population. It is a necessary and indispensable part of climatological investigations. The methods involved are straightforward and provide a valuable tool for many applied studies.

Table 20.3 provides the mean seasonal snowfall at a midwestern city for a 30-year period. Note that the data are given in seasonal rather than calendar dates so that a continuum of related snowfall periods is derived. It is useful to begin the snow year in July so that, for example, the 1939 to 1940 snow season runs from July 1939 to June 1940.

An initial step in the analysis may be to construct a **histogram**—a graphical representation of the data tallied into classes. As in the case of modal classes, the number of classes to be used (i.e., the class interval) can be estimated by finding the range of data and dividing it into a suitable number of classes. The snowfall data in Table 20.3 has a range of nearly 30 inches. Since the optimum number of classes ranges between 6 and 20, if a class interval of 5 were used, this would give at least 6 classes. A more formal method of deriving class intervals is given by

$$K = 1 + 3.3 \log n,$$

where K is the number of classes and n the number of variables. Since divisions in units of 2, 5, 10, and so on are mostly used, the choice of a class interval of 5 is appropriate. Note that in using class intervals, it is necessary to have values where there can be no question as to which group a given observation belongs. The selected modal classes are given with the representative histogram in Figure 20.2.

Although the histogram is useful, the climatologist is more interested in estimates of probabilities over several class intervals that are more conveniently obtained from the cumulative distribution. Also, the cumulative distribution provides better estimates of the probabilities since the arbitrary division into class intervals tends to waste some of the information on the population given by the climatological series. To obtain the cumulative distribution, the data are first put in order as in Table 20.4.

The f's in the table are the cumulative relative frequencies, or estimates, of the cumulative population probabilities and are obtained by the formula $f = m/(n + 1)$, where m is the mth value in order of magnitude of the climatological series and n is the number of terms in the climatological series (in this case, 30). The division by $(n + 1)$ instead of n gives a better estimate of population probabilities, especially at the ends of the distribution. The f calculation is simple. For example, for the m value of 1 (4.5 inches), $f = 1/31 = .032$; for the $m = 30$ value (34.4 inches), $f = 30/31 = .968$. The f's give the probabilities that precipitation is less

Table 20.3 Snowfall, in inches, at Terre Haute, Indiana

1939–40	27.2	1949–50	8.0	1959–60	19.0
1940–41	14.3	1950–51	34.4	1960–61	23.5
1941–42	24.2	1951–52	21.4	1961–62	24.6
1942–43	20.8	1952–53	16.9	1962–63	17.1
1943–44	22.4	1953–54	7.0	1963–64	31.2
1944–45	14.3	1954–55	9.7	1964–65	29.5
1945–46	16.6	1955–56	17.0	1965–66	6.9
1946–47	8.3	1956–57	14.6	1966–67	12.1
1947–48	20.0	1957–58	4.5	1967–68	33.6
1948–49	7.5	1958–59	15.3	1968–69	7.5

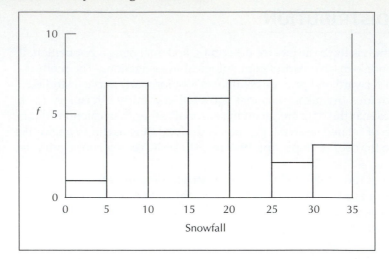

Figure 20.2
Tallied modal data (from Table 20.3) and their representation as a histogram.

Table 20.4 Cumulative Distribution of Snowfall (inches) at Terre Haute, Indiana

m	X	f	m	X	f	m	X	f
1	4.5	.032	11	14.3	.338	21	21.4	.677
2	6.9	.065	12	14.6	.387	22	23.4	.710
3	7.0	.097	13	15.3	.419	23	23.5	.774
4	7.5	.145*	14	16.6	.452	24	24.2	.774
5	7.5	.145*	15	16.9	.484	25	24.6	.806
6	8.0	.194	16	17.0	.516	26	27.2	.839
7	8.3	.226	17	17.1	.548	27	29.5	.871
8	9.7	.258	18	19.0	.581	28	31.2	.903
9	12.1	.290	19	20.0	.613	29	33.6	.935
10	14.3	.338	20	20.8	.645	30	34.4	.968

*Equal values are given the average rank of the positions they occupy.

than any value shown in the table. For example, the probability that X is less than 12.1 inches is 0.290 and that it is greater than 12.1 inches is $1 - f = 0.710$. Note that when probabilities are estimated for a continuous random variable such as snowfall, it is a misunderstanding of sampling principles to use the wording *equaled* or *exceeded* or *less than or equal to* because the probability of any exact value occurring is zero. The probability that it is between 12.1 and 22.4 inches is $0.710 - 0.290 = 0.420$. Thus, the cumulative distribution gives all the information available from histograms and much in addition for it uses every value of the climatological series individually to obtain the probability estimates. The sample cumulative distribution may also be put to graphical form by plotting f on the ordinate against X on the abscissa and connecting the points by lines, as in Figure 20.3.

RELATIONSHIPS

Hypothesis Testing

An important aspect of climatic analysis is establishing relationships between climate and other variables. In such an analysis, it is customary to formulate two hypotheses.

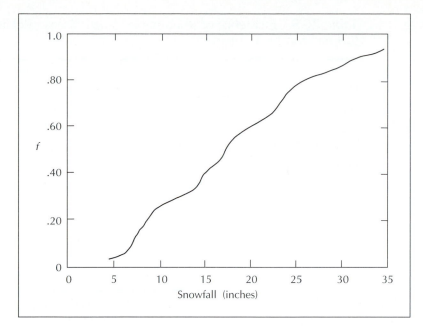

Figure 20.3
Cumulative distribution of data given in Table 20.4.

The first, the null hypothesis (H_o), is that there is a high probability that the observed relationships are due to chance. The alternate hypothesis (H_1) is that there is a statistical relationship between the two data sets being analyzed. Although this is a straight-forward idea, to decide on acceptance and rejection of a null hypothesis, there has to be a decision-making procedure. Such a procedure draws on the idea of significance, where the probability level for rejection of H_o is the significance level, designated α.

If there is a 95 percent confidence level (meaning that the particular result only has a 5 percent chance of occurring through random variation), then $\alpha = .05$. For a 99 percent confidence, the 0.1 level of significance is used. If the decided level is attained, then H_o may be rejected and the alternate hypothesis accepted. The designation of a confidence level is left to the needs of the user. The 99 percent and 95 percent levels are most frequently used in studies, with the former most commonly employed when using large data sets.

Correlation Coefficient

In the analysis of the impact of climate on people, many misconceptions have arisen. Some have resulted from the misinterpretation of incomplete data, others as a result of total lack of rigorous statistical analysis. Of particular importance in the analysis is the sorting out of relationships. For example, are the number of deaths that occur at a given season related to high and low temperatures? Do colds occur more when it is wet and rainy? Such potential relationships can be examined using **correlation coefficients**; that used here is the Pearson Product Moment Correlation Coefficient (r).

A 1967 health report suggested that rheumatic heart disease incidence in children is related to climate. It was noted that, although no clear relationship can be found between summer and winter temperatures and the incidence of the disease, many more cases were reported in northern states than in southern states. To test this idea, latitude and rheumatic heart disease in children were correlated (Table 20.5), and it proves useful to state null and alternative hypotheses:

Null hypothesis: The correlation coefficient (r) = 0. There is no statistical relationship.

Alternative hypothesis: The correlation coefficient is not equal to 0. There is a relationship.

Table 20.5   Incidence of Rheumatic Heart Disease Related to Location by Latitude

Locality	Approx. Lat.	Rheumatic Heart Disease per 100 Children
Wyoming	45°	4.5
Madison, WI	43°	2.2
New Haven, CT	41°	2.5
Eureka, CA	41°	2.1
Denver, CO	40°	1.6
Redlands, CA	34°	0.4
Dublin, GA	32°	1.0
Southern AZ	32°	0.5
Pensacola, FL	30°	0.3
Dade County, FL	25°	0.4

To compute r, a table can be constructed in which the variables are represented by X (latitude) and Y (rheumatic heart disease) and where x is $X - \bar{X}$ (latitude minus mean latitude) and y is $Y - \bar{Y}$ (incidence of disease minus mean incidence). The summation values are used in the formula to calculate r.

$$r = \frac{\Sigma xy}{\sqrt{\Sigma x^2 \Sigma y^2}}.$$

Substituting values from Table 20.5,

$$r = \frac{67.60}{\sqrt{(388.1)(15.92)}} = \frac{67.6}{78.6} = 0.86.$$

As illustrated in Figure 20.4, derived correlation values of +1 and −1 indicate perfect positive and negative correlations, respectively. If $r = 0$, then no statistical relationship occurs. Clearly, a value of 0.86 is a high positive value and suggests that the two sets of data are related. The statistical significance that permits the null hypothesis to be rejected can be calculated by a test statistic—t. This is given by

$$t = r\frac{n - 2}{\sqrt{1 - r^2}},$$

where r is the derived correlation coefficient and n the number of pairs in the correlation. In our example,

$$t = 0.86.\frac{8}{\sqrt{.261}} = 4.76.$$

Prepared tables and an extract are included in Table 20.6 and are available to evaluate this result.

The df in the table refers to degrees of freedom. A mathematical concept, df, refers to a number of restrictions placed on the data for one degree of freedom is lost for each data group (n) correlated. Thus, in this example, where $n = 10$, the degrees of freedom become ($n - 2$) = 8. From Table 20.6, it can be seen that, for 8 degrees of freedom (df), the critical value of t at the .05 level is 1.860. Since our de-

Table 20.6 Selected Extract—Degrees of Freedom

df	a = .10	a = .05	a = 0.25
7	1.415	1.895	2.365
8	1.397	1.860	2.306
9	1.383	1.833	2.262
10	1.372	1.812	2.228

rived value of t is greater (4.76) than this, the null hypothesis is rejected and the alternate accepted. There is a significant statistical correlation.

Although the prior analysis provides the method for completing the correlation coefficient, is the result meaningful in terms of the data being analyzed? To relate a set of events to gross latitude may not be meaningful for so many other factors are omitted. In effect, perhaps the analysis completed provides a statistical manipulation of limited real-world consequences. Statistical significance does not imply cause and effect.

Rank Correlation

Thus far, the analytic techniques outlined have involved actual data—the parameters of the sampled population. Another correlation technique uses a nonparametric test, meaning that the test depends on the distribution of the data but not on specific values in that distribution. It is called the *rank correlation coefficient* (r). As previously noted, if the derived value of r is +1, then it represents a perfect linear correlation; a value of -1 is a perfect negative relationship. If $r = 0$, then no correlation exists.

Contrary to popular belief, desert air is not totally dry and does contain moisture. Table 20.7 provides the maximum average monthly temperatures and average

Table 20.7 Mean Monthly Temperature and Moisture Data for a Desert Station

Month	Mean Max. Temp. °F	Mean Rel. Hum. %	Mean Vapor Pressure. mm.
Jan.	69 (12)	37	5.43 (12)
Feb.	75 (10)	34	7.56 (9)
Mar.	83 (8)	35	10.13 (8)
Apr.	92 (7)	27	10.38 (7)
May	99 (5)	23	10.98 (6)
June	110 (3)	25	16.51 (1)
July	113 (1)	16	11.51 (4)
Aug.	111 (2)	19	12.90 (2)
Sept.	105 (4)	24	12.42 (3)
Oct.	94 (6)	28	11.48 (5)
Nov.	80 (9)	38	5.79 (11)
Dec.	71 (11)	38	7.39 (10)

monthly relative humidity values (recorded at 1300 hours) at the Saharan station (In-Salah, Algeria). There appears to be a relationship between the two for as temperatures rise so the relative humidity decreases. But such a relationship must be anticipated because relative humidity is a temperature-dependent atmospheric moisture measurement. With no change in the amount of water vapor in the air, the relative humidity decreases as the temperature rises. To more accurately assess whether desert air moisture is related to temperature, and hence time of the year, a more conservative moisture measure is required. Table 20.8 lists the mean monthly vapor pressure (mm). Vapor pressure denotes the amount of moisture in the air irrespective of temperature. To find if a relationship exists between temperature and vapor pressure, rank correlation can be used. For the two data sets that are to be correlated, each is given a rank. The numbers in parentheses in Table 20.8 show that, for temperature and vapor pressure, a rank of 1 is given to the highest value, 2 to the second highest, and so on. If there were two or more identical values, they would be given the average rank they occupy. For example, if there were two values in the fourth rank, each would be given a value of $4 + 5/2 = 4.5$. The next rank would be 6.

To quantify the relationship, the rank correlation coefficient is given by

$$R = 1 - \frac{6\sum(x - y)^2}{n(n^2 - 1)},$$

where $(x - y)$ is the difference in rank (see Table 20.8) and n is the number of observations. Drawing on the ranks given in the table, r may be calculated. The value for $(x - y)^2$ is substituted into the formula as is the value for n—in this case, 12.

$$= 1 - \frac{6(22)}{12(144 - 1)}$$

$$= 1 - .007 = 0.92.$$

This value represents a high positive correlation between the mean maximum monthly temperatures and mean monthly vapor pressure. When a value such as this is derived, its closeness to a value of 1 (perfect correlation) clearly indi-

Table 20.8   Ranked Data (see Table 20.7)

Month	x (temp. rank)	y (vapor press. rank)	$x - y$	$(x - y)^2$
Jan.	11	12	0	0
Feb.	10	9	1	1
Mar.	8	8	0	0
Apr.	7	7	0	0
May	5	6	−1	1
June	3	1	2	4
July	1	4	−3	9
Aug.	2	2	0	0
Sept.	4	3	1	1
Oct.	6	5	1	1
Nov.	9	11	−2	4
Dec.	11	10	1	1
				Sum = 22

cates that a significant statistical relationship exists. It was previously noted that as the value of r diminishes, it becomes necessary to carry out tests to determine whether a statistical relationship exists. The statistical relationship does not indicate, however, the climatic reasons for the correlation. What the relationship really tells is that the dominant air at this station comes from different sources over the period of a year.

REGRESSION ANALYSIS

Regression analysis involves a technique by which, using a limited amount of data, it is possible to predict the value of one variable given any corresponding value of another, related value. The relationship or correlation between two variables can be represented in graphical form, where one variable is plotted against the other. Figure 20.4 suggests some of the patterns that may emerge.

Figure 20.4a shows that when July precipitation increases, there is a corresponding rise in the unidentified crop yield. This represents a positive linear relationship between the two variables. A straight line drawn on the graph does not pass through all the plotted values, but represents a fairly good fit. Were the line to pass through the points, a perfect linear relationship would exist. Figure 20.4b values indicate that temperatures decrease with altitude. A negative linear relationship exists. Again a line can be drawn through the plotted values to approximate the relationship. In Figure 20.4c, points are scattered all over the graph; there is no relationship between the values. The lines drawn on the graphs—the *best fit* lines—are positioned so that the distance between all points and the line is minimum. It is a regression line. Because the line has a location and slope within the $X - Y$ coordinates, it can be represented by an equation in the form

$$Y = a + bX,$$

where Y and X are the selected variables and a and b are constants that determine the location and slope of the line. The constant a is the Y intercept, the value of Y at the point where the line crosses the Y axis with $X = 0$, and b is the slope of the regression line—the change in Y that corresponds to one unit change in X. Again, use of the method is best seen in examples.

Variations from normal circulation patterns result in agricultural problems. One experiment was carried out that analyzed the yields of a number of crops grown in a lysimeter near Columbia, South Carolina. Lysimeter data provide evapotranspiration rates and these were compared to the crop productivity. The data are tabulated in Table 20.9a.

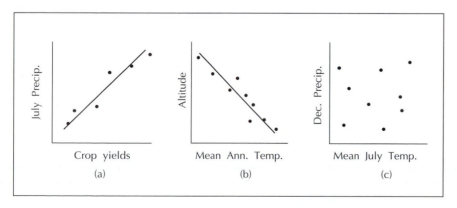

Figure 20.4
Statistical relationships between two variables. (a) A high positive linear relationship, (b) a high negative linear relationship, and (c) no linear relationship exists.

Table 20.9a   Regression Analysis Data (The Relationship Between Evapotranspiration Rates and Crop Yield at a Selected Station)

Evapotranspiration (in/yr)	Dry Matter Yield (1000 lbs/yr)	Evapotranspiration (in/yr)	Dry Matter Yield (1000 lbs/yr)
29.6	6.7	26.8	5.9
29.8	6.3	24.0	3.8
28.4	6.0	29.6	6.8
25.0	4.0	30.2	6.8
24.6	3.9	31.9	8.4
24.0	3.2	32.0	8.8
21.0	0.8	28.0	6.6
28.0	6.4	28.6	6.6
27.9	6.2	28.2	6.5
27.4	5.9	27.9	5.7

Table 20.9b   Calculation of Regression Equation (for Data Given in a)

X (Evapotranspiration)	Y (Yield)	X^2	XY
29.6	6.7	876.16	198.32
29.8	6.3	888.04	187.74
28.4	6.0	806.56	170.74
⋮	⋮	⋮	⋮
27.9	5.7	778.41	159.03
$\sum X = 552.9$	$\sum Y = 115.3$	$\sum X^2 = 15430.75$	$\sum XY = 3281.25$

Table 20.9c   Solution for a and b in Regression Equation

$$S_{XX} = \sum X^2 - \frac{(\sum X)^2}{n} \qquad\qquad S_{XY} = \sum XY - \frac{(\sum X)(\sum Y)}{n}$$

$$= 15430.75 - (552.9)^2/20 \qquad\qquad = 3281.25 - (552.9 \times 115.3)/20$$

$$= 145.82 \qquad\qquad\qquad\qquad\qquad = 73.78$$

So $\qquad\qquad\qquad\qquad b = S_{XY}/S_{XX} = 73.78/145.82 = .643$

To derive a: $\qquad\qquad\quad a = \bar{Y} - b\bar{X} = 5.76 - (.643 \times 27.64) = -12.0$

Given now that $\qquad\quad a = -12.0$ and $b = .643$, the regression equation for Y is
$$Y = -12.0 + .643X$$

Example: Were the annual evapotranspiration (X) equal to 20 inches, then the yield (in 000s lbs/yr) would be given by

$$Y = -12 + (0.634 \times 20)$$
$$= 0.86$$

A glance at the data seems to indicate that the higher the evapotranspiration, the higher the yield. A positive linear relationship seems to exist. Using regression analysis, a prediction equation of yield in terms of evapotranspiration can be derived. The equation takes the form

$$Y = a + bX$$

and, using the least squares formulas,

$$b = S_{xy}/S_{xx}, \quad \text{and}$$
$$a = \bar{Y} - b\bar{X},$$

where

$$S_{xx} = \Sigma x^2 - \frac{(\Sigma x^2)}{n} \text{ and } S_{xy} = \Sigma XY - \frac{(\Sigma X)(\Sigma Y)}{n},$$

and $\bar{Y}$ and $\bar{X}$ are mean values of Y and X, respectively.

The data from the previous tabulation are recalculated as shown in Table 20.9b and the calculation is shown in Part c of the table. The derived regression equation, $Y = -12.0 + 0.643X$, provides yields (in 000s lbs/year) for a given annual evapotranspiration. For example, were the evapotranpiration (X) equal to 20 inches, then the potential yield (Y) would be

$$Y = -12 + (0.643 \times 20)$$
$$Y = 0.86.$$

In employing this method, unless a perfect relationship exists, two best fit lines can be drawn. One is Y on X, the other X on Y. The former is used in the provided example, the latter allows X to be predicted from Y.

The regression analysis completed here uses the method of least squares, which fits a curve to the data by minimizing the sum of the squares of deviations of data points from the fitted curve. Residuals, the variation of data values from the best fit line, provide significant information. Note that, although simple regression is of value, some atmospheric processes are not linear, and what may appear a poor relationship with linear regression is actually very good with curvilinear. Such is true when the *rates* are considered instead of amounts. Multiple regression is also widely used. This uses more than two variables and usually requires application of suitable computer programs.

CLIMATIC TRENDS

The term *climatic change* is one that has a number of meanings, and in analyzing historic climatic data it is important that these be recognized. Figure 20.5 provides a guide to the terms used. The examples use temperature data to illustrate the terms. In this analysis, the major concern is with the identification of a temperature **trend** using data gathered at a single weather station using basic methods of analysis that provide a guide to potential changes. Table 20.10 provides the January mean temperature (°F) for a 46-year period at Terre Haute, Indiana.

Lists of data such as in Table 20.10 are difficult to read meaningfully, and a constructed graph aids in visual interpretation of the climate over time. Figure 20.6 is a graph of the data given in the table. The general impression derived from the graph is that there appears to be a decline in average January temperatures since about 1950. There are a number of ways the data can be

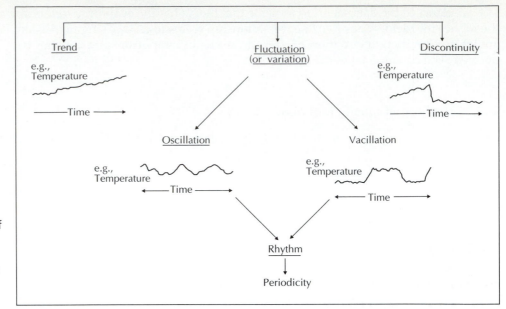

Figure 20.5
An explanation of some of the terms applying to climate variations. Temperature is used as an example in the inset graphs.

Table 20.10 Average January Temperature for Terre Haute, Indiana, 1933–1978

Year	°F	Year	°F
33	40.5	56	27.5
34	35.4	57	23.9
35	32.8	58	29.8
36	23.3	59	25.1
37	33	60	32.1
38	30.5	61	25.1
39	36.8	62	23.8
40	16.8	63	19.8
41	32.1	64	32.2
42	29.8	65	28.1
43	30.2	66	13.6
44	35.2	67	30.2
45	26.4	68	24.1
46	21.7	69	25.1
47	35.5	70	18.3
48	21.9	71	24
49	34.1	72	26.2
50	37.5	73	30.2
51	29.8	74	29.1
52	33.5	75	32.2
53	33.8	76	24.7
54	31.1	77	9.8
55	28.9	78	15.2

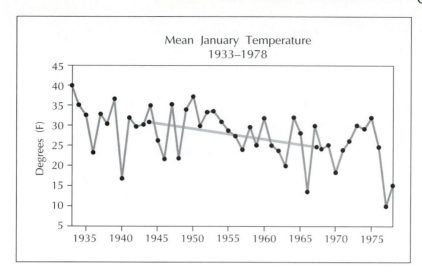

Figure 20.6
Graphical representation of data given in Table 20.10.

treated to clarify the observation. One method uses semi-average data. These are derived by finding the mean value for the first half of the period (1933–1955) and that for the second half (1956–1978). The two values are plotted on the graph and joined by a line. This semi-average line again seems to indicate a downward trend in the second half of the data period. The semi-average means can be tested to determine whether there is a statistically significant difference between the two means. The decision maker is Student's t, which is used for evaluating the difference between two population means. It is not dealt with in this brief description.

Running Means

A frequently used method to depict climatic trends is **running means**. As noted earlier, when all data are plotted on a graph, it is frequently difficult to see any patterns. Use of running means helps smooth the data series. This method involves the calculation of a number of successive means and grouping them to find the group mean. For a 5-year running mean (using the Terre Haute data), the following method is used:

a. Years   1933
                1934
                1935   Mean = 33.06 (Mean 1)
                1936
                1937

b. Drop the first value (for 1933) and add the next value (1938)
                1934
                1935
                1936   Mean = 31.06 (Mean 2)
                1937
                1938

c. Continue for each 5-year group, such that the next mean is derived by dropping 1934 data and adding 1939. This would give Mean 3. The 5-year running means for the data are shown in graph form in Figure 20.7.

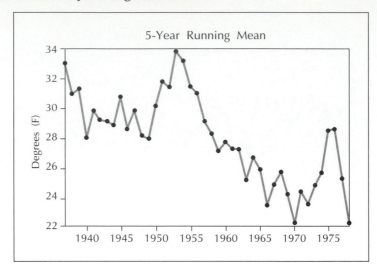

Figure 20.7
Five-year running mean
for representative snowfall
data.

The graph is much easier to interpret for trends than that of single annual means. Although data are lost in the construction of the graph, smoothing the data in this way enables a clearer view of the cooling trend discerned earlier. Note, however, that caution should always be exercised when attempting to discern trends. Smoothing data might ignore the rapidity of a change and mask physical causes of importance.

An interesting observation may be made about the Terre Haute data used in this example. It can be noted that the value of the second semi-average is lower than the first. It is also noted that there is a marked downward trend of temperature over the period analyzed. This was a widespread character of U.S. data from the 1940s to the 1970s. At the time, it resulted in many popular articles describing the cooling of the globe. Of course, subsequent to this period, there has been a substantial warming and major concerns of global warming.

FURTHER ANALYSIS

The outline provided in this chapter suggests some of the simpler statistical techniques that are available to the applied climatologist. The coverage is neither complete nor definitive and is presented to illustrate the basic meaning of some methods. The current availability of computer programs for statistical analysis certainly enhances calculations, but unless the derived results are understood they have little meaning. Only when investigators have an understanding of the methods used should they attempt to complete more complex analytic methods that require extensive data manipulation.

A next step, for example, would be use of multiple regression analysis, where a dependent variable is predicted using more than one independent variable. Such a method might be used to relate all climatic factors to, for example, corn yields. Factor analysis provides a valuable tool for a variety of problems. It is a technique that collapses a set of intercorrelated variables onto a smaller number of basic dimensions or composite variables. In analyzing periodic phenomena, harmonic analysis may be used. This provides values of a dependent variable that are repeated at equal time intervals of an independent variable. A study of published work in climatology could reveal the use of such methods together with many others.

SUMMARY

Statistical analysis of climatic data is basic to a study of climatology. To summarize data using a single value measures of central tendencies, the mean, mode, and median are used. To derive information about dispersion about the central value, the range may be used, but more information is derived from the standard deviation. The frequency distribution permits some estimates of probabilities of the data series.

Relationships between climate and other variables are examined using the null hypothesis and various correlation techniques. A method for examining actual data is the Product Moment correlation; for nonparametric data, it is the Rank correlation. Relationships between variables can also be studied using regression analysis. By construction of best fit lines, a prediction equation may be derived.

Analysis of historic climate data may be completed in a number of ways. In the example given here, simple trend analysis and running means are examined. Such analysis can be extended and tested using more sophisticated techniques.

KEY TERMS

Central tendency	Mean	Regression analysis
Coefficient of variation	Median	Running means
Correlation coefficient	Mode	Skewness
Frequency distribution	Quantiles	Standard deviation
Histogram	Range	Trend

REVIEW QUESTIONS

1. How does the mean differ from the mode? the median?

2. Give an example of the meaning of skewness.

3. How could the standard deviation provide information about the severity of a dry spell?

4. Draw a simplified figure illustrating a (a) histogram, and (b) cumulative distribution curve.

5. Explain the use of the null hypothesis.

6. Give some examples of data that might be meaningfully correlated.

7. What type of data would be used in rank correlation?

8. What is a positive linear correlation?

9. Show how running means are derived.

10. What is a climatic trend?

Appendix / Metric Units

°C		°F	°C		°F	°C		°F	°C		°F
−40.0	−40	−40	−17.2	+1	33.8	5.0	41	105.8	27.2	81	177.8
−39.4	−39	−38.2	−16.7	2	35.6	5.6	42	107.6	27.8	82	179.6
−38.9	−38	−36.4	−16.1	3	37.4	6.1	43	109.4	28.3	83	181.4
−38.3	−37	−34.6	−15.4	4	39.2	6.7	44	111.2	28.9	84	183.2
−37.8	−36	−32.8	−15.0	5	41.0	7.2	45	113.0	29.4	85	185.0
−37.2	−35	−31.0	−14.4	6	42.8	7.8	46	114.8	30.0	86	186.8
−36.7	−34	−29.2	−13.9	7	44.6	8.3	47	116.6	30.6	87	188.6
−36.1	−33	−27.4	−13.3	8	46.4	8.9	48	118.4	31.1	88	190.4
−35.6	−32	−25.6	−12.8	9	48.2	9.4	49	120.2	31.7	89	192.2
−35.0	−31	−23.8	−12.2	10	50.0	10.0	50	122.0	32.2	90	194.0
−34.4	−30	−22.0	−11.7	11	51.8	10.6	51	123.8	32.8	91	195.8
−33.9	−29	−20.2	−11.1	12	53.6	11.1	52	125.6	33.3	92	197.6
−33.3	−28	−18.4	−10.6	13	55.4	11.7	53	127.4	33.9	93	199.4
−32.8	−27	−16.6	−10.0	14	57.2	12.2	54	129.2	34.4	94	201.2
−32.2	−26	−14.8	−9.4	15	59.0	12.8	55	131.0	35.0	95	203.0
−31.7	−25	−13.0	−8.9	16	60.8	13.3	56	132.8	35.6	96	204.8
−31.1	−24	−11.2	−8.3	17	62.6	13.9	57	134.6	36.1	97	206.6
−30.6	−23	−9.4	−7.8	18	64.4	14.4	58	136.4	36.7	98	208.4
−30.0	−22	−7.6	−7.2	19	66.2	15.0	59	138.2	37.2	99	210.2
−29.4	−21	−5.8	−6.7	20	68.0	15.6	60	140.0	37.8	100	212.0
−28.9	−20	−4.0	−6.1	21	69.8	16.1	61	141.8	38.3	101	213.8
−28.3	−19	−2.2	−5.6	22	71.6	16.7	62	143.6	38.9	102	215.6
−27.8	−18	−0.4	−5.0	23	73.4	17.2	63	145.4	39.4	103	217.4
−27.2	−17	+1.4	−4.4	24	75.2	17.8	64	147.2	40.0	104	219.2
−26.7	−16	3.2	−3.9	25	77.0	18.3	65	149.0	40.6	105	221.0
−26.1	−15	5.0	−3.3	26	78.8	18.9	66	150.8	41.1	106	222.8
−25.6	−14	6.8	−2.8	27	80.6	19.4	67	152.6	41.7	107	224.6
−25.0	−13	8.6	−2.2	28	82.4	20.0	68	154.4	42.2	108	226.4
−24.4	−12	10.4	−1.7	29	84.2	20.6	69	156.2	42.8	109	228.2
−23.9	−11	12.2	−1.1	30	86.0	21.1	70	158.0	43.3	110	230.0
−23.3	−10	14.0	−0.6	31	87.8	21.7	71	159.8	43.9	111	231.8
−22.8	−9	15.8	0.0	32	89.6	22.2	72	161.6	44.4	112	233.6
−22.2	−8	17.6	+0.6	33	91.4	22.8	73	163.4	45.0	113	235.4
−21.7	−7	19.4	1.1	34	93.2	23.3	74	165.2	45.6	114	237.2
−21.1	−6	21.2	1.7	35	95.0	23.9	75	167.0	46.1	115	239.0
−20.6	−5	23.0	2.2	36	96.8	24.4	76	168.8	46.7	116	240.8
−20.0	−4	24.8	2.8	37	98.6	25.0	77	170.6	47.2	117	242.6
−19.4	−3	26.6	3.3	38	100.4	25.6	78	172.4	47.8	118	244.4
−18.9	−2	28.4	3.9	39	102.2	26.1	79	174.2	48.3	119	246.2
−18.3	−1	30.2	4.4	40	104.0	26.7	80	176.0	48.9	120	248.0
−17.8	0	32.0									

Table 2 Wind-Conversion Table (wind speed units: 1 mile per hour = 0.868391 knot = 1.609344 km/h = 0.44704 m/s)

Miles per hour	knots	Meters per second	Kilometers per hour	Miles per hour	knots	Meters per second	Kilometers per hour
1	0.9	0.4	1.6	51	44.3	22.8	82.1
2	1.7	0.9	3.2	52	45.2	23.2	83.7
3	2.6	1.3	4.8	53	46.0	23.7	85.3
4	3.5	1.8	6.4	54	46.9	24.1	86.9
5	4.3	2.2	8.0	55	47.8	24.6	88.5
6	5.2	2.7	9.7	56	48.6	25.0	90.1
7	6.1	3.1	11.3	57	49.5	25.5	91.7
8	6.9	3.6	12.9	58	50.4	25.9	93.3
9	7.8	4.0	14.5	59	51.2	26.4	95.0
10	8.7	4.5	16.1	60	52.1	26.8	96.6
11	9.6	4.9	17.7	61	53.0	27.3	98.2
12	10.4	5.4	19.3	62	53.8	27.7	99.8
13	11.3	5.8	20.9	63	54.7	28.2	101.4
14	12.2	6.3	22.5	64	55.6	28.6	103.0
15	13.0	6.7	24.1	65	56.4	29.1	104.6
16	13.9	7.2	25.7	66	57.3	29.5	106.2
17	14.8	7.6	27.4	67	58.2	30.0	107.8
18	15.6	8.0	29.0	68	59.1	30.4	109.4
19	16.5	8.5	30.6	69	59.9	30.8	111.0
20	17.4	8.9	32.2	70	60.8	31.3	112.7
21	18.2	9.4	33.8	71	61.7	31.7	114.3
22	19.1	9.8	35.4	72	62.5	32.2	115.9
23	20.0	10.3	37.0	73	63.4	32.6	117.5
24	20.8	10.7	38.6	74	64.3	33.1	119.1
25	21.7	11.2	40.2	75	65.1	33.5	120.7
26	22.6	11.6	41.8	76	66.0	34.0	122.3
27	23.4	12.1	43.5	77	66.9	34.4	123.9
28	24.3	12.5	45.1	78	67.7	34.9	125.5
29	25.2	13.0	46.7	79	68.6	35.3	127.1
30	26.1	13.4	48.3	80	69.5	35.8	128.7
31	26.9	13.9	49.9	81	70.3	36.2	130.4
32	27.8	14.3	51.5	82	71.2	36.7	132.0
33	28.7	14.8	53.1	83	72.1	37.1	133.6
34	29.5	15.2	54.7	84	72.9	37.6	135.2
35	30.4	15.6	56.3	85	73.8	38.0	136.8
36	31.3	16.1	57.9	86	74.7	38.4	138.4
37	32.1	16.5	59.5	87	75.5	38.9	140.0
38	33.0	17.0	61.2	88	76.4	39.3	141.6
39	33.9	17.4	62.8	89	77.3	39.8	143.2
40	34.7	17.9	64.4	90	78.2	40.2	144.8
41	35.6	18.3	66.0	91	79.0	40.7	146.5
42	36.5	18.8	67.6	92	79.9	41.1	148.1
43	37.3	19.2	69.2	93	80.8	41.6	149.7
44	38.2	19.7	70.8	94	81.6	42.0	151.3
45	39.1	20.1	72.4	95	82.5	42.5	152.9
46	39.9	20.6	74.0	96	83.4	42.9	154.5
47	40.8	21.0	75.6	97	84.2	43.4	156.1
48	41.7	21.5	77.2	98	85.1	43.8	157.7
49	42.6	21.9	78.9	99	86.0	44.3	159.3
50	43.4	22.4	80.5	100	86.8	44.7	160.9

Glossary

Absolute humidity   The ratio of the mass or weight of water vapor per unit volume of air; for example, grams per cubic meter.

Acid precipitation   Precipitation that is more acidic than natural precipitation, which has an average pH of about 5.5.

Adiabatic   Having to do with heating or cooling in gases due strictly to their expansion and contraction.

Advection   Mass motion in the atmosphere; in general, horizontal movement of the air.

Aerology   The study of the free atmosphere through its vertical extent.

Aerosols   Solid or liquid particles dispersed in a gas; dust and fog are examples.

Air mass   A large body of air that is characterized by homogeneous physical properties at any given altitude.

Agroclimatology   The scientific study of the effects of climate upon crops.

Albedo   The reflectivity of the earth, generally measured as a percentage of incoming radiation.

Analemma   A graphical representation of the ephemeris of the sun.

Angstrom   A unit of length equal to 10^8 cm, used in measuring electromagnetic waves.

Anticyclone   An area of above-average atmospheric pressure characterized by a generally outward flow of air at the surface.

Aphelion   The point on the earth's orbit that is farthest from the sun.

Arctic hurricane   An intense low-pressure system that develops along the edge of the pack ice in the Arctic. Arctic hurricanes produce high winds and heavy precipitation and represent a major hazard to shipping along their path.

Astronomical twilight   The period after sunset or before sunrise ending or beginning when the sun is 18° below the horizon.

Atmosphere   The mixture of gases that surrounds the earth.

Atmospheric circulation   The motion within the atmosphere that results from inequalities in pressure over the earth's surface. When the average of the entire globe is considered, the motion is referred to as the general circulation of the atmosphere.

Aurora   A display of colored light seen in the polar skies; called aurora borealis in the Northern Hemisphere and aurora australis in the Southern Hemisphere.

Beaufort wind scale   A system for estimating wind velocity (named for its inventor).

Biome   A well-demarcated environment that contains a complex of organisms defining the ecology of the region. Tundra and rain forest are examples.

Black body   A substance or body that is a perfect absorber and a perfect radiator.

Blocking   The retardation of eastward-moving pressure systems by stagnation of a high-pressure system.

Blizzard   High winds accompanied by blowing snow; usually associated with winter cold fronts in mild latitudes.

Bora   A cold, dry wind blowing down off the highlands of Yugoslavia and affecting the Adriatic coast.

Boundary layer   The layer of air above the earth's surface wherein the motion of the air is influenced by friction with the earth's surface features. Also called the Atmospheric Boundary Layer or the Friction Layer.

Bowen ratio   The ratio of the heat energy used for sensible heat to the heat energy used for latent heat, for a given surface.

Buoyant   Less dense than the surrounding medium and thus able to float.

Calorie   A measurement equal to the amount of heat needed to raise 1 g water 1°C; equal to 4.19 joules.

Carbon monoxide   A poisonous gas with no color and little odor, formed when there is incomplete combustion of fossil fuels.

Carcinogens   A substance or agent producing or inciting cancerous growth.

Castellanus   Towering clouds resembling castle turrets.

Centrifugal force   The apparent force exerted outward on a rotating body or on an object traveling on a curved path.

Chinook   A warm, dry wind blowing down off the Rocky Mountains of western North America.

Chlorofluorocarbons   A class of compounds whose major ingredients are chlorine, fluorine, and carbon. These compounds are instrumental in the destruction of the stratospheric ozone layer.

Cholera   Any of several diseases that result in severe gastrointestinal symptoms.

Cirrostratus   A thin, whitish veil of cloud that forms at high altitudes.

Climate   All of the types of weather that occur at a given place over time.

Climatic regime   The set of annual cycles associated with various climatic elements; for example, the thermal regime is the seasonal patterns of temperature, and the moisture regime is the seasonal patterns of precipitation.

Cloud streets   Long, thin lines of clouds forming in the trade winds when winds are steady and of low velocity over helical currents.

Cold front   The leading edge of a cold air mass that displaces a warmer air mass.

Condensation   The change of state from a gas to a liquid.

Conduction   Energy transfer directly from molecule to molecule. Conduction takes place most readily in solids in which molecules are tightly packed.

Continentality   A measure of the remoteness of a land area from the influence of the ocean.

Convection   Mass movement in a fluid or vertical movements in the atmosphere.

Convergent   Moving toward a central point of an area; coming together.

Cordillera   A group or system of mountain ranges, including the intervening valleys.

Coriolis force   An apparent force caused by the earth's rotation. The Coriolis force is responsible for deflecting winds clockwise in the Northern Hemisphere and counterclockwise in the Southern Hemisphere.

Crepuscular rays   Streaks of light that appear to emanate from the sun shortly before or after sunset. The rays result from the sun shining through a break in the clouds and illuminating particulate matter in the atmosphere.

Cyclone   Any rotating low-pressure system.

Deflation   The lifting and removal of earth particles by the action of the wind.

Dengue fever   An acute infectious disease characterized by the sudden onset of headache, severe aching of the joints, and rash. Dengue fever is caused by a virus transmitted by a mosquito and is found primarily in the tropics.

Desertification   The spread of desertlike conditions in a given location.

Determinism   The philosophical doctrine that every event has a cause and thus that no events are intrinsically uncertain.

Dew point   The temperature at which saturation would be reached if an air mass were cooled at constant pressure without altering the amount of water vapor present.

Die back   In the case of vegetation, to die from the top down. When a population dies back, its numbers decrease.

Diurnal   Occurring in the daytime or having a daily cycle.

Divergence   The condition that exists when the distribution of winds within a given region results in a net horizontal outflow of air from the region. At lower levels, the resulting deficit is compensated for by a downward movement of air from aloft; hence, areas of divergent winds are unfavorable to cloud formation and precipitation.

Doldrums   An area of very ill-defined surface winds associated with the equatorial convergence zone.

Doppler radar   A type of radar that detects the relative direction of motion; used in locating rotation in clouds and tornadoes.

Drainage basin   A part of the surface of the earth that is occupied by a drainage system of rivers and streams.

Drought   A time of unusually low water supply, due to reduced rainfall, reduced snowmelt, or low levels of streams.

Dust devil   A small cyclonic circulation, or dust swirl, produced by intense surface heating. Dust devils are most common in arid regions and resemble miniature tornadoes.

Easterly wave   A weak, large-scale convergence system that is part of the secondary circulation of the tropics.

Eccentricity   The deviation from an established pattern. In astronomy, the difference in the orbit of a body from that of a circle.

Effluents   The outflow of water or waste liquid from an underground source or conduit.

Electromagnetic spectrum   The range of energy that is transferred as wave motions and that does not require any intervening matter to make the transfer. Electromagnetic waves travel at the speed of light (186,000 mi/s).

Electromagnetic waves   Waves characterized by variations of electric and magnetic fields.

El Niño   An event, most pronounced in the south Pacific Ocean, that reverses the normal flow of water and wind.

ENSO   The El Niño southern oscillation. The greater change in the atmosphere in the southern hemisphere, which includes the El Niño.

Evapotranspiration   The total water loss from land by the combined processes of evaporation and transpiration.

Exosphere   The outermost region of the atmosphere from which particles may escape to space. The first interaction of solar radiation with the atmosphere occurs in the exosphere.

Famine   The scarcity of food over a large region, resulting in malnutrition or starvation for very large numbers of people.

Fluid   A substance capable of flowing easily.

Foehn   A central European wind that is the same wind as the chinook wind of North America; also called a leste wind.

Friction layer   The atmospheric layer above the earth's surface in which wind speed and direction are influenced by friction.

Front   The boundary between two different air masses.

Fujita scale   A scale that rates the relative severity of tornadoes based on the extent of damage incurred.

Gas law   The law which states that the pressure exerted by a gas is proportional to its density and absolute temperature.

Geostrophic wind   A wind aloft flowing parallel to the pressure gradient, with the pressure gradient and the Coriolis force in balance.

Geosynchronous   As applied to a satellite, having an orbit such that the satellite remains fixed over the same point at the earth's equator.

Gradient wind   A wind that blows parallel to curved isobars such that centrifugal, Coriolis, and pressure-gradient forces are in balance.

Greenhouse effect   The process by which the heating of the atmosphere is compared to a common greenhouse. Sunlight (shortwave radiation) passes through the atmosphere to reach the earth. The energy reradiated by the earth is at a longer wavelength, and its return to space is inhibited by atmospheric carbon dioxide and water vapor. This process acts to increase the temperature of the lower atmosphere.

Hadley cell   A convectional cell, operating as part of the general circulation, located approximately between the Tropic of Cancer or the Tropic of Capricorn and the equator.

Hantavirus   A pulmonary syndrome characterized by flulike symptoms such as fever, muscle aches, shortness of breath, and coughing. Hantavirus is carried by rodents.

Harmattan   A dry, dust-laden wind blowing south from the Sahara Desert.

Heat index   A value that describes the combined effect of high temperature and high humidity on the body. The apparent temperature.

Heat island   An area of higher air temperature, compared with the surrounding area, in an urban region.

Heat wave   Any unseasonably warm spell, which can occur anytime of the year.

Hectare   A metric unit of area equal to 2.47 acres.

Heterosphere   The upper of a two-part division of the atmosphere, based upon its composition, in which gases are no longer uniform.

Histogram   Graphical representation that uses rectangles to show the frequency distribution of selected class intervals.

Homosphere   The lower of a two-part division of the atmosphere, based upon its composition, in which a uniform mixture of gases is found.

Horse latitudes   Former name for belts of high pressure located at about 35°N and 35°S.

Hurricane   A tropical cyclone that develops in the Atlantic Ocean.

Hydrologic cycle   The processes by which water is cycled through the environment. The hydrologic cycle involves changes of state (solid, liquid, and gas) of water as well as the transport of water from place to place.

Hygroscopic   The attribute of being able to absorb water from the atmosphere.

Hypothermia   The state resulting from a dangerous fall in the body temperature of humans or animals.

Ice shelves   Large masses of floating ice attached to land. They are found most extensively around the Antarctic continent.

Infrared radiation   Radiation in the range longer than red. Most sensible heat radiated by the earth and other terrestrial objects is in the form of infrared waves.

Intertropical convergence zone   The seasonally migrating, low-pressure zone, located approximately at the equator, wherein the northeast and southeast trade winds converge. Composed largely of moist and unstable air, the intertropical convergence zone provides copious precipitation. Also referred to as the intertropical front (ITF).

Inversion   A reversal of the normal atmosphere regime; for example, an increase in temperature with height.

Ionosphere   A zone of the upper atmosphere characterized by gases that have been ionized by solar radiation.

Isobar   A line on a map or chart connecting points of equal barometric pressure.

Jet stream   A high-speed flow of air that occurs in narrow bands of the upper air westerlies.

Katabatic wind   Any air blowing downslope as a result of the force of gravity.

Laminar flow   A flow in which a liquid or gas moves smoothly in parallel layers.

Langley   A measure of radiation intensity equal to 1 g-cal/cm².

La Niña   A strengthening of the normal circulation over the southern Pacific Ocean.

Latent energy   Energy temporarily stored or concealed, such as the heat contained in water vapor.

Leveche   A dry, dust-laden wind blowing from the Sahara Desert into Spain.

Malaria   A disease produced by a parasitic protozoan that multiplies in and destroys red blood cells. Transmitted by mosquitos, malaria is endemic in the tropics.

Maunder minimum   A term applied to a time of minimum sunspot activity that occurred from about 1645–1700.

Mean   A measure of central tendency—the sum of individual values divided by the number of items summed.

Median   A measure of central tendency—the value of the middle item when items are arranged according to size.

Melanoma   A tumor of high malignancy that starts in a dark mole and metastasizes rapidly and widely. Also referred to as malignant melanoma.

Meridional circulation   Air flowing in a circulation pattern that is essentially aligned parallel to meridians of longitude.

Mesocyclone   A very large thunderstorm in which rotation has developed. The rotation often results in the formation of a tornado.

Mesosphere   The layer of the atmosphere above the stratosphere in which temperatures drop fairly rapidly with increasing height.

Metabolism   The set of chemical processes that sustains organisms.

Meteorology   The study of phenomena of the atmosphere, with a view toward understanding and predicting weather and climate.

Methane   A colorless, odorless, flammable atmospheric gas that absorbs Earth radiation quite efficiently.

Microclimate   The climate of a small area, such as a forest floor or small valley.

Middle-latitude cyclone   A low-pressure system with frontal boundaries occurring in the middle latitudes.

Milankovich cycles   A series of long-term changes in Earth's climate resulting from changes in planetary orbital parameters.

Millibar   A unit of pressure equal to 1000 dynes/cm².

Mistral   A cold, dry gravity wind blowing down off the Alps and affecting the French and Italian Riviera.

Mode   A measure of central tendency—the most frequent value that occurs in a set of data.

Monsoon   Atmospheric circulation typified by a change in prevailing wind direction from one season to another.

Nimbostratus   A low, dark gray cloud layer.

Nor'easter   A large low-pressure system in the atmosphere that travels north along the east coast of the United States. It may result in heavy precipitation and flooding along the coast. In winter, nor'easters produce large amounts of snowfall.

Obliquity   The angle between the plane of the earth's equator and the orbit of the earth around the sun.

Occluded front   A warm mass of air trapped when a cold front overtakes a warm front.

Opaqueness   The degree to which light will not pass through a substance.

Orographic precipitation   Precipitation that results from the lifting of air over some topographic barrier such as a coastline, hills, or mountains.

Ozone   Oxygen in the triatomic form (O_3); highly corrosive and poisonous.

Ozone layer   The layer of ozone, 25 kilometers above the earth's surface, that absorbs ultraviolet radiation from the sun.

Particulates   Solid particles found in the atmosphere.

Perihelion   The point in the earth's orbit around the sun when the earth is closest to the sun.

Periodic   Occurring or appearing at regular intervals, such as the sun's rising and setting.

Permafrost   Ground that is permanently frozen at some depth. The surface may melt in the summer, but the soil remains frozen beneath the surface.

Persistence   The continuation of a set of existing conditions.

Photochemical   Having to do with a chemical change that either releases or absorbs radiation.

Photoperiod   The period of each day when direct solar radiation reaches the earth's surface; approximately sunrise to sunset.

Photosynthesis   The process by which sugars are manufactured in plant cells, using water and carbon dioxide in the presence of sunlight.

Plague   A disease that reached pandemic proportions in the past. The plague is carried by fleas and has a high fatality rate. In the Middle Ages, it was known as the Black Death.

Pluvial   Pertaining to precipitation.

Polar front   The frontal zone of a storm separating air masses of polar origin from air masses of tropical origin.

Polar stratospheric clouds   Very high, thin clouds that form in the stratosphere, particularly over the Antarctic continent.

Potential evaporation   The maximum evaporation rate from a water surface; depends on the vapor pressure of the water.

Pressure gradient   The amount of pressure change occurring over a given distance.

Radiation   The process by which electromagnetic radiation is propagated through space.

Rain forest   A forest ecosystem that develops in regions with high rates of precipitation year-round. Rain forests consist mainly of evergreen vegetation, with trees dominant.

Rayleigh scattering   The scattering of solar radiation by particles in the earth's atmosphere.

Relative humidity   The ratio of the amount of water present in the air to the amount of water vapor the air can hold, multiplied by 100.

Remote sensing   Process of acquiring data or information about an object or phenomenon when the instrumentation performing the measurement is not in physical contact with the object.

Respiration   The physical and chemical processes by which an organism supplies itself with oxygen.

Ridge   An elongated area in which pressure is higher than the surrounding region.

Rossby waves   Waves in the middle and upper troposphere of the middle latitudes with wavelengths of 4000–6000 kilometers; named for C. G. Rossby, the meteorologist who developed the equations for parameters governing the waves.

Saffir-Simpson Scale   A relative scale that utilizes wind velocity, storm surge, and atmospheric pressure to categorize the strength of tropical lows.

Sand drift   The movement of sand over an area of dunes in a desert. Sand drift is due to the heating of the sand by the sun and is primarily a daytime phenomenon. The sand moves along the surface and is responsible for a large part of the movement of the dunes.

Santa Ana   A chinook wind occurring in southern California and northern Mexico.

Sastrugi   Ripples produced in snow by persistent gravity winds in Antarctica.

Saturation vapor pressure   The maximum amount of water vapor that the atmosphere can hold at a given temperature.

Savanna   Tropical grassland interspersed with trees and shrubs.

Sensible temperature   The sensation of temperature the human body feels, in contrast to the actual heat content of the air recorded by a thermometer.

Sirocco   A hot, dry wind blowing north across the Mediterranean Sea.

Solar constant   The mean rate at which solar radiation reaches the earth.

Southern oscillation   The shifting of atmospheric circulation over the southern Pacific region.

Specific gravity   The ratio of a unit mass of a substance to a unit mass of water.

Specific heat   The amount of heat needed to raise 1 g of a substance 1°C.

Specific humidity   The ratio of the mass or weight of water vapor in the air to a unit of air that includes the water vapor, such as grams of water vapor per kilogram of wet air.

Standard atmosphere   A conventional vertical profile of temperature, pressure, and density within the atmosphere.

Stationary front   A cold or warm front that has ceased to move; the boundary between two stagnant air masses.

Stefan-Boltzmann Law   A law which states that the amount of radiant energy emitted by a black body is proportional to the fourth power of the absolute temperature of the body.

Stratopause   The upper boundary of the stratosphere.

Stratosphere   A thermal division of the earth's atmosphere located between the troposphere and the mesosphere. The primary zone of ozone formation.

Sublimation   The transition of water directly from the solid state to the gaseous state without passing through the liquid state, and vice versa.

Subsidence   Descending or setting, as in the air.

Supercell   An extremely large thunderstorm, often reaching as high as the tropopause and sometimes extending into the stratosphere.

Surface ozone   Ozone that forms near the ground as a result of sunlight acting on incompletely combusted hydrocarbons. Surface ozone occurs most often and most intensely in urban areas of the industrial world.

Synoptic climatology   A study of climatology that relates local and regional climates to atmospheric circulation patterns.

Taiga   The coniferous forests of Siberia; also, other coniferous forests of the Northern Hemisphere.

Terminal fall velocity   The velocity at which a particle will fall through a fluid when the acceleration due to gravity is balanced by friction.

Terrestrial   Pertaining to the land, as distinguished from the sea or air.

Thermal   A small-scale rising current of warm air.

Thermodynamics   The science of the relationship between heat and mechanical work.

Thermosphere   The atmospheric zone that includes the ionosphere and exosphere.

Thunderstorm   A convective cell characterized by vertical cumuliform clouds.

Tornado   An intense vortex in the atmosphere with abnormally low pressure in the center and a converging spiral of high-velocity winds.

Trade winds   Two belts of winds that blow almost constantly from easterly directions and are located on the equatorward sides of the subtropical highs.

Transpiration   The process by which water leaves a plant and changes to vapor in the air.

Tropical cyclone   A large, rotating, low-pressure storm that develops over tropical oceans, called a hurricane in the Atlantic and a typhoon in the Pacific Ocean.

Tropopause   The upper boundary zone of the troposphere, marked by a discontinuity of temperature and moisture.

Troposphere   The lower layer of the atmosphere marked by decreasing temperature, pressure, and moisture with height; the layer in which most day-to-day weather changes occur.

Trough   An elongated area of low pressure relative to the pressure of the surrounding area.

Tundra   The treeless plains of the Arctic; a climatic region exhibiting tundra vegetation.

Turbulent flow   A flow in which irregular and seemingly random fluctuations occur.

Twilight   The period before sunrise and after sunset in which refracted sunlight reaches the earth.

Typhoon   See tropical cyclone.

Ultraviolet radiation   Radiation of a wavelength shorter than violet. The invisible ultraviolet radiation is largely responsible for sunburn.

Upslope   Moving uphill, as does a breeze or fog.

Upwelling   A vertical current of water in the ocean. Upwellings are common along the eastern sides of the oceans in midlatitudes.

Van Allen belts   Two zones of charged particles existing around the earth at very high altitudes and associated with the earth's magnetic field.

Vapor pressure   The partial pressure of the total atmospheric gaseous mixture that is due to water vapor; also called vapor tension.

Virga   A thin veil of rain seen hanging from a thunderstorm, but not reaching to the ground. The droplets are evaporated before they reach the ground.

Viscosity   The internal friction in fluids that offers resistance to flow.

Volcanic explosivity index   A measure of the role of volcanoes in climatic change.

Vortex   A whirling or rotating fluid with low pressure in the center.

Walker circulation   The atmospheric circulation over the southern Pacific Ocean. The Walker circulation follows the general model of the Hadley cell, with westward drift of the trade winds along the surface near the equator and counterflow aloft.

Warm front   A zone along which a warm air mass displaces a colder one.

Waterspout   A tornado occurring at sea that touches the surface and picks up water.

Water year   A 12-month period used in hydrology that corresponds to the annual precipitation or flow of streams. In the United States, the water year runs from October 1 to September 30. October 1 is the time of year when the water supply in the soil and streams is usually at its lowest.

Wavelength   The linear distance between the crests or the troughs in a wave pattern.

Weather   The state of the atmosphere at any one point in time and space.

Williwaw   A bora wind in Alaska, Greenland, or coastal Antarctica.

Windchill   The physiological effects resulting from the combined effect of low temperature and wind.

Wind rose   A class of diagram used to show wind speed and direction.

Yellow fever   An acute infectious disease characterized by the sudden onset of fever, jaundice, albuminuria, and, often, hemorrhage. Yellow fever is caused by a virus transmitted by a mosquito found primarily in the tropics.

Zebra mussel   A freshwater mollusk accidently introduced into the Great Lakes of North America. The zebra mussel competes with native fish for food and proliferates rapidly. It has become a problem as well in that it clogs inlet pipes taking water from the lake.

Zenith   A point in space directly above a person's head; a point on a line passing through the zenith.

Zonal circulation   The approximate flow of air along a parallel of latitude.

Zonda   A hot, dry wind blowing down the east side of the Andes Mountains of South America. A zonda is formed in the same manner as the chinook in the United States and Canada.

Bibliography

Addison D., *Weatherwise*, © 1994, Building for Comfort and Safety, June/July, pp. 14–18.

Aguado E., and Burt J. E., *Understanding Weather and Climate*, © 2001, Prentice Hall, Upper Saddle River, N.J.

Ahrens C. D., *Essentials of Meteorology*, 3e., © 2001, Pacific Grove, CA., Brooks/Cole.

Arya S. Pal., *Air Pollution Meteorology and Dispersion*, New York, © 1998, Oxford University Press.

Barry R. G., and Corley R. J., *Atmosphere, Weather, and Climate*, 7e., © 1998, New York, Routledge.

Bird E. C. F., *The Effects of Rising Sea Level on Coastal Environments*, © 1993, Chichester, U.K., John Wiley.

Bligh W., *A Narrative of the Mutiny on Board the HMS Bounty*, © 1838, Kent, England, Hodder and Stoughton, Ltd.

Boucher K., *Global Climates*, © 1975, Halstead Press, New York.

Brasseur Guy P., Orlando John J., and Tyndall Geoffrey S., *Atmospheric Chemistry and Global Change*, © 1999, Oxford University Press, New York.

Christopherson R.W., *Geosystems*, 4e., © 2000, Prentice Hall, Upper Saddle River, N.J.

Clawson D. L., *World Regional Geography: A Development Approach*, 7e., © 2001, Prentice Hall, Upper Saddle River, N.J.

Cole F. W., *Introduction to Meteorology*, 3e., © 1980, John Wiley & Sons, New York.

Dana R. H., Jr., *Two Years before the Mast*, © 1840, Harper and Rowe, New York.

Gedzelman S. D., *The Science and Wonders of the Atmosphere*, © 1980, John Wiley and Sons, New York.

Goody Richard,*Principles of Atmospheric Physics and Chemistry*, © 1995, Oxford University Press, New York.

Goudie A.S., *Environmental Change*, © 1983, Oxford University Press, New York.

Griffiths J. F., and Driscoll D. M., *Survey of Climatology*, © 1981, Merrill Publishing Co, Columbus, OH.

Hanwell J., *Atmospheric Processes*, © 1980, George Allen & Unwin, London, England.

Hidore J. J., *A Workbook of Weather Maps*, © 1975, Wm. C. Brown Co, Dubuque, IA.

Hidore J. J., *Weather and Climate*, © 1985, Park Pressm, Champagne, IL.

Hidore J. J., *Global Environmental Change*, © 1996, Prentice Hall, Upper Saddle River, N.J.

Intergovernamental Panel on Climatic Change (IPCC), *Summary for Policy Makers*, © 2001, Online, **www.ipcc.ch/**.

Kump L. R., Kasting J. F., and Crane R. G., *The Earth System*, © 1999, Prentice Hall, Upper Saddle River, N.J.

Lutgens F. K., and Tarbuck E. J., *The Atmosphere*, 8e., © 2001, Prentice Hall, Upper Saddle River, N.J.

Mackenzie F. T., *Our Changing Planet*, 2e., © 1998, Prentice Hall, Upper Saddle River, N.J.

McKnight T. M., *Physical Geography: A Landscape Appreciation*, 6e., © 1999, Prentice Hall, Upper Saddle River, N.J.

Miller A., Thompson J. C., Peterson R. E., and Haragan D. R., *Elements of Meteorology*, 4e., © 1983, Merrill, Columbus, OH.

Moran J. M., and Morgan M. D., *Meteorology: The Atmosphere and the Science of Weather*, 5e., © 1997, Prentice Hall, Upper Saddle River, N.J.

Morgan M. D., and Moran J. M., *Weather and People*, © 1997, Prentice Hall, Upper Saddle River, N.J.

National Aeronautics and Space Administration, NASA, **www.nasa.gov/**.

National Climate Data Center, **www.noaa.gov/ncdc.html**.

National Oceanographic and Atmospheric Administration, **www.noaa.gov/**.

National Research Council, *Acid Deposition: Long-term Effects*, © 1986, National Academy Press, Washington, D.C.

Oliver J. E., and Hidore J. J., *Climatology*, Columbus, © 1984, Merrill Publishing Co, OH.

Oliver J. E., *Climatology: Selected Applications*, © 1981, John Wiley and Sons, New York.

Oliver J. E., and Fairbridge R. W., *Encyclopedia of Climatology*, © 1987, Van Nostrand Reinhold, New York.

Reynolds R. W., *A Real-Time Global Sea Surface Temperature Analysis*, © 1988, Journal of Climate, 1, pp. 75–86.

Thornthwaite C. W., *An Approach Toward a Rational Classification of Climate*, © 1948, Geographical Review, 38, pp. 55–94.

Turco Richard P., *Earth Under Siege: From Air Pollution to Global Change*, © 1996, Oxford University Press, New York.

Turekian K. K., *Global Environmental Change*, © 1996, Prentice Hall, Upper Saddle River, N.J.

Van Loon Gary W., and Duffy Stephen J., *Environmental Chemistry*, © 2000, Oxford University Press, New York.

Warrick R. A., Barrow E. M., and Wigley T. M., *Climate and Sea Level Change*, © 1993, Cambridge University Press, New York.

Wayne Richard P., *Chemistry of Atmospheres*, 3e., © 1999, Oxford University Press, New York.

Index